Studies on Entrepreneurship, Structural Change and Industrial Dynamics

The ‘Studies on Entrepreneurship, Structural Change and Industrial Dynamics’ series showcases exceptional scholarly work being developed on the still unexplored complex relationship between entrepreneurship, structural change and industrial dynamics, by addressing structural and technological determinants of the evolutionary pathway of innovative and entrepreneurial activity.

The series invites proposals based on sound research methodologies and approaches to the above topics. Volumes in the series may include research monographs and edited/contributed works.

This is a SCOPUS-indexed series.

Conceição Rego • Maria Raquel Lucas
María Isabel Sánchez-Hernández
Luísa Cagica Carvalho • Adriana Backx Noronha
Editors

Transitioning to a Circular Economy

From Education and Technological Change to Green Horizons

Springer

Editors
Conceição Rego
Department of Economics
University of Évora
Évora, Portugal

Maria Raquel Lucas
Department of Management
University of Évora
Évora, Portugal

María Isabel Sánchez-Hernández
Faculty of Economics and Business
University of Extremadura
Badajoz, Spain

Luísa Cagica Carvalho
Department of Economics and Management
Instituto Politécnico de Setúbal
Setúbal, Portugal

Adriana Backx Noronha
School of Economics, Business
and Accounting
University of São Paulo
São Paulo, Brazil

ISSN 2511-2023 ISSN 2511-2031 (electronic)
Studies on Entrepreneurship, Structural Change and Industrial Dynamics
ISBN 978-3-031-77663-2 ISBN 978-3-031-77661-8 (eBook)
https://doi.org/10.1007/978-3-031-77661-8

This Springer imprint is published by the registered company Springer Nature Switzerland AG
The registered company address is: Gewerbestrasse 11, 6330 Cham, Switzerland

Acknowledgments

The editors are grateful for the support of FCT—Fundação para a Ciência e a Tecnologia, I.P., within the scope of the project UIDB 04007/2020.

Contents

Introduction

Conceição Rego, Maria Raquel Lucas, María Isabel Sánchez-Hernández, Luísa Cagica Carvalho, and Adriana Backx Noronha

Abstract This book is an important read for anyone interested in understanding and promoting the transition to a circular economy. Divided into three parts, it addresses pathways for social development and education for sustainability, in addition to the Sustainable Development Goals (SDGs), while also presenting construction and energy strategies within the context of the circular economy. With chapters that include case studies and perspectives from various regions, mainly Europe and Latin America, this book offers insights into educating for sustainability, tackling technological challenges, and implementing green practices. Readers will find various discussions on the role of education in promoting sustainable practices, the

C. Rego
Department of Economics, University of Évora, Évora, Portugal
e-mail: mcpr@uevora.pt

M. R. Lucas
Department of Management, University of Évora, Évora, Portugal
e-mail: mrlucas@uevora.pt

M. I. Sánchez-Hernández
Faculty of Economics and Business, University of Extremadura, Badajoz, Spain
e-mail: isanchez@unex.es

L. Cagica Carvalho (✉)
Department of Economics and Management, Instituto Politécnico de Setúbal, Setúbal, Portugal
e-mail: luisa.c.carvalho@esce.ips.pt

A. Backx Noronha
School of Economics, Business and Accounting, University of São Paulo, São Paulo, Brazil
e-mail: backx@usp.br

C. Rego et al. (eds.), *Transitioning to a Circular Economy*, Studies on Entrepreneurship, Structural Change and Industrial Dynamics,
https://doi.org/10.1007/978-3-031-77661-8_1

importance of social innovation, and the emerging business opportunities from the green transition. This book highlights the need for circular economic models that prioritise reduction, reuse, and recycling, providing valuable insights for policy-makers, educators, researchers, and professionals.

By incorporating this book into courses, educators can offer students an understanding of circular economy concepts, sustainable development, and ecological transformation, preparing them to face the environmental challenges of the twenty-first century with innovative and sustainable solutions.

Keywords Entrepreneurship · Technological change · Circular economy · Green transition · Education

The environmental, social, and economic events that have characterised our days, in the first decades of the twenty-first century, illustrate the need for us, individually and collectively, to change our ways of being and our preferences. The socio-productive models, based on linear economy structures, "get-consume-use-throw away", whose temporal origins can be traced back to the Industrial Revolution, which over time have been the basis for the development of the global economy, are no longer considered sustainable today, for social, economic, and environmental reasons, fundamentally due to the excessive exploitation of resources. Instead, there is a reasonable consensus that knowledge has led to an evolution towards Circular Economy (CE) organisational models where the 5Rs—Rethink, Refuse, Reduce, Reuse, and Recycle—predominate.

This transition, however, is neither quick nor cost-free in the various countries and continents. The way of life that has been implemented, especially in the most developed countries—and which serves as a model for the world—which combines quality of life with high and growing income and high levels of consumption, also presupposes a large availability of energy resources, which come overwhelmingly from fossil fuels, the exploitation and use of which is highly polluting and damaging to the environment. Changing the paradigm of what it means to live with quality in the various countries of the world is not an easy task, nor does it mean the same thing. The asymmetries in development between the various countries and continents are very high, so that what is already commonplace in terms of access to goods and services that provide well-being and a healthy life in developed countries, in many cases, is not yet within the reach of residents in developing countries. Therefore, the effort to evolve towards more ecologically balanced lifestyles, to be fair, cannot be made in the same way in all countries.

The Sustainable Development Goals (SDGs), the ecological transition, or the assumptions of Circular Economy (CE) force societies and public and private organisations to overcome challenges and take risks. Aiming to develop new products and services that are economically viable and ecologically efficient, by minimising resource extraction, maximising reuse, increasing efficiency, and developing new business models, the transition to a more sustainable CE, with respect for the SDGs, is a new social and business revolution and one of the dominant challenges for management and education. If, on the one hand, it forces organisations to review production systems and accommodate new constraints, on the other hand, it allows them to add value because citizens are increasingly aware of and committed to organisations, products, and services resulting from processes that are considered more ethical, inclusive and socially responsible and sustainable. For CE practices and the promotion of more sustainable levels of development to become a reality, it is essential to act in several ways: educating citizens, monitoring the progress of transition processes, and providing political decision-makers with information and knowledge to formulate policies that are efficient and fair.

Despite being widely encouraged, CE, sustainable practices, and the SDGs are still not widely experienced. Education for sustainability and CE is crucial to the progress of these paradigms that facilitate sustainable development and increased employability. Education in general and higher education, in particular, are among the social activities that contribute most to improving countries' development and their quality of life and to changing people's attitudes towards fundamental causes. In general, populations with higher levels of formal education are more attentive to the challenges of contemporary society, more proactive in decision-making and in changing consolidated habits and ways of life, bringing them more into line with more sustainable societies. Educating for digital, inclusive, and ecological transitions is the first step in ensuring that no-one is left behind. All over the world, in each place at its own pace and with its own priorities.

This book deals with challenges regarding technological innovation, ecological transition, and new ways of promoting sustainable business and what this translates into citizens' quality of life. This is a wide-ranging publication that aims to be a fundamental reference, building on research in the fields of circular economy and green and inclusive transition in relation to technological change and environmental sustainability.

Nowadays, having a good quality of life is much more than having a high GDP. Quality of life and happiness are based on a large set of issues beyond economic values. Social and environmental balance are crucial in order that no one and no place is left behind, regarding sustainability, below the Sustainable Development Goals (SDGs) framework. The topical focus of this book is identifying paths of deep social development through circular economy approach, green and inclusive transition, and digital transformation. The contributions in the book, from a large diversity of countries, help us learn more about the current challenges of circular economy and the energy efficiency and renewable energy, waste of new and old productions in the context of circular economy but also of sustainable consumption, social engagement, and social innovation. Education as a tool for more environmental sustainability is also highlighted.

This widespread and opportune publication, which complements other previous publications of the editors (Carvalho et al., 2018, 2019; Sánchez-Hernández et al., 2021a, 2021b; Rego et al., 2024), intends to be an essential reference source, building on the available literature in the fields of green transition and sustainability, in the literature devoted to the green transition, technological change, education role, and business dynamics. This book addresses a range of approaches including, conceptual, theoretical, and practical case studies and projects related to these developments. This book contains 15 chapters divided into 3 sections. The first section regards the advanced debate on the paths for sustainability based on different ways of social development, and it includes five chapters. The second section offers insights related to the education for sustainability based on Sustainable Development Goals (SDGs) and includes four chapters. The third section will discuss some main sectorial challenges in the areas of construction and energy, considering strategies for circular economy (CE). The next paragraphs detail the book structure, meaning its sections and respective chapters.

1 Part I. Social Development: Paths for Sustainability

What paths can lead us to a more sustainable society where we are happier? This question is answered in various ways in this section. The articles, by authors from different European countries, highlight the importance of the immaterial dimensions of life for happiness, based on more ecologically balanced models of social organisation and production. Here, it stresses the importance of non-material elements of life such as social and relational capital, work-life balance, mental health, and distributive justice. The happiness framework offers a new perspective on the green transition, showing that cutting on certain types of production and consumption causes greenhouse gas emissions and unhappiness. One of the ways of trying to live better is what neo-nomads do. For this, we will explore neo-nomadism and its sustainability, as well as review how countries have regulated this phenomenon and its associated practices. On the other hand, it is fundamental to understand the conceptual framework, particularly related to social aspects of the circular economy, once circular economy and social enterprise need to be combined. From this point of view, social circular economy is that enhancing personal well-being can go hand in hand with improving society and the environment. Action consistent with the principles, to achieve greater sustainability, implies going beyond what public decision-makers define. This is why we present an exploratory study that examines the relationship between circular economy, sustainability, and social innovation using qualitative data from an intentionally sustainable community. In this regard, we emphasise also the process of integration in the labour market of People with Intellectual Disabilities (PwID), which appears to be a tool for a more inclusive society, with more cohesion and development.

Chapter "Using the Happiness Framework as a Catalyst for the Green Transition" written by Mota argues that the usage of the happiness framework is a way to overcome some of the dilemmas of the green transition. The chapter suggests that the relevance of non-material dimensions for happiness is an opportunity for the green transition, because investing in such dimensions might compensate for some consumption losses imposed by the green transition, even if those losses are minimised by entrepreneurship focused on environmentally friendly technological change and circular economy mechanisms.

Chapter "Neo-nomads and Sustainability" developed by Lezcano-Calderón and Palos-Sánchez uses the Scopus and Web of Science databases to conduct a literature review on socioeconomic and environmental sustainability linked to neo-nomadism and the regulations enacted by countries. The main conclusions revolve around the scarce examination of the sustainability of neo-nomadic practices and the implications it has for territories.

Chapter "Rolling with the Resistance to the Circular Economy? A Contribution to Sustainable Consumption and Social Engagement" written by De Coninck, Anderluh, Mesbahi, Serralha, Paz and Carriço gives an overview of different views on the social circular economy, including their strong points and weaknesses. And explore the skills needed for the green transition. Since a circular economy transition can only be accomplished by combining the material, strategic, and consumer perspectives, this contribution also provides a discussion of research results on consumer perspectives on the circular economy.

Chapter ""It's Only Waste If You're Not Innovative": Understanding Micro-Scale Circular Economy Practices as Social Innovations" proposed by Nogueira, Pinto, and Marques analyses how a community's circular economy practices developed at the micro level can be understood as social innovation and their relevance for sustainable development. The findings help to understand the role of social innovation in promoting circular practices and sustainable development. The practices analysed have a strong DIY—Do It Yourself—component, stemming from the creativity of the members of this community. The study highlights the challenges circular economy practices face, such as the production of aesthetically pleasing items and the dissemination of the practice.

Chapter "Universal Design Applied to Inclusive and Sustainable Economy for All" composed by Oria-Segura and Muñoz-Medina establishes that support for the group of People with Intellectual Disabilities (PwID) can and must be an element of social cohesion and economic development of the territories. The chapter describes the experience of the development and promotion of employment with training that has put PwID in contact with the real world of work.

2 Part II. Education for Sustainability and SDG

The second part of this book is devoted to the questions of Education for Sustainability and SDG. Education is a transformative tool. With teachers, students, and schools where creativity and dreams can go hand in hand, it is possible to contribute to building fairer and more sustainable societies. The contributions for this section come from Europe and Latin America (Brazil and Colombia). First, it will be presented a *Consortium* of Higher Education Institutions created from a European Union Call for European Universities; all of them are located in peripheral regions, in their countries, and intend to work together, developing teaching and research activities from the common ground of sustainability and promotion of SDGs. From Brazil comes a case of designing new forms of executive education courses, through co-creation, focused in the framework of Circular Economy and Sustainable Business. From Colombia, a research team brings us the reflection about the ways in which sustainability and SDGs issues are included in the teaching-learning field. The authors also make suggestions on how to improve the pedagogical approach to these issues.

Chapter "EU GREEN: How Peripheral Regional Universities Created a European Alliance Based in SDGs" proposed by Delicado, Yuste, Pérez-Aloe, Pelosi, and Pires describes how nine peripheral universities located far away from the centre of their countries launched a consortium that focused on a particular set of sustainability-related practices. This chapter describes the method and tools used to relate the different stages and spheres of the creation process of the alliance that will be based on sustainability and address the major societal challenges identified in the 17 SDGs of the UN 2030 Agenda.

Chapter "Designing New Forms of Business Education at Circular Economy and Sustainable Business: An Experience Report on Stakeholders' Engagement Through Co-creation" written by Siloto, Lima, Pereira, and Barcelos addresses the problem of how to design a business education course that focuses on the circular economy through co-creation. This experience report takes place in Macaé/RJ, Brazil's oil capital and deeply dependent on productive activities linked to fossil fuels.

Chapter "Teaching Sustainable Development Goals from Theory to Praxis in Universities Based on the Theory of Graphs" suggested by Isaza Zapata, Portilla Rosero, and Calle Piedrahita discusses the teaching of the Sustainable Development Goals (SDGs) in universities and present a study that revealed that instructors predominantly approach the teaching of SDGs from a theoretical standpoint, emphasising the need to raise awareness about the pressing real-world challenges our planet faces in the absence of intervention. While the SDGs are integrated into various subjects' curricula, little effort is made to actively contribute to sustainability.

Chapter "Environmentally Sustainable Conscious Students When Starting Up" developed by Calle Piedrahita, Colorado Agudelo, Blanquised Rivera, Zapata Jiménez, and Malavera Pineda aims to determine the characteristics of entrepreneurial students with regards to environmental sustainability. The main finding of this chapter suggested that students in training demonstrated a critical and positive awareness of environmental sustainability. Additionally, those who work exhibited a greater acceptance of engaging in activities that promote environmental sustainability.

3 Part III. Construction and Energy as CE Strategies

The third part of this book is devoted to the sectors of construction and energy, regarding CE strategies. These sectors of activity are among those that simultaneously generate the most economic value and cause the most externalities in the ecological balance of societies. They therefore deserve special attention. The chapters in this section have contributions from European and Brazilian researchers. The first two chapters of this section deal with issues related to the construction industry and the problems associated with construction and demolition waste. These chapters are underpinned by the need to make these activities more ecologically balanced, in line with the principles of the CE and the green transition. The following chapters discuss the problem of energy efficiency and the creation of energy from ecologically more efficient systems in line with CE practices. Thus, these chapters present, on the one hand, consumers' perceptions of the information available on the energy efficiency of housing at the time of making a rental decision; on the other hand, a literature review based on the main technologies associated with the treatment of photovoltaic panel waste will be described; finally, a comparative study between Brazil and Italy, starts from the framework of the Russia-Ukraine conflict to emphasise the need to promoting a shift from fossil fuels to sustainable renewable energy. In this scenario, the use of waste shows a huge potential to enhance this change, being crucial to know the main barriers and opportunities to circular economy adoption in European and Brazilian agroindustry.

Chapter "Architecture for Deconstruction: A Methodological Approach to Eco-construction" designed by Marques related the context of construction industry and the demolition process in the context of circular economy concerning barriers to design architecture for deconstruction there are sound economic reasons to recognise demolition as economical resource.

Chapter "Circularity of Household-Generated Construction and Demolition Waste: Management Principles for Green Transition in Latvia" proposed by Cudečka-Puriņa, Mavļutova, Vesere, Osipova, Atstāja, Purviņš, Kuzmina and Butkevičs provides a holistic view of the construction waste problem, revealing the lack of a unified approach towards waste sorting and further reuse possibilities. The goal of the study is to investigate the principles of circular economy in building construction and demolition waste management and to find possible solutions for their development in household-generated waste management for green transition in Latvia.

Chapter "Impact of Home Energy Efficiency Labeling Investment on Rent Housing Decision of Students in Spain: A Choice Experiment Approach" written by González González, Crespo-Cebada, Díaz-Caro, and Mirón Sanguino aims to assess students' preferences about the rental housing market and energy efficiency certificate. The main results show that the energy certificate is a relevant attribute in the choice of the apartment to rent; however, when the market segmentation is carried out, different groups are observed.

Chapter "Photovoltaic Module Waste and the Circular Economy: A Comprehensive Review" created by Jiménez-Islas, Pérez-Romero, Álvarez-García, and Río-Rama discusses how the technologies associated with the disposal of photovoltaic module waste and its relationship with the circular economy, in addition proposed, to explain evolution of scientific publications associated with the field of study, bibliometric indicators.

Chapter "Renewable Energy Production with a Circular Economy Approach: Entrepreneurial Brazilian and Italian Case Studies" developed by Finco, Shikida, Bentivoglio, Vian, and Chiaraluce aims at identifying and analysing the main barriers and opportunities to circular economy adoption in European and Brazilian agroindustry based on case studies. In detail, the chapter focuses on the Brazilian case of sugarcane ethanol and an Italian frozen vegetable company because they are strategic sectors, they produce considerable quantities of waste, and both produce renewable energy from waste.

This book allows an enlarged perspective, theoretical and empirical, on an issue that is still quite undeveloped and unknown, providing a set of perspectives related with conceptualisation, framework, adoption, operationalisation, and new trends on the CE.

References

Carvalho, L., Rego, C., Lucas, M. R., Scnchez-Hernández, M. I., & Backx Noronha, A. (Eds.). (2018). *Entrepreneurship and structural change in dynamic territories: Contributions from developed and developing countries* (Structural change and industrial dynamics). Springer.

Carvalho, L., Rego, C., Lucas, M. R., Sánchez-Hernández, M. I., & Backx Noronha, A. (2019). *New paths of entrepreneurship development. The role of education, smart cities, and social factors* (Structural change and industrial dynamics). Springer.

Rego, C., Lucas, M. R., Hernandez, M. I., Carvalho, L., & Backx Noronha, A. (2024). *Entrepreneurship, technological change and circular economy for a green transition. Research contributions for a more productive environment* (Studies on entrepreneurship, structural change and industrial dynamics). Springer.

Sánchez-Hernández, M. I., Carvalho, L., Rego, C., Lucas, M. R., & Backx Noronha, A. (2021a). *Entrepreneurship in the fourth sector: Entrepreneurial ecosystems and sustainable business models* (Studies on entrepreneurship, structural change and industrial dynamics). Springer.

Sánchez-Hernández, M. I., Carvalho, L., Rego, C., Lucas, M. R., & Backx Noronha, A. (2021b). *Social innovation and entrepreneurship in the fourth sector. Sustainable best-practices from across the world* (Studies on entrepreneurship, structural change and industrial dynamics). Springer.

Part I
Social Development: Paths for Sustainability

Using the Happiness Framework as a Catalyst for the Green Transition

Gabriel Leite Mota

Abstract This chapter argues that the usage of the happiness framework is a way to overcome some of the dilemmas of the green transition. It explains how society's focus on GDP as the metric of progress can be challenged by the science of happiness, which stresses the importance of non-material elements of life such as social and relational capital, work–life balance, mental health, and distributive justice. The happiness framework offers a new perspective on the green transition, showing that cutting on certain types of production and consumption that cause greenhouse gas emissions and unhappiness would generate benefit instead of harm. Additionally, cuts on certain segments of production and consumption would cause fewer welfare losses than what is predicted when GDP is considered as the metric of progress. Lastly, preserving nature can have a direct positive impact on the enjoyment of life, something that is not counted in GDP. This chapter suggests that the relevance of non-material dimensions for happiness is an opportunity for the green transition, because investing in such dimensions might compensate for some consumption losses imposed by the green transition, even if those losses are minimized by entrepreneurship focused on environmentally friendly technological change and circular economy mechanisms. Finally, this chapter emphasizes that the green transition is a global project that can only be successful if a large part of society is involved and engaged, with a focus on sustainability, equity, and well-being.

Keywords Happiness · Economics of happiness · Green transition · GDP · Circular economy

---

G. Leite Mota (✉)
Instituto Superior de Serviço Social do Porto, Centro de Investigação em Ciências do Serviço Social and Instituto de Sociologia da Universidade do Porto, Porto, Portugal
e-mail: gabriel.mota@isssp.pt

C. Rego et al. (eds.), *Transitioning to a Circular Economy*, Studies on Entrepreneurship, Structural Change and Industrial Dynamics,
https://doi.org/10.1007/978-3-031-77661-8_2

1 Introduction

The Industrial Revolution of the eighteenth century marked a tipping point in human evolution: the emergence of the exponential upward slope in the time series of economic progress. Due to that, humanity gained power to expand in numbers, life expectancy, and living conditions. A revolution that begun in Great Britain later spread to other countries and continents, eventually generating the world we now live in. But what seemed like a continuous route of progress, revealed itself a dangerous path. The environmental challenges that we now face are a direct consequence of that economic progress, and a fundamental question has to be answered: how can we keep moving forward economically, while not destroying the ecological equilibrium that supports earth's liveability?

For that to happen, we need what has been called the *green transition*: a process that changes the way we produce and consume the things we need, so that earth's liveability is not endangered. And although some steps have been taken in that direction, both from governments and entrepreneurs, like the spread of circular economy, sharing economy, the increment of alternative and renewable energy sources, or several technological advances that now allow some countries to move on with their decoupling process (increasing GDP while reducing greenhouse gas emissions), several obstacles remain. In particular, the need for an effective coordinated global action, finding an operative way to redistribute the costs of this transition, achieving popular support for such changes (when many even deny global warming), and generating technological breakthroughs in a sufficiently short time are problems still without a definite solution.

Throughout this chapter, we argue that we might be able to find an answer to this problem by framing the green transition using the concept of happiness and the empirical evidence that the science of happiness has already produced.

In **Sect. 2** of this chapter, we explain with detail the economic, political, and intellectual processes that have been shaping the world after the Industrial Revolution, namely, how those forces created a culture we might designate as "the religion of GDP" where everything that matters in society can be, more directly or indirectly, reconducted to the process of economic growth. Such bias towards economic growth, disregarding possible hazards of that path, originated the planetary consequences we now witness that call for an ecological emergency intervention.

In **Sect. 3,** we discuss the difficulties faced when trying to change the course of progress. In particular, we shed light on the challenges and dilemmas the green transition has to deal with. If we opt for a drastic cut on greenhouse gas emissions, we will have to deal with the necessary reduction on GDP, an economic recession no one wants. If we bet on the technological solution, we face the risk of not having the innovations on time. If we try to combine the two, we might face the opposition from those entrenched on their viewpoint, having the difficulty to reach a compromise. In practice, the green transition has to: deal with the opposition from those with vested interests on greenhouse gas emissions based economic growth and the force of climate change denialism with its traction on political discourse; temper the argumentative appeal of those defending technological innovations as a panacea, namely, for environmental risks; enlighten the implementation issues of degrowth, deacceleration and a global redistributive policy; and achieve a worldwide collective action and mentalities' change under the pressure of urgency.

In **Sect. 4,** we synthesize the main results from the happiness literature and make the case that framing the green transition within the happiness perspective is a solution for some of the most important dilemmas previously stated. We stress that focusing on the notion of progress as happiness, rather than economic growth, is a game changer: not only environmental preservation contributes directly to happiness, the empirical relation, with diminishing marginal returns, between GDP and happiness supports changes in the composition of growth, or even economic degrowth, that might not endanger happiness. At the same time, given the non-GDP determinants of happiness (the immaterial factors), progress can be achieved without economic growth, that is, with no environmental negative externalities.

We **conclude** this chapter with the main lessons to take from using happiness as a catalyst for the green transition.

2 How We Got Here: The Religion of GDP

If it is always legitimate to say that the present is a consequence of the past, that's even more the case regarding contemporaneity. We live in the most globalized and interdependent era of all times, and the world has never been economically richer. And all that is a consequence of the Industrial Revolution of the eighteenth century. A combination of intellectual (enlightenment), political (liberalism), and economic forces (namely, capitalism) produced a long-lasting change on the living conditions of humans on planet earth. Not only have the living conditions changed but the world vision also changed, with people starting to expect continuous gains in the material conditions of life from one generation to the next.

What has begun as an economic revolution in the Great Britain soon spread to neighbouring countries, starting a transformative movement that is still in progress. The profound changes China and India are witnessing in the twenty-first century can be interpreted as a late consequence of the Industrial Revolution. The fact is, those countries are being able to rescue from poverty millions of people, producing one of the most impressive escapes from poverty in the history of humanity. And that thanks to globalization and the industrialization of such countries.

This economic and social process that has begun three centuries ago hasn't spotted yet. And looking to all the countries and people in the world that currently do not benefit from it (in many parts of Africa, Latin America, and Asia), it is expectable that many more are eager to participate and enjoy the advantages of strong economic growth. But that, as we shall discuss in the next section, might be a problem.

If the eighteenth century started the process, the twentieth century marked the acceleration and consolidation of that way of living. Of course, as already mentioned, not equally between countries, as some of them are still apart from those transformations. Nevertheless, it was during the twentieth century that economic progress became more exponential (prominently after the end of the Second World War) and that the economic growth as a metric of success became institutionalized. As more and more countries joined international trade, industrialization, specialization, and mass production, the *innovation*, the *economies of scale,* and the *network effects* all collaborated to the exponential economic growth the world has witnessed during that period.

While these transformations were taking place, a parallel institutional movement consolidated. Even if it can be argued that the works of Adam Smith, and fellow thinkers of that time, established the foundations of the liberal capitalism that now governs the world, the twentieth century was crucial for that to happen. In particular, a combination of an *economic doctrine* (neoclassical economics, spread as universal and converted into mainstream), the inauguration of *management degrees* (where students were trained to foster firms' productivity and profits), the launch of the *Gross Domestic Product*,[1] and the *political and public adherence* to the notion that more economic growth is always better (so that we should compare countries' civilizational success based on its capacity to generate growth) paved the way to the political, ideological, and practical consolidation of economic growth as the way to go.

It is not by accident that all major international organizations created during the twentieth century had a focus on economic growth (e.g. International Monetary Fund, World Bank, United Nations, Organisation for Economic Co-operation and Development, European Union). Let it be to define the most powerful nations on earth (as in the G20 or the G7/G8 reunions), let it be to identify poor countries and provide aid, economic growth was established as the benchmark of success. Even nowadays, and despite all the efforts some of those institutions are making to redefine the concept and metrics of progress (as in the OECD Better Life Initiative, the UN World Happiness Report, or the EU "beyond GDP" initiative) the practice remains largely unaltered: economic policies are still being designed and assessed based on its effects on GDP, Public Debt, Public Deficit, Inflation, Unemployment or Balance of Trade—traditional macroeconomic indicators—irrespective of the relative consequences each of those economic variables have on happiness. The case of the EU is acute, as the European Central Bank has, as its only mandate, to control inflation (and prevent financial instability), when studies such as Di Tella et al. (2001) have shown that unemployment matters more to happiness than inflation (supporting the double target of the US Federal Reserve's mandate). Furthermore, the EU treaties define obligations based on the ratios of Public Debt and Public Deficit to GDP, not to public happiness.

Moving forward, it is worth looking with more detail into the factors that were behind the consolidation of economic growth as a panacea. We can divide those factors into three categories:

[1] GDP—a metric of economic production that its creator, Simon Kuznets, back in 1934, had the caution to clearly state that it was not a metric of national well-being.

- Economic forces
- Political forces
- Intellectual forces

Economic Forces

One must recognize that the process originated during the Industrial Revolution would have not survived, until the present, unless *some good was being produced* and some *endogeneity was in motion*.

Starting with the *good*, when we talk about economies, we are talking about production and consumption of goods and services. Goods and services that are, supposedly, to our enjoyment and comfort. That is, when an economy grows, more goods and services are being produced and consumed.[2] And that increased consumption should materialize into more welfare—see Stevenson and Wolfers (2008) on the positive relation between economic performance and happiness. That is the very simple logic through which economic growth can be defended. And, in fact, since the eighteenth century until today, there is an enormous number of things that were created that contributed a lot to our comfort and enjoyment. To name just a few: basic sanitation, health care infrastructures, roads, cars, public transports, modern houses with electricity, heating and cooling equipment, freezers, mass production of food (that originated an abundance and diversity of calories with no parallel in human history), or even mass production of clothes (that allowed people to be protected from the weather, comfortably and stylishly).

When the Industrial Revolution started, the priority was to produce the basic goods and services that, once available, would generate large gains in welfare. But once basic needs were satisfied, other needs could be attended and basic needs could be served with better care, more sophistication, and made available to more and more people. The logic would be: richer countries would benefit from the continuation of economic growth because ever more needs could be satisfied, and all needs would be served better; poor countries would follow the leaders, and walk the same path the rich countries already navigated, first attending basic needs, later serving more sophisticated necessities, should economic growth be enough.[3] That is, in a nutshell, the fundamental argument still in use for those (mostly economists) who claim economic growth is a civilizational conquest that should not be jeopardized by climate change. Quite the opposite, they argue, it is in the technological innovation, an intrinsic dimension of economic growth, that humanity will be able to find ways to adapt and prosper in a world with a different ecological balance, or to revert future ecological unbalances. In fact, it is only because so many needs have been attended and so many people around the world have been a beneficiary of economic growth that this process is being sustained for so long.

---

[2] Even if we save and invest, that also represents consumption, only in the future.

[3] Interestingly, Adam Smith or John Stuart Mill (as most of the classical economists) believed that basic needs would rapidly be satisfied by economic progress, so that only superficial wants would be able to foster future economic growth, envisioning, even if not saying it explicitly, that economic growth could start losing its power to generate welfare overtime.

Looking now to what we call the *endogeneity factors* of economic growth, we must investigate the economic forces that tend to continuously feed economic growth. And we shall mention the existence of *economies of scale*, the gains from *specialization and learning by doing*, the *networks of interdependence* and the incentives of the *capitalist game*, and different forces with positive correlations, which only reinforce endogeneity.

One of the main reasons for the economic growth generated by the Industrial Revolution was the capacity that many firms had to reduce their unitary costs of production once they had started increasing their size. These *economies of scale*, which is precisely the ability to produce more goods at a lower cost per unit by increasing the scale of production,[4] were facilitated by the use of machines, interchangeable parts, and mass production techniques that enabled manufacturers to produce goods more efficiently, and at a lower cost, leading to increased profitability and growth. This is the force that has been behind the rise of modern corporations (mostly during the twentieth century), that were able to take advantage of larger production facilities and foster the specialization of labour. Furthermore, as corporations grew larger and more complex, they were able to invest in research and development, leading to further innovation and growth. This is an ongoing and self-reinforced process, with big contemporary corporations continuing to push the boundaries of what is possible through economies of scale. There are reasons why becoming big is good for profit and for growth, something that pushes medium and small companies to try getting bigger. And when they succeed, the economy grows with them.

At the same time, there has been *specialization and learning by doing*. These interrelated concepts are part of the central argument that Smith (1776:1904) has put forward in his seminal work on why nations get rich.[5] In practice, specialization enables workers to focus on a particular task and become more skilled and efficient, while learning by doing allows workers to gain expertise and improve their performance through repeated practice. Both contribute to the increment of firm's productivity, which is a key determinant of economic growth as soon as many companies start to become specialized. Additionally, specialization and learning by doing have also played a critical role in driving innovation. As workers became more skilled and efficient in their specialized tasks, they were better able to identify areas for improvement and suggest innovations that could lead to further productivity gains. Paradigmatically, the development of new machines and production techniques in the early days of the Industrial Revolution was largely driven by workers who had become experts in their tasks. More contemporarily, as firms grew larger and more complex, they were able to invest in research and development, leading to further innovation and growth. The use of specialized teams, such as marketing or research and development, has allowed corporations to develop new products and services

[4] See Stigler (1958) for a classic explanation of this process or Kwack and Sun (2005) for a recent empirical example of this force at paly.

[5] See Becker and Murphy (1992) for a more recent appraisal of the same question.

more efficiently, leading to increased profitability and growth. Again, the success of the firms who had enjoyed the benefits of this process served as a benchmark that others seeking profit wanted to copy. Once many copy this, all workers start to become more skilled and efficient, and by driving innovation, productivity, and profitability, sustained economic growth happens.

Another reason why economic growth tends to reinforce itself is the existence of *networks of interdependence* that specialization and trade foster. There can only be specialization if all specialized firms and workers can trade their production with each other, so that they can gain access to the consumption bundle they desire. But once these networks are in place, it is virtually impossible to refuse to participate in the process. The recent Covid-19 pandemic was a clear example of the level of interdependence the world now lives in, with even the most closed nations suffering both from the viral contamination and ruptures in the supply chain. The unprecedented levels of economic integration we now have around the world (the world has never been more globalized) are a consequence and a cause of specialization and trade, all directly linked to productivity gains. It is only because nations and firms engage in this worldwide process of specialization, segmentation of the production process, supply chains, and trade that we can generate the current level of wealth.[6] This network dynamic, where being apart leads to failure, being part leads to adaption, learning, and specialization, is another force conducing economic agents (firms and individuals) in the direction of economic growth. Furthermore, interdependence fosters the sharing of knowledge (we need to speak a common economic and operational language), which also favours productivity gains and, aggregately, economic growth.

Operating as a unifying force for all the above mentioned, and as a fundamental determinant of economic growth, is the power of *capitalistic incentives*. What we might call the capitalist game has been a dominant force shaping the world since the Industrial Revolution. Very basically, the incentives of the capitalist game push economic agents into seeking profit through market competition. The means of production being privately owned determines that capitalists want to take the most advantage possible by putting them in productive activities (in different sectors), always seeking to maximize profit. The competition for clients in the market forces enterprises to seek innovation and productivity (to save costs or increase revenues), generating a circle of continuous innovation and productivity, which generates economic growth. Some of the most important theoretical advocates of capitalism (as those from the Austrian School of Thought) attribute to this economic system (with the price mechanisms it conveys and the ingenuity of entrepreneurship) the main share in explaining the economic progress since the Industrial Revolution. Without the need of central command or State orientation, capitalism forces agents into a restless search for novelty (moved by the fear of going bankrupt and the ambition to

[6] See Hidalgo and Hausmann (2009) for an illustration of this idea using an indicator of the complexity of exports as a determinant of economic growth.

gain more) culminating in a never-ending process of economic growth—see Shleifer (1998).

Summarizing, we could say "what works, will work", an expression to symbolize this endogeneity dimension. Particularly, learning by doing, getting bigger, specializing, and exchanging with others are attitudes that, individually and jointly, promote productivity, profit, and economic growth. Seeking gains, countries and individuals who engaged in the capitalist game of competition, profit, specialization, and growth, got successful. Those who didn't still find themselves struggling with high levels of poverty, low life expectancy, and basic levels of knowledge and skills, which pushes them into vicious circles of poverty, the opposite of the virtuous circles of growth here discussed.

Political Forces

The Industrial Revolution was not only a technological transformation, but it was also a societal transformation. After millennia of existence in different kinds of feudalistic societies, the eighteenth century in Europe was the time to break up with that past. Although that cut had different implementations across countries (namely, the British reformism versus the French revolutionism), and occurred at different moments, the end of feudalism was a tendency that spread out to the world. With the few exceptions of absolute monarchies in the middle east, and the "feudalist republic" of North Korea, all countries of the world already abandoned feudalism. And that has a lot to do with the process initiated by the enlightenment and the Industrial Revolution.

In fact, since the beginning of the Industrial Revolution, an ongoing movement of economic liberalism took place. Despite marked differences between countries, and over time, the tendency was set in motion. More and more space was given to private enterprises and entrepreneurs that would try the chance for making money via profit gains. The gradual openness to interregional and international trade, the increased diversity of firms in terms of size, products and technologies and the emergence of a new social class, the bourgeoisie, all contributed to the reinforcement of economic growth and the logic of capitalistic competition. This new social class, by becoming richer and richer, gained political and institutional power, replaced clergies and nobleman in some strategic social positions, and was able to shape the course of society more and more in the direction of economic growth. Not by accident, it was the bourgeoisie who later founded and funded management degrees and business schools to prepare human resources for their companies and to boost the capitalistic ideology (contributing to the intellectual forces favouring economic growth). At the same time, workers looked up to the bourgeoisie, fantasying with the possibility of becoming one of them, many times not opposing the advancement of the capitalists' interest, irrespective of poor working conditions enjoyed by the proletariat during the first stages of the Industrial Revolution.

When we reached the twentieth century, and several countries transitioned to democratic regimes, this cultural environment influenced the political process. In particular, the shaping of public policies. Through voting pressure, with voters penalizing governments when economic performance is bad and re-electing governments when the economy is booming—see Fox and Phillips (2003) and Cerda and Vergara (2022)—and through lobbyists' influence, each fighting for their profits, eventually contributing to economic growth—see Mork (1993), Julio (2014), and De Bruycker and Beyers (2019)—politicians were guided towards the promotion of economic growth as the most important thing to be done. Even if several policy goals were defined, and different people claimed for different objectives, the notion that more economic growth was a good thing, and less of it a bad, never left the room of the democratic game. And politicians wanting to get to, or staying in power, soon were co-opted by this vision even if they had different perspectives on this topic.

And it was also during the twentieth century that another decisive step was taken towards the consolidation of economic growth as the grail of policy: the birth, in 1934, by the hands of Simon Kuznets, of the Gross Domestic Product (GDP). In the aftermath of the Great Depression, the USA decided to create a statistical indicator to measure the aggregate production of the nation, valued at market prices. It was seen as crucial to have information on how the economy was going to prevent the booms and busts of the, then, recent past. But what was created solely as an index of economic production,[7] a figure for economists to care about, soon became the most important number of a nation, a statistic that all wanted to augment and reach the top of the world ranking. A figure known by politicians and the lay people (even if not aware of what it really meant) gained so much importance, nationally and internationally, that all the forces that were already at play pushing society towards economic growth ended up with just one more powerful ally. Even today, GDP continues to maintain its centrality in the policy domain. And although it might not be a universal remedy to all earthly issues, it is still seen as a sine qua non condition by all countries.

Intellectual Forces

Understanding why economic growth became so well-established in the contemporary world demands that we also look at the intellectual/cultural dimension. Despite all the other above-mentioned forces at play, the cultural environment has been especially relevant to maintain economic growth as the civilizational priority. And for that, we can pinpoint three main contributors: *neoclassical economics*; *management science*; and *liberal individualism*.

Although many outside the economic discipline do not realize the importance of economics to the consolidation of market economies and the societal centrality of economic growth, the methodological and intellectual direction that this discipline took during the transition from the nineteenth to the twentieth centuries was crucial. Such transition originated what is called Neoclassical economics, an economic theory based on the assumptions of rationality, self-interest, and the market as the most

[7] It is now famous this quotation: "The welfare of a nation can, therefore, scarcely be inferred from a measurement of national income" Kuznets (1934), where the creator of the index promptly and explicitly declares that GDP is not to be used as a metric of well-being or development, by no means the ultimate goal of national policy.

efficient resource allocation mechanism. With those assumptions, it provided a framework for understanding economic growth as the result of maximizing individual utility through market transactions, with individuals and firms seeking to maximize their profits and wealth by engaging in exchange transactions in a competitive market. The resulting allocation of resources was considered efficient and led to economic growth.[8] At the same time, neoclassical economics emphasized the importance of technological progress and innovation generating more efficient production methods, which led to lower costs and higher productivity. This, in turn, leads to higher output, income, and economic growth.

In neoclassical economics, economic growth is viewed as a means to achieve higher levels of human welfare and well-being, with economic growth seen as a way to increase the material standard of living, reduce poverty, and improve social welfare, leading to greater happiness and fulfilment for individuals and society as a whole. Finally, it promoted the idea that economic growth was necessary for the survival and success of a nation in a competitive global economy, providing for the security and well-being of its citizens. Overall, neoclassical economics generated the intellectual and philosophical grounds for economic growth to become the most desirable societal goal by emphasizing the importance of individual utility maximization, technological progress, human welfare, and national competitiveness. This view has been influencing economic policy and development strategies in many countries, particularly in the post-World War II period.

Another intellectual force at play pushing society in the direction of economic growth was the creation of *management science* and the development of management degrees and schools. Sprung from the initiative and funding of business owners (notably Edward Tuck) who wanted to form more capable managers and generate a pro-business culture, and the early systematic contributions of Frederick Taylor, Henri Fayol, or Max Weber, management paved the way to more and more productive enterprises. As soon as the theories of good organization began to be applied, companies started to collect the fruits of better management. And the fruits were productivity, profit, and growth. As firms that hired managers became more successful, others followed the lead, spreading management throughout the economy and reinforcing the possibilities for economic growth. Nowadays, management courses are amongst the most taught discipline in the world, with business departments at universities typically being the richest. The creation and maintenance of a pro-business culture was a very successful endeavour that eventually moved beyond the walls of profit-oriented companies and spread its influence—the logic of efficiency and productivity—to the public sector and non-profit organizations. That is, not only economic growth was boosted by the private sector, but it also came from the public and third sectors.

[8] This idea is formalized in the Fundamental Theorems of Welfare Economics which state that Pareto optimal allocation of resources can be obtained through free markets and competition.

A third intellectual force worth mentioning here is *liberal individualism*. Coincident with the Industrial Revolution was the birth of liberal philosophies.[9] Liberal philosophies also marked a revolution, this time a cultural one, moving societies away from organic conceptions where the ruler of the state was the key figure, and the one deciding what was good for society and individuals, to a notion where individuals gained that centrality. Under liberal individualism, every person is the ultimate judge of her own choices and the decider of what's good. Individuals are conceived as objective entities with subjective preferences that no one can deny. This kind of conception of society and individuals was transported to economics with the notion that the central economic entity is the economic agent, a rational individual (let it be a consumer or a producer) who seeks to maximize utility, a subjective reality that only reveals itself through market choices and prices. Price mechanisms and market competition are economic devices that match perfectly with, and reinforce, liberal individualism. In this framework, moving away from capitalism and market mechanisms is moving away from the possibility of individuals to express their preferences and pushing production in the direction of their needs. It was also in this context that economic thought moved away from Classical Political Economy (and the notion of surplus and its distribution, with economy being seen as producing above the needs of society, the question being the distribution of such surplus) to neoclassical economics where the economy was conceived as producing below wants (which were infinite), so scarcity, not surplus, was key—see Xenos (1987) and Martins (2013). As such, economic growth was the only way to go: use the market to know what individuals want; and produce as much as possible to try to close the gap between the infinite desires and limited productive capacity. Additionally, liberal individualism also fits well with the notion of entrepreneurship and that of the entrepreneur as the hero of capitalism that Joseph Schumpeter popularized, while economic growth continues to be the holy grail.

Closing the circle, not only are Neoclassical economics and management well-aligned in their focus on economic growth but both are also compatible with liberal individualism, where economic growth is interpreted in terms of augmented freedom of choice and augmented freedom to produce and consume, resulting in augmented well-being.

3 The Dilemmas of the Green Transition

The idea that greenhouse gas emissions generate the heating of the earth temperature, and that a consequent climate change is ongoing, is a known scientific fact—see Parmesan and Yohe (2003).

[9] The utilitarianism proposed by Jeremy Bentham was one of the most important and most influential—see Bentham (1789).

Nevertheless, many deny that scientific evidence. An no matter how scientist and politicians pass the message - that the quantity of greenhouse gases freed during the production and or/consumption of the goods and services we use is too much for earth's ecological liveability - many tend to refute that evidence—see Bohr (2016). And while some simply refute it all "there is no planetary warming, economic production cannot disrupt ecological balances" (mostly lay people, but many politicians as well— Donald Trump or Jair Bolsonaro are clear examples), others accept the first part "there is global warming caused by human action" but refuse the second part saying "we don't need to reduce economic growth in order to preserve favourable ecological balances" (many mainstream economists fit in this category).

It is the midst of this intellectual environment that the green transition must occur. And, theoretically, there are three fundamental alternatives to pursue the path of an economic transition that can preserve the ecological balances compatible with the existence of humans on the planet: significantly cut the production and consumption responsible for pollution and resources' exhaustion—*the cutting solution*; operate a set of technological breakthroughs that allow humanity to continue with their consumption habits without endangering earth's liveability—*the technology solution*; activate some combination of the two—*the mix solution*.

In all cases, there are a bunch of dilemmas that are not easy to overcome. If we go with the *cutting solution*, we need to find a way not to cause the frustration of expectations, massive poverty, hunger, and premature deaths that a drastic economic recession could cause.[10] Additionally, we would need to deal with the trade-off between present and future: do we maintain, as economists do, that the present is more important than the future (there is a positive rate of interest), so not too many current losses are acceptable to sustain possible future gains, or we adhere to the *strong longtermism* thesis of Greaves and MacAskill (2021) and claim that the future is more important than the present, so huge suffering in the present can be justified by the prospects of enormous future happiness? Furthermore, how can we convince the population to accept the costs of the cutting solution? If we choose the *technology solution*, then we need to deal with the issue of speed and earth's limited resources. Regarding speed, technology would have to be developed, but in a truly short time frame, so that the ongoing global temperature warming could be reverted on time, which is a longshot.[11] Additionally, technologies that save, eliminate, or even capture greenhouse gases do not solve the problem of finitude of resources. The limited quantity of fresh water (or even water in total) or natural elements used in the production of our modern life determine a cap in production possibilities, no matter how far technological progress might go (and remember that we need to think how to serve all people on earth with abundance, not just serve the people in the rich countries as now happens). Furthermore, some strategies already put in

[10] One might argue that the effects of climate change will be even worse in all those aspects. But the timing issue remains: the cutting solution would impose those costs now, the "not doing nothing" solution would impose those costs only in the future, even if nearby.

[11] On this respect, some say we need to boost economic growth, not hamper it, to maximize the probability of a major technological breakthrough, a kind of "all-in" solution.

place in this direction have shown little capacity to overcome global warming. Finally, when trying to combine these two avenues (which is mostly what we have been doing since the climate crisis began), we need to face the dilemma of reaching a compromise between people with different viewpoints, many of which are entrenched in their positions claiming the other view is nothing but a delusion. That is, for instance, how environmentalists tend to see economists' faith on technology, and how economists tend to see the environmentalists' proposals. Adding to all that, the green transition must be fair—there must be environmental justice as in Mohai et al. (2009)—which means that the costs and the benefits that occur as consequence of these societal transformational processes must be equitably distributed between rich and poor countries, rich and poor people, men and women, old and young and between ethnicities, which is something much easily said than done.

Returning to the *cutting solution*, it is important to recognize that such a proposal messes with the installed powers of oil industries, the vested interests of many companies, institutions, and governments, and disrupts the way of living of the typical person, who will fight back against those who try to alter things as they are now. Economic growth is so embedded in our lives, and the notion that more of it is good, less of it is bad, that any proposition to stop or reduce it is easily deemed radical, extremist, or dangerous. Not casually, and despite the intellectual effort that is being made by that current of thought—see Kallis (2011)—the degrowth movement faces all kinds of resistances from economists, politicians, or even the public opinion, all framing this line of reasoning, and their proposals, as extremists or unfeasible—see van den Bergh (2011). At the same time, and even if we accept parts of the degrowth argumentation, there is a very sensitive issue regarding the concept of sufficiency. According to degrowth theories, we should move away from satisfaction of wants to the practice of sufficiency. That is, we should stop giving the consumer the possibility to buy everything he/she desires and start focusing on achieving sufficiency for all. As appealing as it might sound from an ecological perspective, an operable and agreed upon definition of sufficiency is something extremely hard to achieve—see Sekulova et al. (2013).[12] And a philosophical question remains: sufficiency might be analyzed in ontological terms—is there something sufficient for an individual or society, or are we animals of infinite needs and wants? If we stick with liberal individualism, we will reject the notion of sufficiency and deem it an unacceptable intromission on freedom of choice. The cutting solution has also to deal with another, but related, line of opposition: the moral argument that what rich countries did, travelling the pollution road, should be allowed to poor countries. Otherwise, the argument goes, rich countries would be condemning poor countries to an eternal poverty. And China or India are usually given as examples of countries that can now fight poverty, only because they are polluting a lot.[13]

[12] Despite these difficulties, IPCC has been doing some work on that direction. Nevertheless, achieving such a definition is no guarantee that it will become operative.

[13] But see O'Neill et al. (2018) on how planetary resources might not be enough to guarantee material abundance to all, despite technological progress, which will force us do adopt sufficiency practices.

Instead of seeing in the environmental crisis evidence that our economic model is unsustainable, some prefer to look at it as a reason to foster and accelerate economic progress. Those who have strong faith in technology prefer the *technology solution*. For technology believers, innovation has already proved capable of solving exceedingly complicated economic and societal problems. In fact, the figures of the current world population and the average life expectancy of people are a direct result of technological progress (both material and organizational). Healthcare, food availability, shelter, and comfort are available for an unprecedented number of individuals. And all in a short period of time after the Industrial Revolution. In fact, the material abundance of the present was seen as impossible by the most sophisticated thinkers at the beginning of the Industrial Revolution. The famous Malthusian prediction, that population growth would always function as a mechanism to impede the augmentation of per capita income, was revealed to be totally mistaken. And Thomas Malthus not having foreseen the force of technological progress is currently being used as a caricature of the degrowth supporters, labelled as neo-Malthusians, by failing to understand the power of technology (present and future) to solve the current environmental problems.

In line with the fitness of this approach with the status quo, the implementation of the technology solution is ongoing. And several steps have been taken, some of which we can group into the following categories: *circular economy*; *alternative energy sources*; *sharing economy*; and *radical innovations*.

Circular economy encompasses the efforts that are being done in several economic sectors trying to generate closed circles of production-consumption-recycle-production, generating minimal waste and maximum efficiency, allowing people to enjoy the same utility from consumption without generating the waste and pollution produced before. These kind of solutions rely on the incrementation of the practices of reuse, recycle and repair, betting that is possible to find a way to maintain consumption without harming the environment—see Ghisellini et al. (2016). While this strategy is being spread to different sectors and with increasing efficiency, it remains unclear how could circular economy produce the fundamental change the climate emergency demands—see de Jesus and Mendonca (2018).

Another way to take advantage of the technological progress to save the environment is generating energy using *alternative sources of energy*, namely, renewables—see Gielen et al. (2019). Wind, solar, tides, waves, and organic decomposition, all are energy sources that don't generate greenhouse gas emissions (at least after the infrastructures have been produced and installed, and discounting repairing and substitutions). Furthermore, they have the advantage of being available all over the world (some types more abundant in certain geographies), not discriminating between poor and rich nations. Nevertheless, the progress in this area is still limited, not only because the energy generated from renewables, all over the world, still accounts for a tiny fraction of the current energy needs but because this energy suffers from intermittency, something that batteries could solve, might batteries not be harmful for the environment.

This bet on renewable energies is being fostered throughout the world, even in the Middle East by the barons of fossil fuels, with several countries subsidizing research, development, production, and distribution of renewables, which has granted the advances that we now have but do not guarantee that this bet is the solution for global warming.

Simultaneously, efforts are being made to transform the way we use things, through what is called the *sharing economy*. Again, as in the circular economy, the main aim is to obtain gains in efficiency by taking advantage of identified inefficiencies, mostly in consumption.[14] While the circular economy has more a "production perspective", sharing economy has more a "consumption/distribution perspective". One of the key contributions coming from the sharing economy is idea of maintaining (or gaining) utility in consumption while efficiently adjusting the consumers' access to goods and services, only making them available when the consumption is needed. Furthermore, once some platform is established, individuals can act as producers or consumers that meet through the platform, satisfying their needs, or making profit, dispensing traditional chains of distribution and production. There are several practical examples of the sharing economy, from transport to music, hospitality, or even blockchain opportunities (see Esmaeilian et al. (2020) for examples), but doubts are being casted regarding the effectiveness of this strategy in stopping climate change—see Martin (2016).

No matter what strategy the technology believers put into practice, betting on the *technology solution* is betting on a radical innovation that is yet to come, and that is very unlikely to materialize with the necessary urgency. The truth is the data on climate change and greenhouse gas emissions are telling us that we need to stop the flux and find a way to capture the stock. Even if tomorrow the nuclear fusion was made commercially available, it would take lots of time to transform the installed systems to be compatible with that new source of clean power. And that process would produce greenhouse gases. That's the reason why many are very sceptical about true decoupling possibilities and green growth—see Ward et al. (2016) and Hickel and Kallis (2020).

In practice, what the world has been doing is a combination of the two extreme strategies discussed, the *mix solution*.

Using international forums, as the Intergovernmental Panel on Climate Change (IPCC), gathering world leaders trying to reach a minimal understanding on a global collective action to fight climate change, or using the UN Sustainable Development Goals as tools to improve climate awareness, many throughout the world are trying to gain support for the actions needed. Some countries are moving fast (using circular economy, renewable energies, supporting entrepreneurial initiatives on environmentally friendly technologies, or promoting conscious consumption) while others are still very dependent on fossil fuels and do not join instinctively the environmental agenda.

Another route already taken is being pushed by economists (theories coming from environmental economics) trying to conjugate market incentives and

[14] The example of cars is often given, because when we own a car, most of the times the car is parked somewhere (with a cost), to be used only in a fraction of that time, and normally by one person only (when four could be transported). Sharing that car would be optimal.

self-interest to channel actions into climate sustainability. One clear example of such endeavour comes from the design and implementation of emissions markets—such as the EU Emission Trading System—see Ellerman and Buchner (2007)—where participants are endowed with a fixed volume of emissions that they can opt not to emit and sell those possibilities in the market to other players, making a revenue. With that, and no further state control, participants have incentives to move into non-polluting forms of production, which fosters technological innovation and the reduction of emissions.

The problem with this mixed solution is twofold. First, it faces the opposition of those who see their standpoint as the sole solution—as Hobson and Lynch (2016) claiming that green growth, circular economy and sharing economy are all delusional in fighting climate change, or orthodox economists deeming degrowth as radical an unscientific; second, it has also to deal with the question of distributive justice—how to distribute benefits (riches and technologies) and costs (pollution and cuts on consumption and jobs) between countries and people and obtain environmental justice.[15]

4 The Happiness Lessons and Solutions for the Green Transition

The fundamental assumption behind the bet on economic growth was that somehow, the complex process of economic progress would always be capable of producing well-being. Supported on the philosophies, theories, and empirical evidence already discussed in the previous sections, that bet was considered a success.

If it is true that the world still struggles with a series of problems, those supporting economic growth always look at those issues as a result of low economic growth, never as an excess of it. Let it be poverty, crime, diseases, wars, underdevelopment, unemployment, or unhappiness, those are questions that will be solved once we promote economic growth, develop markets, competition, and capitalism.

On the other side, authors belonging to the degrowth movement, environmentalists, many development economists and some welfare economists, and many philosophers, soon started to oppose the notion that economic growth is a panacea, quite the opposite, sometimes the very root of problems. Poverty, wars, underdevelopment, financial instability, and inequality are pointed out by these opponents as problems caused by capitalism and the obsession with economic growth, not issues those mechanisms could resolve. And when we add the climate crisis, economic growth soon appears as the main culprit.

It is in this line of inquiry that the economics of happiness emerges. In Easterlin (1974), the author was able to correlate subjective well-being with GDP only to find

[15] See Menton et al. (2020) on how this problem is difficult, even with the help and good intentions of SDG.

out that the established idea that more economic growth is always better was empirically false. Looking at cross-sectional data and over time, he was able to identify this pattern: richer nations tend to be happier than poor ones, but rich nations, while becoming richer, do not get happier. Furthermore, after a certain per capita income, the cross-sectional happiness disparities between poorer and richer nations also tend to disappear. These results clearly challenged the orthodoxy on the mighty of economic growth but went unnoticed for a long time. Only the arrival of the twenty-first century changed the course of events, with the emergence of the economics of happiness integrated in the larger movement of happiness studies, recuperating and reconfirming Richard Easterlin's analysis—see Bruni et al. (2021) for a state of the art review.

Having in mind the results the economics of happiness has been producing (and more broadly the results from the science of happiness coming from different disciplines) we can frame the dilemmas of the green transition with a new perspective. And, hopefully, facilitate such transition.

Here, we opt to pinpoint three central ideas that spring from the happiness literature and that we find the most relevant to frame and foster the green transition.

First, the notion that the relation between *GDP and happiness is non-linear*, with diminishing marginal rates of return. Second, the fact that *immateriality matters* to happiness, so a considerable set of determinants of happiness can be activated without the need of economic growth. Third, the direct relation that we can find between *environment and happiness* and what that teaches us regarding the environmental protection.

The *non-linearity between GDP and happiness* is now a consolidated empirical fact. And despite all the resistance from the part of those entrenched in the dogma of economic growth, it is unscientific to claim that more economic growth is always better and is always a good thing. And this raises a fundamental question: if more economic growth might not be capable of enhancing happiness but is, undeniably, the main source of greenhouse gas emissions that are jeopardizing earth's liveability, why pursue that route?[16] It would only be defensible under the "all-in" assumption that only rapid economic growth can produce the technological breakthrough that would solve climate warming (although not resolving the limited resources issue, and being a longshot). Once we frame economic growth within the happiness literature, casting doubts on the importance of economic growth gains scientific support. And that can be used as an argument to advance the green transition, claiming that changing focus from economic growth to happiness means that we don't need to conciliate growth and green transition, we need only to conciliate happiness with the green transition. And that might dispense economic growth and the consequent greenhouse gas emissions.

Another argument in this direction appears if one thinks in the opposite direction, that is, what happens if GDP diminishes? Traditionally, that would mean a welfare

[16] See Kubiszewski et al. (2013) framing this question from the Genuine Progress Indicator perspective, showing how GDP has detached from GPI decades ago—GPI has flattened while GDP continues to grow.

loss (recall, more GDP is good, less GDP is bad). But, again, the economics of happiness tells a different story. Not only increments in GDP can be unable to foster happiness, but some realities that augment GDP harm happiness. That is the case of long working hours—see Ratzel (2012) where working for too long augments production but steals time for leisure and family, which brings unhappiness; the case of commuting—see Stutzer and Frey (2008) where the long time spent travelling from home to work and around is not compensated (happiness-wise) by the wage earned; or the perversities of conspicuous consumption, where the pleasure of having something is hampered by what others have, forcing us into a rat race of GDP growth with flatted happiness—see Winkelmann (2012). These more perverse situations, where economic growth runs in the opposite direction of happiness give strong arguments to temper GDP. In these situations, GDP growth is simultaneously hurting the environment and the happiness of individuals. In these situations, diminishing GDP, by cutting on those realities, would have a double positive effect: would augment happiness and would decrease greenhouse gas emissions.

Finally, there is another strong effect of the non-linear, with diminishing marginal returns, relation between happiness and GDP: redistributing from the rich to the poor is conducive to enhancements on average happiness. This a conclusion already advanced by the classical utilitarians, but one that got lost in time, despite neoclassical economics' claims of using a utilitarian background. As a corollary, there is room for wealth transfers from the rich countries in the direction of the poor ones, something that can also be used to boost the green transition. As mentioned in previous sections, one line of argument in defence of economic growth states that rich countries have a moral obligation to let poor ones follow the pollution growth track as that opportunity was open to rich countries at the time of the Industrial Revolution. Using climate warming as an argument to stop their economic progress would be unfair. But using this result from the economics of happiness, we have a new possibility. Transfers of wealth can be done, allowing the poor people to gain access to comforts only available in the rich countries, not having the necessity to add that production (and pollution) to the world. Transfer income from the rich countries - that will have a small happiness loss - to the poor ones - that will enjoy large happiness gains - and the world average happiness increases, greenhouse gas emissions stay the same and morality is granted. The same exercise can be done between poor and rich individuals within a country—see Pacek and Radcliff (2008) on how the welfare states, and its redistributive policies, manage to increase happiness.

The second main topic emerging from the happiness literature, and that might serve the purpose of the green transition, is the fact that *non-material realities matter a lot to happiness*. There are dimensions of the life of individuals and societies that are not materialized and are not things that we can produce, sell, and buy in the market, even in the service sector. As such, those realities do not enter GDP calculations, but are central to us. Those dimensions matter a lot to our well-being and are not independent of economic growth and marketization, but stand in a non-linear relation with them. The happiness literature shows us that a huge part of human well-being depends upon the quantity and quality of interpersonal relations—see

Becchetti et al. (2008). Furthermore, that quantity and quality are sometimes crowded out by economic growth and market mechanisms—see Graafland (2010). What this means is that if we take into consideration these dimensions, we have a path to augment happiness with no need for economic growth (or even decreasing it). Although the enjoyment of interpersonal relations might be facilitated by goods and services that are a consequence of economic progress (such as dining out with friends or enjoying the comforts of our family staying at the beach resort), the core of interpersonal relations is the interaction between individuals, not the material context of that relation. Operating based on only producing the context, without caring for the core, is a risk that contemporary societies have been taking, given its bias towards economic growth. Changing that focus allows us to concentrate on the conditions to foster quality and quantity of interpersonal relations, which can be done in harmony with the green transition.

Another very relevant immaterial dimension is trust. Let it be trusted in others (strangers or friends and family) or in the institutions, the social capital is a capital that does not depend on the economic growth (typically, it goes the other way around, more social capital facilitates economic growth) and is a vital component of societal happiness—see Helliwell (2006). Finding ways to boost trust within society is a powerful tool to build happiness with no environmental costs. A more trustworthy society might even be more available to enter the green transition as its high levels of institutional trust might facilitate public opinion listening to scientists and their proposals and become more environment aware—see Hao et al. (2020).

Alongside, comes another immaterial reality, job quality. Individuals tend to spend most of their active life at the job, and job is not only about gaining the money to buy the things we need. Job is also about forming an identity, feeling useful for society, and finding a meaning for life. Not by accident, the literature on happiness gives jobs a central role. Let it be the lack of it—unemployment being one of the main determinants of unhappiness as in Clark and Oswald (1994)—let it be the quality of it—good job quality increases job satisfaction and job satisfactions determines life satisfaction as in Cannas et al. (2019). Although some determinants of job satisfaction might be correlated with economic growth, others are not, namely, those more related with the quality of interpersonal relations at work that, once more, might be contaminated and degraded by an excessive focus on competition, profits, and growth. Betting on strengthening these dimensions can be done without increasing growth, generating a green transition compatible with happiness.

All these immaterial dimensions relevant to happiness have a flexible relation with economic growth, and might be used to support the green transition. Focusing on the advancement of immateriality has no environmental costs (might even have benefits) and produces psychological gains. An example would be to change the mainstream purpose of life from getting rich (which implies to be richer than others, which is socially self-defeating and threatens the environment), to be kind (which is socially self-sustaining and environment friendly) or to embrace neo-humanism as in Sarracino and O'Connor (2023).

The third lesson we can take from the happiness literature to help with the green transition comes from the *direct relationship that subjective well-being has with the environment*. Some recent studies have shown that being in direct contact with nature has a direct positive effect on our physical and mental well-being—see MacKerron and Mourato (2013), Wolsko et al. (2019), and Burns et al. (2021). Destroying landscapes or endangering ecological balances reduces the opportunity for this kind of pleasure. At the same time, and apart from all economic losses caused by climate swings and extreme events, there is a direct effect on well-being—let it be the fear of the coming of a non-liveable planet—eco-anxiety as in Panu (2020)—or the negative emotions triggered by living in a degraded environment—see MacKerron and Mourato (2009). In both cases, what we can take is that the environment has a value for happiness that must be accounted for. When discussing the merits or demerits of the green transition and the costs and benefits of degrowth strategies, it should be put forward the direct costs of endangering the environment in the minds of people alongside the missed happiness from having natural spaces destroyed.

The happiness literature is a vibrant scientific endeavour with a strong flow of results. Regarding the relations between economic growth and happiness, between the environment and happiness, and all the interrelations between them, this branch of science can be used to enter the discussion of the green transition helping in the design of policies and in passing the message to the public opinion of what are the real losses and gains from shifting our current way of living (based on the notion of infinite economic growth and net welfare gains from it) to another more compatible with planetary ecological boundaries without the need to incur in wellbeing losses.

5 Conclusion

This chapter proposes that the happiness literature can serve as a framework for the green transition, solving or diminishing some of the dilemmas such transition presents.

It was explained that we live in a world that has been shaped, since the Industrial Revolution, by the conception and practice of economic growth as the most important national achievement, and the indispensable route for well-being and success. Such cultural environment was fed by endogenous forces that maintained the bet on economic growth throughout time. The combination of liberal individualism, capitalism, neoclassical economics, management science, endogenous economic dynamics, and political pressures generated a circumstance where putting economic growth into perspective was considered nonsensical and radical.

When climatologists began to alert us with the fact of climate change, and the connection between economic activity and global warming has been established, a cultural shock emerged. How could it be that the cause of all good is now the cause of a potential Armageddon?

The reaction to this problem varies depending on the analyst, but we could discriminate three main approaches: the *cutting solution* advocated by the degrowth movement, where a drastic cut to consumption, production, and gas emissions is required to save humanity, changing the mindset from satisfaction of wants to sufficiency; the *technology solution*, where a bet is made on the ingenuity of entrepreneurs to come up with a radical innovation capable of producing abundant clean energy or the creation of carbon capture devices that would allow us to maintain and enlarge our consumption patterns and protect the environment, using circular economy, sharing economy and renewable energy sources, in the meantime, as a mitigation strategy; or the *mix solution*, which is what is in practice, with the combination of institutional efforts (such as the IPCC or the SDG), theoretical advances in economics (environmental economics and the impacts of emissions markets), alongside the incentives for circular and sharing economy, renewables and cuttings on spurious consumption and production, jointly trying to mitigate the environmental damages, although evidence is telling us that that is not enough.

Given these difficulties, the natural resistance to a change of paradigm, and the intellectually and politically crisped environment around these topics, the science of happiness offers some results that could educate the debate and make the green transition softer.

It was suggested that we could use the happiness literature to identify GDP components that are destroying happiness. In such cases, a GDP reduction is a good thing, generating more happiness and less ecological damages. Furthermore, we could use the happiness literature to identify determinants of happiness that are not accounted for in GDP (as relational and social capital). In these cases, more happiness is possible with no economic growth, that is, more happiness is possible with no ecological footprint. Additionally, we could use happiness literature to evaluate the direct importance of nature to happiness, explaining that preserving nature is a good thing, not a cost or a wasted opportunity to grow.

Finally, fighting poverty can be done with more redistribution, not with more economic growth. Such redistributive policy would generate higher average happiness and less pressure to produce/pollute more.

The main lesson should be that once we shift the goal of humankind from economic growth (GDP) to happiness, many of the most important dilemmas presented by the green transition vanish and a compatibility between progress and earth's liveability is restored.

References

Becchetti, L., Pelloni, A., & Rossetti, F. (2008). Relational goods, sociability, and happiness. *Kyklos, 61*(3), 343–363.

Becker, G. S., & Murphy, K. M. (1992). The division-of-labor, coordination costs and knowledge. *Quarterly Journal of Economics, 107*(4), 1137–1160. https://doi.org/10.2307/2118383

Bentham, J. (1789). *The principles of moral and legislation*. Prometheus Books.

Bohr, J. (2016). The 'climatism' cartel: Why climate change deniers oppose market-based mitigation policy. *Environmental Politics, 25*(5), 812–830. https://doi.org/10.1080/09644016.2016.1156106

Bruni, L., Smerilli, A., & De Rosa, D. (Eds.). (2021). *A modern guide to the economics of happiness*. Edward Elgar.

Burns, A. C., Saxena, R., Vetter, C., Phillips, A. J. K., Lane, J. M., & Cain, S. W. (2021). Time spent in outdoor light is associated with mood, sleep, and circadian rhythm-related outcomes: A cross-sectional and longitudinal study in over 400,000 UK biobank participants. *Journal of Affective Disorders, 295*, 347–352. https://doi.org/10.1016/j.jad.2021.08.056

Cannas, M., Sergi, B. S., Sironi, E., & Mentel, U. (2019). Job satisfaction and subjective well-being in Europe. *Economics & Sociology, 12*(4), 183–196. https://doi.org/10.14254/2071-789x.2019/12-4/11

Cerda, R., & Vergara, R. (2022). Economic growth and political approval ratings: Evidence from Latin America. *Political Behavior, 24*. https://doi.org/10.1007/s11109-022-09798-y

Clark, A. E., & Oswald, A. J. (1994). Unhappiness and unemployment. *Economic Journal, 104*(424), 648–659.

De Bruycker, I., & Beyers, J. (2019). Lobbying strategies and success: Inside and outside lobbying in European Union legislative politics. *European Political Science Review, 11*(1), 57–74. https://doi.org/10.1017/s1755773918000218

de Jesus, A., & Mendonca, S. (2018). Lost in transition? Drivers and barriers in the eco-innovation road to the circular economy. *Ecological Economics, 145*, 75–89. https://doi.org/10.1016/j.ecolecon.2017.08.001

Di Tella, R., MacCulloch, R. J., & Oswald, A. J. (2001). Preferences over inflation and unemployment: Evidence from surveys of happiness. *American Economic Review, 91*(1), 335–341.

Easterlin, R. A. (1974). Does economic growth improve the human lot? Some empirical evidence. In R. Easterlin (Ed.), *Happiness in economics* (pp. 5–41). Edward Elgar. Accessed from http://search.ebscohost.com/login.aspx?direct=true&db=eoh&AN=0758622&site=ehost-live&scope=site

Ellerman, A. D., & Buchner, B. K. (2007). The European Union emissions trading scheme: Origins, allocation, and early results. *Review of Environmental Economics and Policy, 1*(1), 66–87. https://doi.org/10.1093/reep/rem003

Esmaeilian, B., Sarkis, J., Lewis, K., & Behdad, S. (2020). Blockchain for the future of sustainable supply chain management in industry 4.0. *Resources Conservation and Recycling, 163*, 15. https://doi.org/10.1016/j.resconrec.2020.105064

Fox, G., & Phillips, E. N. (2003). Interrelationship between presidential approval, presidential votes and macroeconomic performance, 1948-2000. *Journal of Macroeconomics, 25*(3), 411–424. https://doi.org/10.1016/s0164-0704(03)00046-6

Ghisellini, P., Cialani, C., & Ulgiati, S. (2016). A review on circular economy: The expected transition to a balanced interplay of environmental and economic systems. *Journal of Cleaner Production, 114*, 11–32. https://doi.org/10.1016/j.jclepro.2015.09.007

Gielen, D., Boshell, F., Saygin, D., Bazilian, M. D., Wagner, N., & Gorini, R. (2019). The role of renewable energy in the global energy transformation. *Energy Strategy Reviews, 24*, 38–50. https://doi.org/10.1016/j.esr.2019.01.006

Graafland, J. J. (2010). Do Markets crowd out virtues? An Aristotelian framework. *Journal of Business Ethics, 91*(1), 1–19. https://doi.org/10.1007/s10551-009-0072-6

Greaves, H., & MacAskill, W. (2021). *The case for strong longtermism*. Global Priorities Institute Working (5).

Hao, F., Liu, X. S., & Michaels, J. L. (2020). Social capital, carbon dependency, and public response to climate change in 22 European countries. *Environmental Science & Policy, 114*, 64–72. https://doi.org/10.1016/j.envsci.2020.07.028

Helliwell, J. F. (2006). Well-being, social capital and public policy: What's new? *Economic Journal, 116*(510), C34–C45.

Hickel, J., & Kallis, G. (2020). Is green growth possible? *New Political Economy, 25*(4), 469–486. https://doi.org/10.1080/13563467.2019.1598964

Hidalgo, C. A., & Hausmann, R. (2009). The building blocks of economic complexity. *Proceedings of the National Academy of Sciences of the United States of America, 106*(26), 10570–10575. https://doi.org/10.1073/pnas.0900943106

Hobson, K., & Lynch, N. (2016). Diversifying and de-growing the circular economy: Radical social transformation in a resource-scarce world. *Futures, 82*, 15–25. https://doi.org/10.1016/j.futures.2016.05.012

Julio, P. (2014). The politics of growth: Can lobbying raise growth and welfare? *Journal of Macroeconomics, 42*, 263–280. https://doi.org/10.1016/j.jmacro.2014.08.007

Kallis, G. (2011). In defence of degrowth. *Ecological Economics, 70*(5), 873–880. https://doi.org/10.1016/j.ecolecon.2010.12.007

Kubiszewski, I., Costanza, R., Franco, C., Lawn, P., Talberth, J., Jackson, T., et al. (2013). Beyond GDP: Measuring and achieving global genuine progress. *Ecological Economics, 93*, 57–68. https://doi.org/10.1016/j.ecolecon.2013.04.019

Kuznets, S. (1934). *National Income, 1929–1932. Senate Document No. 124, 73rd Congress, 2nd Session*. U. S. Senate.

Kwack, S. Y., & Sun, L. Y. (2005). Economies of scale, technological progress, and the sources of economic growth: Case of Korea, 1969–2000. *Journal of Policy Modeling, 27*(3), 265–283. https://doi.org/10.1016/j.jpolmod.2004.08.012

MacKerron, G., & Mourato, S. (2009). Life satisfaction and air quality in London. *Ecological Economics, 68*(5), 1441–1453. https://doi.org/10.1016/j.ecolecon.2008.10.004

MacKerron, G., & Mourato, S. (2013). Happiness is greater in natural environments. *Global Environmental Change-Human and Policy Dimensions, 23*(5), 992–1000. https://doi.org/10.1016/j.gloenvcha.2013.03.010

Martin, C. J. (2016). The sharing economy: A pathway to sustainability or a nightmarish form of neoliberal capitalism? *Ecological Economics, 121*, 149–159. https://doi.org/10.1016/j.ecolecon.2015.11.027

Martins, N. O. (2013). Classical surplus theory and heterodox economics. *American Journal of Economics and Sociology, 72*(5), 1205–1231. https://doi.org/10.1111/ajes.12045

Menton, M., Larrea, C., Latorre, S., Martinez-Alier, J., Peck, M., Temper, L., et al. (2020). Environmental justice and the SDGs: From synergies to gaps and contradictions. *Sustainability Science, 15*(6), 1621–1636. https://doi.org/10.1007/s11625-020-00789-8

Mohai, P., Pellow, D., & Roberts, J. T. (2009). Environmental justice. *Annual Review of Environment and Resources, 34*, 405–430. https://doi.org/10.1146/annurev-environ-082508-094348

Mork, K. A. (1993). Living with lobbying - A growth policy co-opted by lobbyists can be better than no growth policy at all. *Scandinavian Journal of Economics, 95*(4), 597–605. https://doi.org/10.2307/3440913

O'Neill, D. W., Fanning, A. L., Lamb, W. F., & Steinberger, J. K. (2018). A good life for all within planetary boundaries. *Nature Sustainability, 1*(2), 88–95. https://doi.org/10.1038/s41893-018-0021-4

Pacek, A., & Radcliff, B. (2008). Assessing the welfare state: The politics of happiness. *Perspectives on Politics, 6*(2), 267–277. https://doi.org/10.1017/s1537592708080602

Panu, P. (2020). Anxiety and the ecological crisis: An analysis of eco-anxiety and climate anxiety. *Sustainability, 12*(19), 20. https://doi.org/10.3390/su12197836

Parmesan, C., & Yohe, G. (2003). A globally coherent fingerprint of climate change impacts across natural systems. *Nature, 421*(6918), 37–42. https://doi.org/10.1038/nature01286

Ratzel, S. (2012). Labour supply, life satisfaction, and the (dis)utility of work. *Scandinavian Journal of Economics, 114*(4), 1160–1181. https://doi.org/10.1111/j.1467-9442.2012.01717.x

Sarracino, F., & O'Connor, K. J. (2023). Neo-humanism and COVID-19: Opportunities for a socially and environmentally sustainable world. *Applied Research in Quality of Life, 18*, 9–41. https://doi.org/10.1007/s11482-022-10112-5

Sekulova, F., Kallis, G., Rodriguez-Labajos, B., & Schneider, F. (2013). Degrowth: From theory to practice. *Journal of Cleaner Production, 38*, 1–6. https://doi.org/10.1016/j.jclepro.2012.06.022

Shleifer, A. (1998). State versus private ownership. *Journal of Economic Perspectives, 12*(4), 133–150. https://doi.org/10.1257/jep.12.4.133
Smith, A. (1776:1904). *An inquiry into the nature and causes of the wealth of nations*. Methuen.
Stevenson, B., & Wolfers, J. (2008). Economic growth and subjective well-being: Reassessing the Easterlin paradox. *Brookings Papers on Economic Activity, 2008*, 1–87. https://www.jstor.org/stable/27561613
Stigler, G. J. (1958). The economies of scale. *Journal of Law & Economics, 1*, 54–71. https://doi.org/10.1086/466541
Stutzer, A., & Frey, B. S. (2008). Stress that doesn't pay: The commuting paradox. *Scandinavian Journal of Economics, 110*(2), 339–366. https://doi.org/10.1111/j.1467-9442.2008.00542.x
van den Bergh, J. (2011). Environment versus growth - A criticism of "degrowth" and a plea for "a-growth". *Ecological Economics, 70*(5), 881–890. https://doi.org/10.1016/j.ecolecon.2010.09.035
Ward, J. D., Sutton, P. C., Werner, A. D., Costanza, R., Mohr, S. H., & Simmons, C. T. (2016). Is decoupling GDP growth from environmental impact possible? *PLoS One, 11*(10), 14. https://doi.org/10.1371/journal.pone.0164733
Winkelmann, R. (2012). Conspicuous consumption and satisfaction. *Journal of Economic Psychology, 33*(1), 183–191. https://doi.org/10.1016/j.joep.2011.08.013
Wolsko, C., Lindberg, K., & Reese, R. (2019). Nature-based physical recreation leads to psychological well-being: Evidence from five studies. *Ecopsychology, 11*(4), 222–235. https://doi.org/10.1089/eco.2018.0076
Xenos, N. (1987). Liberalism and the postulate of scarcity. *Political Theory, 15*(2), 225–243. Accessed from http://www.jstor.org/stable/191676

Neo-nomads and Sustainability

Yendry Lezcano-Calderón and Pedro Ramiro Palos-Sánchez

Abstract Globalization and digital technologies have facilitated the rise of neo-nomads. During the pandemic, companies increased the possibilities to work from home and made working conditions more flexible, as a mechanism to continue their activities. Territories also saw the opportunity to take advantage of neo-nomadism to reduce the economic impact generated by immobilization during the crisis. However, this way of life may have implications for the sustainability of the territories. This study aims to explore neo-nomadism and its sustainability, as well as to review how countries have regulated this phenomenon and its associated practices. This study uses the Scopus and Web of Science databases to conduct a literature review on socioeconomic and environmental sustainability linked to neo-nomadism and the regulations enacted by countries. The main conclusions revolve around the scarce examination of the sustainability of neo-nomadic practices and the implications it has for territories. Furthermore, this review reveals that studies on the regulations enacted by countries do not cover the issue of sustainability.

Keywords Digital nomads · Energy nomads · Sustainability · SDGs · Territories

1 Introduction

In the professional world and in the media, digital nomads are frequently discussed. However, this is an incipient topic of academic research (Aroles et al., 2020). According to Šímová (2022), most research has focused on conceptualization,

Y. Lezcano-Calderón (✉)
Golfito, Costa Rica
e-mail: yendry.lezcanocalderon@ucr.ac.cr

P. R. Palos-Sánchez
Seville, Spain
e-mail: ppalos@us.es

C. Rego et al. (eds.), *Transitioning to a Circular Economy*, Studies on Entrepreneurship, Structural Change and Industrial Dynamics,
https://doi.org/10.1007/978-3-031-77661-8_3

lifestyles, location (in)dependence, and the factors defining and affecting the lives of digital nomads. Other factors such as globalization, digital technology, labor flexibility, transportation, and the need to feel greater freedom have contributed to individuals seeking for alternative places to work and travel (Cook, 2020; Hannonen, 2018; Müller, 2016; Orel, 2019; Reichenberger, 2018; von Zumbusch & Lalicic, 2020). Aroles et al. (2020) mention that the vision of the positioning of digital nomads has changed. At first, it was believed that these individuals were an elite who could afford to travel while they worked, but it has become evident that some are engulfed in precarious conditions. This has forced us to review the family and work implications for societies (Hannonen, 2018), and we argue the need to review the underlying socio-environmental impacts of digital nomads.

Currently, digital nomads have access to various jobs (Aroles et al., 2020, 2022). Due to the significant growth in this sector of workers before the pandemic, it is expected that after this health crisis, digital nomads will skyrocket even more, as any country could be an option to locate the digital nomad workplace. For instance, a study on the link between nomads and the hotel industry showed (Chevtaeva & Denizci-Guillet, 2021) that it is necessary for the tourism sector to take advantage of the characteristics of the region, culture, food, local participation, and other opportunities. Although digital nomadism is becoming institutionalized (Aroles et al., 2020) with more countries taking measures to confront, regulate, and promote this way of life/work, this phenomenon should be studied in depth because it can contribute to the establishment of policies that may equip companies and allow them to promote neo-nomadism. Further research in this field would guide the communities that visualize digital nomads as an opportunity to boost their economy.

Few studies have reviewed the situation of digital nomads in particular countries and cities (Šímová, 2022), and only four studies have addressed the issue of sustainability associated with digital nomadism (Cruz et al., 2021; Foley et al., 2022; Orel, 2020; Schwarz, 2021). Because of the war between Russia and Ukraine, the cost of energy has risen significantly for those individuals who must face intense cold winters, impacting their economies. Therefore, those whose work can be done remotely or by telecommuting are choosing to move to warm places, and they are calling themselves energetic nomads. Although there are still no academic papers that refer to this phenomenon, the truth is that in the world this event is being increasingly discussed. We believe that countries should evaluate how this sector can be used for economic dynamics with a socio-environmental balance.

In this chapter, we will reflect on the importance of neo-nomadism as this new lifestyle evolves from emerging technological facilities and their widespread promotion during the pandemic. This way of living has significant implications in the territories and mobility. Contributions and efforts made by energy nomads to reduce their carbon footprint in their home territory are also worth reviewing. Considering the limited regulations in which countries have addressed this labor issue, we would expect more scholars, experts, and researchers to comprehensively review neo-nomadism to explore and estimate how this lifestyle impacts green transition and sustainability, given the lack of studies that have addressed the environmental and social balance implications of these neo-nomads.

2 Framework

2.1 Nomadic Lifestyle

The lifestyle of the digital nomad implies constant movement, both between countries and between workspaces (Richter & Richter, 2020). Research has focused on the particularities of physical spaces or the characteristics of their actions and motivations to link work and leisure (Reichenberger, 2018). For example, studies have found that digital nomads naturally have virtual space, but they also require an appropriate physical space (Lowry & Moskos, 2005) that must fulfill certain conditions to effectively carry out their work (Cook, 2020). It is unclear what constitutes the physical environments of the nomadic spaces, although most of these are a mixture of work, leisure, social activities, and collaborative work (Orel, 2020).

It is known that digital nomads want to be acknowledged as professionals by their partners or families because of the lifestyle they lead, so they accommodate physical spaces to facilitate their work (Nash et al., 2021). Some nomads prefer places where they can separate their professional and personal lives to increase their productivity, while others prefer locations that allow them to combine leisure and work (Lee et al., 2019). In this regard, Sánchez-Vergara et al. (2023) affirm that digital nomads are not tourists who are motivated by the novelty of a destination but rather a new work culture.

The price of natural gas in the market has historically been volatile, and it has been exacerbated since 2022 following the Russia–Ukraine conflict (Chen et al., 2023; Kröger et al., 2023). This crisis further demonstrates the inequality of spending with respect to income, with individuals who own or rent homes with less thermal protection being the most affected (Kröger et al., 2023). Stemming from an economic issue or energy awareness, some digital nomads prefer to move to more warm places. There are still no scientific articles that refer to these energy nomads.

2.2 Sustainability

Brundtland (1987) established that sustainability entails living standards, which implies the satisfaction of human needs and levels of consumption within ecologically acceptable limits and to which all people could aspire. Since then, many studies have been trying to redefine and conceptualize sustainability. Holden et al. (2014) agree that it is about meeting social needs limited by environmental possibilities. For years, the economy was seen, in a very limited view, as the mechanism to cover social needs. However, rapid growth, inequalities, irrational use of resources, and high energy consumption affect sustainable development (Kostakis et al., 2023).

Nowadays, reflecting over sustainability should lead to constant scrutiny of Sustainable Development Goals (SDGs) (Kostakis et al., 2023) because these goals were set with the solemn purpose of empowering all people to satisfy their basic needs, and becoming knowledgeable about sustainable lifestyles. Therefore, these objectives detail the need to reduce poverty and hunger, achieve gender equality, promote health and well-being for all, promote human rights, understand cultural diversity and peace, promote responsible production and consumerism, protect, and rationally use natural and environmental resources (United Nations, 2015).

2.3 Sustainability Associated with Digital Nomads

It is known that to compensate for consumer behavior, digital nomads try to carry out sustainable activities and contribute to reducing the ecological footprint by not using heating in winter, avoiding the purchase of counterfeit products, vehicles, and electronic books, or choosing natural or recyclable wrappers (Schwarz, 2021). On the other hand, air transport and the need for innovative technology are examples of consumerism associated with digital nomads that threaten sustainability (Orel, 2020; Schwarz, 2021).

Also, researchers have reported that digital nomads prefer local consumption, which contributes to local economic dynamics (Schwarz, 2021). In addition, it is increasingly common that environmental initiatives, sustainable development, and collaborative activities are considered in the local communities where digital nomads decide to settle (Schlosberg & Coles, 2016; Schwarz, 2021). Other studies have revealed that people who travel for tourism have a vision of environmental sustainability (Agyeiwaah et al., 2020) although they have unsustainable behaviors (Agyeiwaah et al., 2021). Therefore, it is important to review in depth whether this is a shared characteristic by nomads, especially digital ones, who may be less mobile and whose main reason for traveling digresses from tourism.

The growing tendency of more and more individuals to become digital nomads and energy nomads responds to a way of balancing good living with work despite limited regulations. This lifestyle has its own mobility dynamics shaped by either migratory obligations (so as not to exceed permitted tourism or visa stays) or personal preferences (deciding to explore new paths and cultures). These mobility dynamics in temporary spaces and the inherent socio-environmental implications represent a research gap that needs to be measured. Thus, the purpose of this study is to provide a more comprehensive study to explore neo-nomadism and sustainability.

3 Methodology

A systematic literature review was selected as it is frequently used because it reduces bias and subjectivity in the selection of items, it allows for the presentation of a synthesis of the results of research conducted in a given period, it aids in the detection of research gaps or contradictory findings, which may suggest future research, and it may finally provide suggestions on new research criteria (Abarca et al., 2020; Palos-Sánchez et al., 2022; Ramey & Rao, 2011; Rao & Ramey, 2011).

Three questions were posed for this research to learn about the relationship between neo-nomads and socioeconomic sustainability, as well as the aspects discovered on environmental sustainability that involve the practice of digital neo-nomadism, and finally, the policies and regulations issued by countries that may have implications on the sustainability of neo-nomads.

3.1 Search Strategy

Systematic literature reviews are frequently used because they reduce bias and subjectivity in the selection of items, they allow for the presentation of a synthesis of the results of research conducted in a given period, they facilitate the detection of research gaps or contradictory findings, which may suggest future research, and they can finally shed light on new research criteria. The databases and search strings used are listed in the subsequent study selection section, along with the criteria used to select the documents and determine which studies were included. This study's methodology is divided into three stages: selecting articles, identifying them, and finally synthesizing them (Thomas, 2017; Zhao et al., 2021).

This chapter is based on a bibliographic review of academic articles from the Web of Science and SCOPUS databases. The justification for using articles from Web of Science-indexed journals with quality indexes (JCR) and a sufficient period of time (Durán-Sánchez et al., 2019; Rojas-Sánchez et al., 2023). The Scopus database is used because it is multidisciplinary (Chadegani et al., 2013), has journal articles classified in different fields (Hassan et al., 2019), has quality indexes (SJR), and covers 20% more articles than Web of Science (Falagas et al., 2008; Lezcano-Calderón, 2023; Rojas-Sánchez et al., 2023).

3.2 Study Selection

In this systematic review, search strings were used in the article's title, keywords, and abstract. For the keywords, various ways of referring to the phenomenon of digital nomadism are used, specifically the words "digital nomad*" OR "neo-nomad*" OR "neo nomad*" OR "nomad* work*" OR "knowmad*", and in conjunction with words derived from the three pillars of sustainability "sustainability" OR "sustainable development" OR "economic development" OR "economic growth" OR "social development" OR "environmental*" and in addition to other more specific words derived from the Sustainable Development Goals "gender" OR "hunger" OR water OR energy* OR poverty OR climate* OR food. This preliminary search yielded 28 and 24 papers, respectively, in SCOPUS and Web of Science.

The search period is determined by the date of publication of sustainability reports, with reports ranging from 2012 to 2023. The Asset Owners Disclosure Project's global climate change risk survey reports date from 2016, while The Sustainable Development Goals: Agenda 2030 were published in 2015. Meanwhile, the Institutional Investors Group on Climate Change issued its investor expectations reports in 2014, and two sustainability frameworks were published in 2013: The International Integrated Reporting Framework by the International Integrated Reporting Council and The Climate Disclosure Standards Board's 2012 Climate Change Reporting Framework is the oldest.

Because the topic of sustainability is so broad and must be approached holistically, we filtered our search to the following fields of study: social science, computer science, engineering, arts and humanities, business, management and accounting, decision sciences, environmental science, earth and planetary science, and energy; however, this did not reduce the number of documents in SCOPUS or WoS.

Furthermore, the search was restricted to English-language documents in the SCOPUS and WoS databases. Finally, to make the study more rigorous, the search was limited to scientific articles. Using these criteria, the number of articles in SCOPUS and Web of Science was reduced to 11 and 12, respectively. Furthermore, eight articles that appeared in both databases were eliminated, and a manual screening revealed that one of them was unrelated to the topic. As a result, 14 articles were included for review in this study.

4 Results

The first results that appear are the scarce research that has been carried out to date. Figure 1 shows that as of 2020 there is a significant increase, but still scarce.

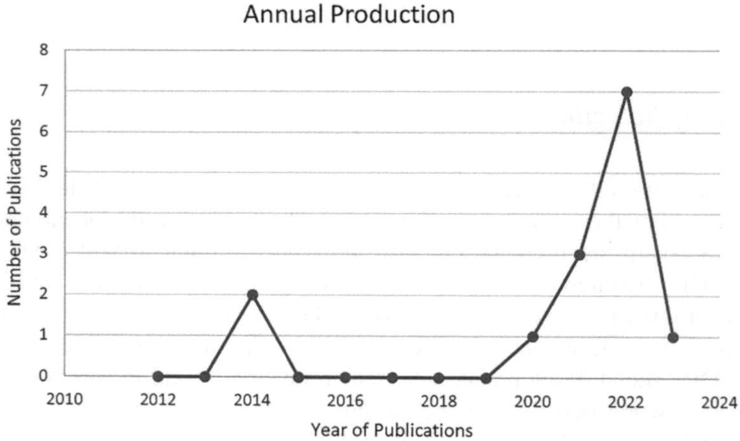

Fig. 1 Annual productions. Own elaboration, 2023

Considering that sustainability implies a balance between economic, social, and environmental factors and that 17 SDGs overlap in these areas, in order to answer the research questions posed, this review has been divided into socioeconomic and environmental aspects. The last part of the review shows the results of the policies found thus far that have implications for the sustainability of neo-nomads.

4.1 Neo-nomadism and Socio-Economic Sustainability

Our research shows that there are no studies that aim to comprehensively investigate the socioeconomic sustainability of neo-nomadic practices, albeit some studies have reviewed some characteristics of neo-nomads that are linked to one or more objectives of sustainable development. For example, during the review, we found some studies related to labor and economic growth. Mourato et al. (2023) reveal that the social responsibility and environmental concern of digital nomads impact positively with entrepreneurship, social entrepreneurial self-efficacy, and innovation. Furthermore, this study showed that the entrepreneurial attitude of digital nomads is positively associated with social entrepreneurial self-efficacy. Finally, it was evidenced that entrepreneurial attitude mediates the relationship between social responsibility and entrepreneurial attitude. In fact, Czarniawska (2014) posited that digital nomads can be both employees and entrepreneurs.

On the other hand, Czarniawska (2014) highlights that digital nomadism offers the possibility of developing work for companies located in different countries precisely because they are carried out remotely. In addition, digital nomadism is an opportunity for globalization and diversification of work. Sánchez-Vergara et al. (2023) show how digital nomads face challenges in maintaining certifications and courses, although not necessarily a formal university degree. However, Czarniawska (2014) shows that working as a digital nomad can mean job instability. Regarding labor issues, Delli Paoli and Maddaloni (2021) found that digital nomadism in Athens cannot be considered a mechanism of upward social mobility for privileged individuals with high status. Instead, it only allows them to sustain their middle-class status while navigating tourism, leisure, and work. Conversely, Parreño-Castellano et al. (2022) reported that the mobility of digital nomads is largely based on the large income differences between countries, and these practices sometimes imply that people lower their labor rates and accept precarious working conditions, leading to downward social mobility. This is often accentuated because the operating costs of companies with on-site workers are transferred to nomad workers, and this is why many times these neo-nomads look for places where their currency or salary can appreciate slightly, to reduce precariousness, which according to Aroles et al. (2022) in their study on the meta-work of digital nomads shows that this precarization occurs mainly at the beginning, when they are attracted by the lifestyle, only to realize that few people receive the promises associated with this phenomenon. Humphry (2014) also reflects that some digital nomad workers have longer work hours compared to regular work. To elaborate, arranging workspace for nomadic work takes more time and generates additional pressure (Aroles et al., 2022; Humphry, 2014), besides the burden of always being available (Aroles et al., 2022).

We found some studies that uncover data related to sustainable development objectives linked to gender equality and reduction of inequalities. Parreño-Castellano et al. (2022) in their study address the features of digital nomads and the policies implemented in Spain during the pandemic, finding that three types of digital nomads can be distinguished. The first are those with higher mobility, mostly young people who view mobility as a way of life. Another type refers to nomadic families, especially couples, although the number of families is increasing, who are forced to adjust their movements according to the school year or vacation periods. The other type of nomads are the corporate and self-employed, whom companies encourage to move for long periods. Likewise, Czarniawska (2014) affirms that this is a job opportunity that grants freedom of movement, especially to young people who are single and have not yet established a family (Delli Paoli & Maddaloni, 2021).

Aroles et al. (2022) show that, for women, the chances of finding safe and friendly places are lower, decreasing the probability of truly achieving a digital nomad lifestyle. Humphry (2014) agrees with other studies that find that mobility practices associated with neo-nomads are linked to gender inequality and discrimination. Consequently, Parreño-Castellano et al. (2022) mentioned that ideal places for digital nomads must ensure good connectivity, good quality of life and work conditions, among other factors such as climate, air quality, safety, tolerance toward women, foreigners and LGBTQ, quality of education and healthcare, cultural resources, daily use of English, low cost of living, outdoor leisure opportunities, and ease of walking mobility (Parreño-Castellano et al., 2022).

The results show that there are few studies related to the objectives of economic development and sustainable communities. For example, the study by Domínguez-Mujica et al. (2022), which analyzes mobility and immobility during the pandemic in the Canary Islands, considers digital nomads as irregular immigrants, which reflects that they move with tourist pretexts when they really intend to settle in a place to develop remote work. In a study on challenges and opportunities for four small developing island states, Foley et al. (2022) show that some island states viewed post-COVID visas for extended stays for different people, including those who may be nomadic professionals, as an opportunity to contribute to economic dynamics. Parreño-Castellano et al. (2022) also show that, during the pandemic, Spanish island territories implemented measures to promote digital nomadism as a way of mitigating a decrease in tourism and boosting the local economy.

Delli Paoli and Maddaloni (2021) study the motives of Italians who migrate to Athens, Greece, and find that there are five types of emigrants, one of them being digital nomads, who intend to move continuously between places and stay for short periods, mixing tourism and work. Likewise, Cruz et al. (2021) show that rethinking the tourism and hospitality sector to provide coworking spaces for nomadic workers is a way to contribute to economic dynamics. Mourato et al. (2023) suggest that we should take advantage of the fact that digital nomads care about the local community and that, naturally, they generate profits in local businesses and revitalize the economy. In sum, when trying to explore the fluid identity of digital diasporas, neo-nomadism re-establishes connections between people who share human, social, and life links regardless of territory, facilitating adaptability and openness toward traditions and cultures (Ajeesh & Rukmini, 2022).

4.2 Neo-nomadism and Environmental Sustainability

Regarding the environmental sustainability associated with the neo-nomad lifestyle, there are fewer studies that address these issues or whose results inform about practices that include sustainable development goals related to resilient, sustainable infrastructure and innovation, sustainable management of water and sanitation, affordable, reliable, sustainable and modern energy, sustainable consumption and production models, climate change and its effects, sustainable use of oceans, seas and marine resources, and sustainable terrestrial ecosystems.

The results show that only five studies have reported findings on the environmental sustainability associated with neo-nomadism. For example, de Carvalho et al. (2022) show that IT infrastructure can facilitate or hinder the work of nomads and be a relevant factor in overcoming emerging difficulties inherent to their work. Cruz et al. (2021) focus on studying the role of furniture in coworking spaces used by digital nomads. This research shows that both design, ergonomics/comfort, flexibility/form, and materials/sustainability are fundamental for these workspaces; in this case, sustainability is associated with the life cycle of products and the choice of their materials. On a personal level, Mourato et al. (2023) show that digital nomads are concerned about the products they consume and that they are concerned about not harming the local environment.

Keseru et al. (2021), in their study on stakeholder preferences for the future of transportation in Europe, present several scenarios, including one on digital nomads. This scenario considers that there are no limits between people's private and work life because they are always connected and available. Finally, Keseru et al. (2021) consider that the future of transportation will be a minimum carbon scenario that reduces demand and use of the services as a way of minimizing carbon emissions.

Foley et al. (2022) warn that attracting digital nomads to small islands can negatively affect the environment, as it involves improving digital infrastructure, electricity, telecommunications, water plants, and sewage, and the latter can mean a demand for water, a scarce resource in these territories.

4.3 Policies and Regulations that May Have Implications for the Sustainability of Neo-nomadism

Sánchez-Vergara et al. (2023) explore how 24 countries are developing policies, supporting business development, and positioning themselves as ideal territories to attract digital nomads. They noted that the hotel sector has begun to transform its business for digital nomads, reflected in the rise of local investment, development, and infrastructure for the community. This study shows that businesses have organized information and improved marketing strategies to target digital nomads and inform them about the different services they can find in the community, while also offering attractive sanitation and security infrastructure for these types of workers so that they value longer stays. This is important to contribute to Goals 6, 9, and 11 of the SDGs (Sustainable Development Goals).

Sánchez-Vergara et al. (2023) show that digital nomads choose a destination considering technological, economic, sustainability, and regulatory policy conditions. This study also shows that the time a digital nomad stays in a country may depend on the time the visa grants, which in some nations can be extended up to 3 years but also be limited to only 5 months, in the case of countries that have at least tried to start regulating this phenomenon. However, the study also shows that there are nomads who, due to the lack of regulation, use other types of visas, some of which allow them to stay a little over 90 days. Implicitly, the allowed period of stay could be associated with the obligation of mobility between countries and, therefore, transportation. This result shows that visa periods may or may not contribute to the achievement of Goal 11 of the SDGs. Sánchez-Vergara et al. (2023) add that obtaining digital nomad visas has different degrees of difficulty depending on the country of origin, predominating classism or privileged nationalities. The study by Aroles et al. (2022) highlights once again how country of origin reflects inequalities, especially when it comes to accessing visas. These results show a contradiction to Goal 10 of the SDGs.

Meanwhile, Keseru et al. (2021) propose a scenario that requires greater regulation of companies in the transport and mobility sector. Given that this sector is characterized by unlimited consumption, individualism, global citizens, and greater flexibility, there is a need for cooperation between the public and private sectors to increase efficiency and reduce carbon emissions. This study ultimately suggests combining these scenarios, regulations of the sectors, and public/private cooperation, to meet the expectations of stakeholders.

5 Discussion and Conclusions

The literature review carried out in this study shows that despite the fact that neo-nomadism has had an important increase, especially during and after the pandemic, the research around this phenomenon has augmented at a lower rate, and the studies concerning the sustainability implied in neo-nomadism practices are limited.

This study shows that there is a contradiction in the contribution of neo-nomadism to work and poverty in terms of SDGs, given that authors state that these people can opt for being entrepreneurs (Czarniawska, 2014; Mourato et al., 2023), employees (Czarniawska, 2014), workforce for companies worldwide thanks to globalization (Czarniawska, 2014; Sánchez-Vergara et al., 2023), yet face work instability (Czarniawska, 2014) and precariousness (Parreño-Castellano et al., 2022). Moreover, Delli Paoli and Maddaloni (2021) affirm that having a life-style mixing work with tourism and leisure does not represent an ascending social mobility mechanism for neo-nomads. More importantly, the study by Parreño-Castellano et al. (2022) sustain that this phenomenon generates descending social mobility, although digital nomads move to more economically viable sites to mitigate precariousness (Aroles et al., 2022).

In the same vein, the study reflects that youngsters are the ones who take advantage of digital nomadism the most, particularly those who are single (Czarniawska, 2014; Delli Paoli & Maddaloni, 2021; Parreño-Castellano et al., 2022), and have small families (Parreño-Castellano et al., 2022). As a result, these work practices contribute to the fulfillment of youth employment, a concern in the SDGs.

In terms of the SDGs that aim at gender equality, our study shows that mobility related to digital nomadism discriminates against women, for there are fewer safe and friendly spaces for them, hindering their freedom to move (Aroles et al., 2022; Humphry, 2014). For this reason, Parreño-Castellano et al. (2022) suggest that it is important that the places willing to accept these people have a good environment and be safe for women and other minority groups, which would also contribute to SDG Goal 11 on safe, inclusive, and sustainable communities.

Likewise, both Humphry (2014) and Aroles et al. (2022) show that nomadism implies high workloads, which include the need of being always available and conditioning the new dwelling places, a topic that companies may not be contemplating and that can affect their health and wellbeing, which may be contrary to what the SDGs are aiming for.

The studies analyzed in our research depict that digital nomads have a behavior similar to regular tourists, albeit there are also irregular immigrants. For this reason, during the pandemic, some insular territories took advantage of the digital nomadism phenomenon to mitigate the effects of the world immobility that restricts the transit of tourists and, in this way, somehow balance the economies in these places (Domínguez-Mujica et al., 2022; Foley et al., 2022; Parreño-Castellano et al., 2022). Cruz et al. (2021) show that tourism and the hotel industry should condition spaces that serve for the digital nomads' work dynamics, and take advantage of the fact that they combine work activities with tourism (Delli Paoli & Maddaloni, 2021). This shows that the phenomenon of digital nomadism can contribute to the goal related to economic growth. However, Foley et al. (2022) admonished that investment in infrastructure can also affect the environment and diminish available natural resources in the case of small islands. It is significant that these studies ensure that the most competitive tourist destinations integrate strategies for sustainability alongside the services offered (Araya et al., 2023; Ozkaya & Demirhan, 2022; Shen et al., 2023). It is important to note that while Goal 10 of the SDGs aims to empower island states, the practices of neo-nomads may undermine the goals related to environmental issues.

Considering that economic growth increases carbon dioxide emissions (CO_2) (Adebayo et al., 2023) and that the main SDGs seek to guarantee environmental protection taking into account development implies that the community participation and social benefits (Yang et al., 2023), as well as low carbon tourism, has become part of the world agenda, which can be used to mitigate the impacts of climate change caused by tourist destinations (Fakfare & Wattanacharoensil, 2023). The infrastructure and the sustainable services, the subsistence means, and the management of tourist places are critical areas (Yang et al., 2023). Low carbon tourism and, therefore, lodging, transportation, and tourism activities providers should work hand in hand with a sustainable administration in the destination, and these crucial factors must be considered in the development of the insular destination (Fakfare & Wattanacharoensil, 2023). Our review shows that the effects of digital nomadic practices on climate change and the low-carbon economy have not been studied. We therefore suggest that these elements be considered in island territories, as well as in other places that receive or plan to receive neo-nomads, for example, coastal sites, rural places, ecologically fragile spaces, and, in general, aspiring sustainable cities and communities.

Our research found that only one study explicitly shows the sensibilization of the products consumed and neo-nomads' environmental awareness (Mourato et al., 2023) and that the social responsibility of this population influences eco-innovation, i.e., the new ways to do business or production processes that contemplate diminishing the environmental risk of pollution and the environmental impact in general in the long run (He et al., 2018).

Moreover, the constant in neo-nomads is mobility, which can confound them with tourists. Nevertheless, although neo-nomads engage in tourism activities or tourism-like activities, our study reveals that there is a gap in the awareness of their mobility and sustainability practices. Yang et al. (2023) suggest that a high level of satisfaction with clients should be kept, creating awareness about the problems regarding sustainability and spreading sustainable tourist practices among them. We conclude that, added to sustainability, it is important to further review the consumption and mobility practices of neo-nomads.

Ajeesh and Rukmini (2022) show that one of the qualities of neo-nomads is that, thanks to the mobility and independence of the territory, these people normally adapt, and accept traditions and different cultures more easily, which can be maximized by the territories. Studies show that the actions, practices, and organizational policies that, through programs, help to improve the local community and its employees economically are part of social responsibility (Jamali et al., 2015). Sánchez-Vergara et al. (2023) affirm that it is essential to promote social relationships, exchanges, and the well-being of nomads, coinciding with Frochot and Batat (2013) in that this contributes to improve the sense of belonging and the connection with the stakeholder and the environment.

Our study shows that the infrastructure is relevant for the neo-nomads, for it can facilitate the labor carried out (Cruz et al., 2021; de Carvalho et al., 2022). Nonetheless, it was not possible to find elements evidencing a sustainability conscience in the establishment of the infrastructure used by the neo-nomads. Despite indications that the sustainability of the building industry and construction is crucial to contribute with the global transition toward a sustainably constructed carbon-neutral environment (Scrucca et al., 2023). Studies also indicate that, historically, the estate sector has been slow to innovate and respond to the pressures of climate change, although recently the use of low-carbon materials and designs has been promoted (Wainwright & Demirel, 2023). Likewise, recent research studies suggest that policy makers, administrators, and even researchers are more and more worried about trying to reach infrastructure, industry, and sustainable innovation objectives according to the SDGs. In this way, the digital gap is closed in the hopes to warrant justice and digital equality.

Wainwright and Demirel's (2023) study and others reflect the importance of the spaces for coworking and coliving for the development of neu-nomadism. However, the literature has not revealed studies dealing with the importance of considering the example of energetic flexibility of the buildings destined to these purposes, aimed at reducing electric energy costs (Patrício et al., 2023). The latter is not limited to a merely economic topic, but to a minimization of environmental impact (Sarabia-Escriva et al., 2023). Considering noncarbohydrate energy sources that produce few or zero CO_2 emissions is an assessment to be carried out (Iqbal et al., 2022). Besides providing more clean energies to the industrial and infrastructure sector, this should be reflected in reforms to the industrial and technological policy to achieve sustainability, including rural areas. Because of this, Adebayo et al. (2023) suggest critical ideas to change fossil energies for renewable energies to achieve the SDGs: those related to accessible, clean energies, and climate-oriented actions. In Europe, the debate over energetic security and the implications of depending on fossil fuels has been reactivated, with scholars/experts calling on establishing policies that guarantee a minimum economic protection (Kröger et al., 2023). For this reason, this study suggests reflecting on affordable nonpolluting energies to permeate corporate activities and digital nomadism into other nations. Also, it is important to analyze if, in the case of energetic nomads, there is a compensation when moving the activities to a country with a lower economic and environmental cost, e.g., moving into a country with high energy costs or less clean energy instead of the new country where they are settled.

Our research shows that the study by Keseru et al. (2021) is the only one that states the topic of transportation and neo-nomads. It suggests considering the decrease in the demand and the use of means of transport for the reduction of carbon emissions in future scenarios. The scarce literature in this regard is concerning due to the fact that, as mentioned by Liao (2023), both the petroleum prices and the geopolitical conflicts over renewable energies have an effect on rising costs, and this can be crucial for companies and the performance of neo-nomads' business. It is also important to remember that urban mobility should be planned with a transition toward a low-carbon-based economy and ecologic transformation. Furthermore, aviation has an impact on climate change due to the increase of CO_2 (Puliafito, 2023). Therefore, being a means of transportation used by neo-nomads, we suggest that enterprises in this sector make efforts to mitigate their carbon footprint and that neo-nomads select companies that implement courses of action for environmental compensation. The "Zero Total Emissions for 2050" by the International Energy Agency (IEA, 2021) and the Conference of the United Nations about Climate Change (COP27) have boosted that enterprises seek to reduce greenhouse gas emissions with the purpose of fulfilling the objectives of The Paris Agreement (Liao, 2023). However, our results show that these issues regarding the practices of neo-nomads and the companies or territories that promote this phenomenon have not yet been investigated.

Lastly, more and more countries seek to attract neo-nomads because they are seen as an option of economic development (Sánchez-Vergara et al., 2023) and, thus, they have established policies and regulations pertaining to the characteristics of these people, their economic activities, and their lengths of stay. Also, this same study highlights that there are better-organized places that have implemented strategies to host clusters of neo-nomads. However, the analysis of these policies revealed that the conditions to opt for these visas can be discriminatory per se and augment inequalities (Aroles et al., 2022; Sánchez-Vergara et al., 2023).

In the same vein, the length of the visa conditions mobility; hence, countries with shorter visa lengths force neo-nomads to move more frequently, and thus this could have major implications in the carbon footprint. However, the absence of this type of regulation can overlap as neo-nomads make their ways into the countries through irregular immigration (Delli Paoli & Maddaloni, 2021), with business or work visas (Sánchez-Vergara et al., 2023), or as regular tourists.

Our study has practical implications, since tourism companies should consider the location, mobilization, and consumption habits of neo-nomads and carry out actions aimed at mitigating the environmental footprint and increasing socio-economic impact. In addition, companies that hire neo-nomadic workers should consider, within their social responsibility, actions to raise awareness and improve sustainability practices.

Companies in the transport sector should take into account the mobility practices of neo-nomads, review the mitigation strategies they endorse, and even show the data and environmental compensation strategies so that digital nomads may more aptly assess criteria for choosing a transport company.

Stemming from our research findings, territories can also review their visa policies for nomads, the periods and criteria of the countries from which they come. They may also establish policies to promote the attraction of neo-nomads to territories that have organized themselves with the establishment of coliving spaces, coworking, and environmentally sustainable infrastructure. Lastly, this study may encourage hotels to make the necessary adaptations to receive neo-nomads, and to contemplate sustainability practices. On the other hand, public administrators should encourage communities that offer conditions for mixing leisure and work to develop sustainable actions.

This research points out that there are territories that have already begun to regulate neo-nomadism in terms of migration and others that have not, and that for the most part have seen neo-nomads as an opportunity to contribute to the economy of the countries. However, it is not yet evident as an option to contribute to such social development and in balance with the environment, while guaranteeing better conditions for the future. Policymakers and institutions can promote better conditions for nomads settling in their territories, for example, security, health and welfare, and inclusive conditions for women. It is also relevant that territories should be concerned about enabling neo-nomads to link with the local community (Chevtaeva & Denizci-Guillet, 2021; Mourato et al., 2023).

Our study has theoretical implications because other researchers can fill the gaps in the literature and study other lines of research that allow them to better understand the practices of neo-nomads which are intrinsically linked to sustainability. Another implication is studying the actions of companies that allow these types of labor contracts and the social responsibility linked to the sustainability of the work of their employees. Finally, other researchers can take this study as a starting point to explore the policies of countries and territories to attract neo-nomads through sustainable practices.

References

Abarca, V. M. G., Palos-Sanchez, P. R., & Rus-Arias, E. (2020). Working in virtual teams: A systematic literature review and a bibliometric analysis. *IEEE Access, 8*, 168923–168940. https://doi.org/10.1109/ACCESS.2020.3023546

Adebayo, T. S., Kartal, M. T., & Ullah, S. (2023). Role of hydroelectricity and natural gas consumption on environmental sustainability in the United States: Evidence from novel time-frequency approaches. *Journal of Environmental Management, 328*, 116987. https://doi.org/10.1016/j.jenvman.2022.116987

Agyeiwaah, E., Pratt, S., Iaquinto, B. L., & Suntikul, W. (2020). Social identity positively impacts sustainable behaviors of backpackers. *Tourism Geographies, 25*(3), 1–22. https://doi.org/10.1080/14616688.2020.1819401

Agyeiwaah, E., Dayour, F., Otoo, F. E., & Goh, B. (2021). Understanding backpacker sustainable behavior using the tri-component attitude model. *Journal of Sustainable Tourism, 29*(7), 1193–1214. https://doi.org/10.1080/09669582.2021.1875476

Ajeesh, A. K., & Rukmini, S. (2022). Transnational voices in the digital age: Exploring the affordances of identity in Pico Iyer's the global soul. *The International Journal of Literary Humanities, 20*(2), 97–106. https://doi.org/10.18848/2327-7912/CGP/v20i02/97-106

Araya, J. D., Hernando, A., Tejera, R., & Velázquez, J. (2023). Sustainable tourism around ecosystem services: Application to a case in Costa Rica using multi-criteria methods. *Land, 12*(3), 628. https://doi.org/10.3390/land12030628

Aroles, J., Granter, E., & de Vaujany, F. (2020). 'Becoming mainstream': The professionalisation and corporatisation of digital nomadism. *New Technology, Work and Employment, 35*(1), 114–129. https://doi.org/10.1111/ntwe.12158

Aroles, J., Bonneau, C., & Bhankaraully, S. (2022). Conceptualising 'meta-work' in the context of continuous, global mobility: The case of digital nomadism. *Work, Employment and Society*. https://doi.org/10.1177/09500170211069797

Brundtland, G. H. (1987). *Report of the world commission on environment and development: "Our common future"*. UN.

Chadegani, A. A., Salehi, H., Yunus, M. M., Farhadi, H., Fooladi, M., Farhadi, M., & Ebrahim, N. A. (2013). *A comparison between two main academic literature collections: Web of science and scopus databases*. https://doi.org/10.48550/ARXIV.1305.0377

Chen, Y., Wei, Y., Bai, L., & Zhang, J. (2023). Can green economy stocks hedge natural gas market risk? Evidence during Russia-Ukraine conflict and other crisis periods. *Finance Research Letters, 53*, 103632. https://doi.org/10.1016/j.frl.2023.103632

Chevtaeva, E., & Denizci-Guillet, B. (2021). Digital nomads' lifestyles and coworkation. *Journal of Destination Marketing & Management, 21*, 100633. https://doi.org/10.1016/j.jdmm.2021.100633

Cook, D. (2020). The freedom trap: Digital nomads and the use of disciplining practices to manage work/leisure boundaries. *Information Technology & Tourism, 22*(3), 355–390. https://doi.org/10.1007/s40558-020-00172-4

Cruz, R., Franqueira, T., & Pombo, F. (2021). Furniture as feature in coworking spaces. Spots in Oporto City as case study. *Res Mobilis: Revista Internacional de Investigación En Mobiliario y Objetos Decorativos, 10*(13), 316–338. https://doi.org/10.17811/rm.10.13-3.2021.316-338

Czarniawska, B. (2014). Nomadic work as life-story plot. *Computer Supported Cooperative Work, 23*(2), 205–221. https://doi.org/10.1007/s10606-013-9189-3

de Carvalho, A. F. P., Saeed, S., Reuter, C., Rohde, M., Randall, D., Pipek, V., & Wulf, V. (2022). Understanding nomadic practices of social activist networks through the lens of infrastructuring: The case of the European social forum. *Computer Supported Cooperative Work: CSCW: An International Journal, 31*(4), 731–769. https://doi.org/10.1007/s10606-022-09442-7

Delli Paoli, A., & Maddaloni, D. (2021). Conceptualizing motives for migration: A typology of Italian migrants in the Athens area. *Journal of International Migration and Integration, 22*(4), 1465–1484. https://doi.org/10.1007/s12134-021-00818-0

Domínguez-Mujica, J., Parreño-Castellano, J. M., & Moreno-Medina, C. (2022). Pandemic mobility/(im)mobility in the Canary Islands: Irregular migrants becoming hotel guests. *Belgeo, 3*. https://doi.org/10.4000/belgeo.56022

Durán-Sánchez, A., Álvarez-García, J., de la Cruz del Río-Rama, M., & Rosado-Cebrián, B. (2019). Science mapping of the knowledge base on tourism innovation. *Sustainability, 11*(12), 3352. https://doi.org/10.3390/su11123352

Fakfare, P., & Wattanacharoensil, W. (2023). Low-carbon tourism for Island destinations: A crucial alternative for sustainable development. *Sustainable Development, 31*(1), 180–197. https://doi.org/10.1002/sd.2382

Falagas, M. E., Pitsouni, E. I., Malietzis, G. A., & Pappas, G. (2008). Comparison of PubMed, Scopus, Web of Science, and Google Scholar: Strengths and weaknesses. *The FASEB Journal, 22*(2), 338–342. https://doi.org/10.1096/fj.07-9492LSF

Foley, A. M., Moncada, S., Mycoo, M., Nunn, P., Tandrayen-Ragoobur, V., & Evans, C. (2022). Small Island developing states in a post-pandemic world: Challenges and opportunities for climate action. *Wiley Interdisciplinary Reviews: Climate Change, 13*(3), e769. https://doi.org/10.1002/wcc.769

Frochot, I., & Batat, W. (2013). *Marketing and designing the tourist experience*. Goodfellow Publishers.

Hannonen, O. (2018). Second home owners as tourism trend-setters: A case of residential tourists in Gran Canaria. *Journal of Spatial and Organizational Dynamics, 6*(4), 345–359.

Hassan, S.-U., Visvizi, A., & Waheed, H. (2019). The 'who' and the 'what' in international migration research: Data-driven analysis of Scopus-indexed scientific literature. *Behaviour & Information Technology, 38*(9), 924–939. https://doi.org/10.1080/0144929X.2019.1583282

He, F., Miao, X., Wong, C. W. Y., & Lee, S. (2018). Contemporary corporate eco-innovation research: A systematic review. *Journal of Cleaner Production, 174*, 502–526. https://doi.org/10.1016/j.jclepro.2017.10.314

Holden, E., Linnerud, K., & Banister, D. (2014). Sustainable development: Our common future revisited. *Global Environmental Change, 26*, 130–139. https://doi.org/10.1016/j.gloenvcha.2014.04.006

Humphry, J. (2014). Officing: Mediating time and the professional self in the support of nomadic work. *Computer Supported Cooperative Work, 23*(2), 185–204. https://doi.org/10.1007/s10606-013-9197-3

IEA (2021). Net Zero by 2050, IEA, Paris. https://www.iea.org/reports/net-zero-by-2050, Licence: CC BY 4.0

Iqbal, M., Chand, S., & Ul Haq, Z. (2022). Economic policy uncertainty and CO_2 emissions: A comparative analysis of developed and developing nations. *Environmental Science and Pollution Research, 30*(6), 15034–15043. https://doi.org/10.1007/s11356-022-23115-4

Jamali, D. R., El Dirani, A. M., & Harwood, I. A. (2015). Exploring human resource management roles in corporate social responsibility: The CSR-HRM co-creation model. *Business Ethics: A European Review, 24*(2), 125–143. https://doi.org/10.1111/beer.12085

Keseru, I., Coosemans, T., & Macharis, C. (2021). Stakeholders' preferences for the future of transport in Europe: Participatory evaluation of scenarios combining scenario planning and the multi-actor multi-criteria analysis. *Futures, 127*, 102690. https://doi.org/10.1016/j.futures.2020.102690

Kostakis, I., Paparas, D., & Tsagarakis, K. P. (2023). Disaggregated energy use and socioeconomic sustainability within OECD countries. *Journal of Environmental Management, 334*, 117475. https://doi.org/10.1016/j.jenvman.2023.117475

Kröger, M., Longmuir, M., Neuhoff, K., & Schütze, F. (2023). The price of natural gas dependency: Price shocks, inequality, and public policy. *Energy Policy, 175*, 113472. https://doi.org/10.1016/j.enpol.2023.113472

Lee, A., Toombs, A. L., Erickson, I., Nemer, D., Ho, Y., Jo, E., & Guo, Z. (2019). The social infrastructure of co-spaces: Home, work, and sociable places for digital nomads. *Proceedings of the ACM on Human-Computer Interaction, 3*(CSCW), 1–23. https://doi.org/10.1080/14616688.2020.1819401

Lezcano-Calderón, Y. (2023). Resources and capabilities of family SMEs: A bibliometric study. *Estudios*.

Liao, S. (2023). The Russia–Ukraine outbreak and the value of renewable energy. *Economics Letters, 225*, 111045. https://doi.org/10.1016/j.econlet.2023.111045

Lowry, D., & Moskos, M. (2005). *Hanging on the mobile phone: Experiencing work and spatial flexibility*. National Institute of Labour Studies.

Mourato, I., Dias, Á., & Pereira, L. (2023). Estimating the impact of digital nomads' sustainable responsibility on entrepreneurial self-efficacy. *Social Sciences, 12*(2), 97. https://doi.org/10.3390/socsci12020097

Müller, A. (2016). The digital nomad: Buzzword or research category? *Transnational Social Review, 6*(3), 344–348.

Nash, C., Jarrahi, M. H., & Sutherland, W. (2021). Nomadic work and location independence: The role of space in shaping the work of digital nomads. *Human Behavior and Emerging Technologies, 3*(2), 271–282. https://doi.org/10.1002/hbe2.234

Orel, M. (2019). Coworking environments and digital nomadism: Balancing work and leisure whilst on the move. *World Leisure Journal, 61*(3), 215–227. https://doi.org/10.1080/16078055.2019.1639275

Orel, M. (2020). Life is better in flip flops. Digital nomads and their transformational travels to Thailand. *International Journal of Culture, Tourism and Hospitality Research, 15*(1), 3–9. https://doi.org/10.1108/IJCTHR-12-2019-0229

Ozkaya, G., & Demirhan, A. (2022). Multi-criteria analysis of sustainable travel and tourism competitiveness in Europe and Eurasia. *Sustainability, 14*(22), 15396. https://doi.org/10.3390/su142215396

Palos-Sánchez, P. R., Baena-Luna, P., Badicu, A., & Infante-Moro, J. C. (2022). Artificial intelligence and human resources management: A bibliometric analysis. *Applied Artificial Intelligence, 36*(1), 2145631. https://doi.org/10.1080/08839514.2022.2145631

Parreño-Castellano, J., Domínguez-Mujica, J., & Moreno-Medina, C. (2022). Reflections on digital nomadism in Spain during the COVID-19 pandemic—Effect of policy and place. *Sustainability, 14*(23), 16253. https://doi.org/10.3390/su142316253

Patrício, J. T., Lopes, R. A., Majdalani, N., Aelenei, D., & Martins, J. (2023). Aggregated use of energy flexibility in office buildings. *Energies, 16*(2), 961. https://doi.org/10.3390/en16020961

Puliafito, S. E. (2023). Civil aviation emissions in Argentina. *Science of the Total Environment, 869*, 161675. https://doi.org/10.1016/j.scitotenv.2023.161675

Ramey, J., & Rao, P. G. (2011). The systematic literature review as a research genre. In *2011 IEEE International Professional Communication Conference*, pp. 1–7. https://doi.org/10.1109/IPCC.2011.6087229.

Rao, P. G., & Ramey, J. (2011). Use of mobile phones by non-literate and semi-literate people: A systematic literature review. In *2011 IEEE International Professional Communication Conference*, pp. 1–10. https://doi.org/10.1109/IPCC.2011.6087228.

Reichenberger, I. (2018). Digital nomads – A quest for holistic freedom in work and leisure. *Annals of Leisure Research, 21*(3), 364–380. https://doi.org/10.1080/11745398.2017.1358098

Richter, S., & Richter, A. (2020). Digital nomads. *Business & Information Systems Engineering, 62*(1), 77–81. https://doi.org/10.1007/s12599-019-00615-1

Rojas-Sánchez, M. A., Palos-Sánchez, P. R., & Folgado-Fernández, J. A. (2023). Systematic literature review and bibliometric analysis on virtual reality and education. *Education and Information Technologies, 28*(1), 155–192. https://doi.org/10.1007/s10639-022-11167-5

Sánchez-Vergara, J. I., Orel, M., & Capdevila, I. (2023). "Home office is the here and now." Digital nomad visa systems and remote work-focused leisure policies. *World Leisure Economic policy uncertainty and Journal, 65*(13), 1–20. https://doi.org/10.1080/16078055.2023.2165142

Sarabia-Escriva, E.-J., Soto-Francés, V.-M., Pinazo-Ojer, J.-M., & Acha, S. (2023). Economic and environmental analysis of domestic hot water systems for single-family homes. *Energy and Buildings, 286*, 112925. https://doi.org/10.1016/j.enbuild.2023.112925

Schlosberg, D., & Coles, R. (2016). The new environmentalism of everyday life: Sustainability, material flows and movements. *Contemporary Political Theory, 15*(2), 160–181. https://doi.org/10.1057/cpt.2015.34

Schwarz, S. (2021). 'Going' green? Nomadic consumers' paradoxical journey toward sustainability. *ACR North American Advances*.

Scrucca, F., Ingrao, C., Barberio, G., Matarazzo, A., & Lagioia, G. (2023). On the role of sustainable buildings in achieving the 2030 UN sustainable development goals. *Environmental Impact Assessment Review, 100*, 107069. https://doi.org/10.1016/j.eiar.2023.107069

Shen, C.-C., Liu, H.-L., & Wang, D. (2023). The influence of different factors of product attachment on Taoist tourism loyalty. *Sustainability, 15*(3), 2123. https://doi.org/10.3390/su15032123

Šímová, T. (2022). A research framework for digital nomadism: A bibliometric study. *World Leisure Journal, 65, 1–17*(2). https://doi.org/10.1080/16078055.2022.2134200

Thomas, J. (2017). *An introduction to systematic reviews* (2nd ed.). Sage.

United Nations. (2015). *Transforming our world: The 2030 agenda for sustainable development*. United Nations. Accessed from https://sdgs.un.org/2030agenda

von Zumbusch, J. S. H., & Lalicic, L. (2020). The role of co-living spaces in digital nomads' well-being. *Information Technology and Tourism, 22*(3), 439–453. https://doi.org/10.1007/s40558-020-00182-2

Wainwright, T., & Demirel, P. (2023). Multiple logics in financialisation? Moving to carbon sustainability in build-to-rent development. *Environment and Planning A: Economy and Space, 55*(1), 22–45. https://doi.org/10.1177/0308518X221123813

Yang, Y., Wani, G. A., Nagaraj, V., Haseeb, M., Sultan, S., Hossain, M. E., Kamal, M., & Shah, S. M. R. (2023). Progress in sustainable tourism research: An analysis of the comprehensive literature and future research directions. *Sustainability, 15*(3), 2755. https://doi.org/10.3390/su15032755

Zhao, Y., Pinto Llorente, A. M., & Sánchez Gómez, M. C. (2021). Digital competence in higher education research: A systematic literature review. *Computers & Education, 168*, 104212. https://doi.org/10.1016/j.compedu.2021.104212

Rolling with the Resistance to the Circular Economy? A Contribution to Sustainable Consumption and Social Engagement

Sarah De Coninck, Alexandra Anderluh, Zahra Mesbahi, Fátima N. Serralha, Maria Catarina Paz, and Nelson Carriço

Abstract Strategical documents on circular economy increasingly mention aspects of social justice, inclusion, and multistakeholder engagement. However, defining the social aspects of the circular economy has been a difficult task. Therefore, circular economy and social enterprise need to be combined, leading to the emergence of a new term: 'social circular economy'. The fundamental premise of the social circular economy is that enhancing personal well-being can go hand in hand with improving society and the environment. First, this chapter gives an overview of different views on the social circular economy, including their strong points and weaknesses. Second, this chapter explores the skills needed for the green transition. Given the increasing focus on systemic approaches to the green transition, specific skills are required from all stakeholders to adequately value our planet and take appropriate action to protect it. The third part of this chapter focuses on the ability to tolerate climate change anxiety and still being able to undertake action. Since a circular economy transition can only be accomplished by combining the material, strategic, and consumer perspectives, this chapter ends with a discussion of research results on consumer perspectives on the circular economy.

Keywords Social circular economy · Resistance · Consumer perspectives · Green skills

S. De Coninck (✉)
University College Leuven Limburg, Diepenbeek, Belgium

Katholieke Universiteit Leuven, Leuven, Belgium
e-mail: sarah.deconinck@kuleuven.be

A. Anderluh · Z. Mesbahi
St. Pölten University of Applied Sciences, Saint Pölten, Austria
e-mail: alexandra.anderluh@fhstp.ac.at; zahra.mesbahi@fhstp.ac.at

F. N. Serralha · M. C. Paz · N. Carriço
Instituto Politécnico de Setúbal, Barreiro, Portugal
e-mail: maria.serralha@estbarreiro.ips.pt; catarina.paz@estbarreiro.ips.pt; nelson.carrico@estbarreiro.ips.pt

C. Rego et al. (eds.), *Transitioning to a Circular Economy*, Studies on Entrepreneurship, Structural Change and Industrial Dynamics,
https://doi.org/10.1007/978-3-031-77661-8_4

1 Introduction

The European Green Deal aims to reach null net carbon emissions by 2050 (European Commission, 2019). In order to reach this goal, economic growth needs to be detached from resource use. In addition to these ecological goals, Europe aims to make this transition *just* and *inclusive*. The key objective is to improve the health and well-being of every citizen. Specific target areas of the European Green Deal with a clear social focus are healthy and affordable food, easily accessible, affordable and healthy public transport, the right to repair goods, sustainable jobs, and appropriate skills training. Apart from the social focus within the European Green Deal, the need for multistakeholder co-creation, including citizens, communities, and organisations, is increasingly recognised. As a result, there is a clear need for systemic approaches and multistakeholder engagement. Scaling up circularity is crucial to the Green Deal, and the Circular Economy Action Plan aims to accelerate the transition towards a circular economy (European Commission, 2020). The main idea is that the EU needs to strive for a regenerative economic model in which more is given back to the planet than is taken. To reach this goal, our consumption footprint needs to be reduced, while the circular material use rate needs to be increased. Co-creation with economic actors, consumers, citizens, civil society organisations, and governments is key to reaching the goals of the green transition. In the Circular Economy Action Plan, the focus is on multistakeholder involvement in multiple ways. First of all, the Circular Economy Action Plan aims to provide high-quality, functional, and safe products for citizens. Moreover, products are aimed to be more circular in the sense that they should last longer and are designed for reuse, repair, and high-quality recycling. Products and sustainable services should also bring about a better quality of life, innovative jobs, and upgraded knowledge and skills. Multi-stakeholder co-creation, including economic actors, consumers, citizens, and civil society organisations, is seen as key to reaching the transition to a circular economy. In sum, the Circular Economy Action Plan wants to make circularity work for people, regions, and cities.

Nonetheless, the current circular economy paradigm is insufficient to achieve a more equitable society and a thriving environment. Therefore, circular economy and social enterprises need to be combined, which leads to the emergence of a new term: 'social circular economy' (Robinson, 2018). Explicitly including the social aspect in the circular economy helps overcome the potential pitfalls of the circular economy (e.g. prioritising economic gains and as a result overlooking societal challenges) and social enterprises (e.g. prioritising social interest and as a result overlooking negative impact on the environment). Consequently, the social circular economy offers a comprehensive systemic framework. It considers the environmental principles of the circular economy, the societal benefits that social enterprises seek, and economic prosperity. The fundamental premise of the social circular economy is that enhancing personal well-being can go hand in hand with improving society and the environment. It provides an operating model ensuring economic benefits while benefitting both society and the environment.

Thus, the social circular economy emerges, which extends the traditional circular economy and emphasises the role of social enterprises and local communities. The first part of this chapter gives an overview of the different frameworks that exist related to the social circular economy. The social circular economy aims to create a more inclusive and equitable system that benefits all stakeholders. Given that multistakeholder involvement requires different skills than typical technical skills, a second part will give an overview of the current frameworks on skills needed for the transition to a circular economy, followed by the third part on how to deal with resistance. For these first three parts, a narrative literature study is conducted, including academic literature and strategic documents.

In the traditional circular economy model, the role of consumers is often overlooked. However, consumers play a crucial role in the economic system by deciding which products to buy and how to consume and dispose of them. Therefore, it is essential to consider the consumer perspective in the transition to the circular economy. For this reason, the final part of the chapter will give an overview of the results of a study on the consumer perspective on the circular economy. As such, the objectives of this chapter are (1) to clarify what social circular economy is, (2) to provide an overview of the skills needed for the transition to a circular economy, (3) to give some insight in how to deal with resistance to the circular economy, and (4) to present insight into the consumer perspective on circular economy.

2 Social Circular Economy

Despite the societal focus of the circular economy becoming increasingly apparent, defining the social aspects of the circular economy has been a difficult task. While various taxonomies exist, a common understanding of the social aspect of circular economy is lacking, and specifying how to measure social indicators remains challenging and little researched. One taxonomy by Padilla-Rivera et al. (2020) identified four thematic areas related to the social aspect of the circular economy, namely, (1) labour practices and decent work, (2) human rights, (3) society, and (4) product responsibility. Alternatively, Mies and Gold (2021) looked at social aspects appearing in key international reports on the circular economy to identify both overarching and actor-specific social dimensions of the circular economy. They identified 6 relevant actors (workers, organisations, customers, local community, society, and stakeholder collaborations), 22 overarching social aspects, and 4 actor-specific social aspects. The range of social aspects identified by the authors goes from infrastructure availability to well-being, satisfaction, social justice, and cultural heritage (Mies & Gold, 2021). Given this wide range of social factors identified, a clear view on what the social aspect of the circular economy is, is still lacking.

As mentioned before, a recent development to address the lack of emphasis on social sustainability in the circular economy is the emergence of the social circular economy. This approach combines the concepts of the circular economy and social enterprise to create a system that benefits people and planet. Social enterprises, which use business principles to achieve positive social and environmental outcomes, play a vital role in this model as they work to create a more equitable and sustainable world. The social circular economy is a business model that prioritises positive societal outcomes and aims to prevent the economy from harming society. However, the circular economy aims to address environmental issues and does not necessarily consider negative social externalities or societal challenges. From the perspective of the United Nations' Sustainable Development Goals (SDGs), the circular economy primarily aligns with SDG12, which is Responsible Consumption and Production. In contrast, the social circular economy could meet not only this goal but also three more SDGs: SDG5 (Gender Equality), SDG8 (Decent Work and Economic Growth), and SDG10 (Reduced Inequalities). Thus, a purely circular economy approach could create different constraints when compared to the social circular economy (Robinson, 2018). The challenge is that, despite the general acceptance of a social dimension to sustainability, there has been a lack of clarity and agreement on its precise implications for sustainability strategies. Social impacts can encompass a wide range of factors related to human well-being.

However, the absence of a clear definition of social sustainability has made it challenging to measure its impacts accurately. This lack of measurement can limit the potential incentives for progress. Therefore, there is a need for more specific and measurable criteria to define and evaluate the social element of circular economy effectively (Old et al., 2022). Moreover, sustainable consumption poses a significant challenge to the social circular economy. Sustainable consumption encompasses the aspects of environmental sustainability, social sustainability, and economic sustainability. The 1994 Oslo Symposium defined sustainable consumption as 'the use of goods and services that respond to basic needs and bring a better quality of life, while minimising the use of natural resources, toxic materials and emissions of waste and pollutants over the life cycle, so as not to jeopardise the needs of future generations' (Oslo Roundtable on Sustainable Production and Consumption, n.d.). Several theoretical frameworks are important for enabling sustainable consumption and relate to the social circular economy. This chapter discusses some of these concepts, such as the social economy (SE), the social and solidarity economy (SSE), and the doughnut economy (DE).

According to the European Commission (2023), the social economy includes various businesses, organisations, and legal entities that share the common goal of systematically prioritising people, generating positive impacts on local communities, and pursuing social causes. Within the SE, citizens can be given meaningful opportunities to affect the patterns of governance, production, and consumption that directly impact their lives, communities, and environment (Smith, 2005). The SE and sustainable consumption are closely linked, as the SE prioritises social and environmental goals and can contribute to developing sustainable consumption practices. The SE has a critical role to play in promoting sustainability, as it is closely related to the circular economy and encourages local production and consumption patterns (Hug, 2022).

The social and solidarity economy encompasses various types of organisations and enterprises that operate based on principles of solidarity and participation.

These entities produce goods and services while simultaneously pursuing social and economic goals. Examples of SSE entities include cooperatives, mutual benefit societies, associations, and social enterprises that operate in the social or economic sector (Fonteneau et al., 2010). Sustainable Production and consumption (SDG12) is among the sustainable development goals most closely linked to the social solidarity economy (United Nations, 2021). Significant overlaps exist between the SSE and SDGs. A case study found that the SSE contributes to SDG12 (responsible consumption and production) and SDG8 (decent work and economic growth) (Villalba-Eguiluz et al., 2020). The SSE has conceptual contributions to the Circular Economy concerning ecological and social sustainability. The SSE focuses on creating a fair and inclusive workplace and can contribute to a social circular economy by providing opportunities for communities to engage in sustainable practices, such as recycling and reusing materials, and creating economic systems that prioritise social and environmental sustainability over profit (Malagón-Vélez, 2021).

The doughnut economy is a new approach that considers both planetary boundaries and social foundations in the development of societies (Luukkanen et al., 2021). It proposes a framework that aims to meet the needs of all within the means of the planet. With this model, progress is not synonymous with growth but is considered the balance between the foundation of basic human needs and the planet's ecological limits (Raworth, 2018). The circular economy and doughnut economy are two different models, but they are complementary in their focus on sustainability. The doughnut economy can complement social circular economy by transforming degenerative economies into regenerative ones and divisive economies into distributive ones by giving more visibility to the cooperative and caring side of human behaviour (Klara Drèze et al., 2022). Combining the social circular economy with additional concepts, such as the doughnut economy, can create a more holistic and equitable economy that benefits both people and the planet (Everett, 2022). Doughnut economics recognises that human behaviour can be nurtured to be cooperative and caring, just as it can be competitive and individualistic, and that economies, societies, and the rest of the living world are complex, interdependent systems that are best understood through the lens of systems thinking (Doughnut Economics Action Lab, 2023). The connection between the doughnut economy and sustainable consumption lies in their shared objective of promoting progress while safeguarding the sustainability of natural resources (Raworth, 2018).

In summary, social circular economy, social economy, social and solidarity economy, and doughnut economics all emphasise the importance of social and environmental sustainability. However, each concept proposes different approaches and translates differently in the implementation. In addition, in product development, the integration of sustainability has primarily focused on environmental considerations, while the social aspect has been relatively neglected. However, environmental and social dimensions must be considered equally for product development to be genuinely sustainable. Unfortunately, methodological support for incorporating the social aspect of sustainability in product development is currently immature. Therefore, there is a need for more significant emphasis on integrating social sustainability criteria into the product development process to achieve truly sustainable outcomes (Lagun Mesquita, 2021).

Thus, achieving a sustainable circular economy requires a business model that effectively balances the three often conflicting environmental, economic, and social sustainability goals. To accomplish this balancing act, co-creating a value network of different stakeholders is crucial. This approach ensures that the needs and perspectives of all stakeholders are considered, resulting in the creation of more sustainable products and services that meet the needs of society and the environment while being economically viable (Leiting et al., 2023). To succeed in a circular economy, companies need to adopt measurement frameworks for decision-making that go beyond traditional economic and environmental considerations and account for potential social impacts. However, existing frameworks, methods, and tools are insufficient in addressing the complexity of the social dimension in a circular economy.

Therefore, there is a need for more holistic approaches that account for the social dimension's complexities in a circular economy (Scarpellini, 2022). Anecdotal evidence indicates that entrepreneurial companies tend to take a more comprehensive and inclusive approach to implementing a circular economy. This approach is motivated by a genuine concern for the well-being of people and the natural environment. The reasons why entrepreneurs choose to adopt circular economy practices can significantly influence the impact and direction of these practices in society. As such, the underlying factors that drive entrepreneurial companies to commit to circular economy principles are crucial (Rovanto & Finne, 2022).

The sharing economy is an example of a business model for social entrepreneurship. This economic model goes beyond the traditional notion of mass production and consumption. Instead, it involves sharing resources rather than owning them. The growth of the sharing economy has been greatly facilitated by advances in information and communication technologies, particularly social networking platforms that connect individuals willing to share their possessions. The key idea behind collaborative sharing systems is that untapped potential exists in goods and services their owners do not fully utilise. Consequently, with the advent of technology-based platforms, the sharing economy has the potential to address social issues effectively. Hence, social enterprises can leverage the principles of the sharing economy to develop their business models (Roh, 2016). However, the actual impact of the sharing economy on sustainability is difficult to determine due to the various factors at play. Although the sharing economy has expanded significantly over the last decade, it currently falls short of delivering equitable and fair benefits for all participants involved. Despite its potential to support a social circular economy, the sharing economy has generated unchecked externalities and unintended consequences, highlighting the need for further development and regulation to ensure that it aligns with sustainability goals (United Nations, Department of Economic and Social Affairs, 2020).

These complexities in implementing the circular economy have slowed the transition process. For a successful transition, stakeholder engagement is crucial.

Consumers are one of the most relevant stakeholders in this context, yet their engagement in the circular economy is still underdeveloped.

3 Skills for the Green Transition

Given the importance of multistakeholder engagement in the transition to a circular economy, and the complex societal issues involved, new and updated skills and competencies aligned with the circular economy principles are required (Giannoccaro et al., 2021; Growth that Matters AB, 2021). As privileged vehicles to implement these skills, higher education institutions play a critical role through teaching activities that form new young generations of business leaders, designers, scientists, educators, and engineers—amongst many other disciplines—and may reskill professionals according to the new requirements of a circular economy (Kirchherr & Piscicelli, 2019).

According to a study published by the Ellen MacArthur Foundation (2021), the direction of travel for higher education institutions to accelerate the transition to a circular economy in cities focuses on five drivers:

1. Teaching the circular economy, so that every student engages with the circular economy when they choose to study at an institution in their city.
2. Applied research that supports the circular ambitions of businesses within the city.
3. Campus management that enables institutions to be circular in their operations.
4. Student-led initiatives that empower students to drive forward circular economy activities both on and off campus within the city.
5. Leadership and influence of faculty that inform business or government strategy.

Despite these efforts, the following misconceptions about what circular economy truly is still exist: (1) it's all about making better use of waste, (2) it's only about recycling more, (3) it's all about efficiency, (4) it's a new world of sustainability, and (5) waste-to-energy is part of a circular economy (Ellen MacArthur Foundation, 2021). A study carried out in two universities in Portugal and Spain indicates that students did not know the concept of the circular economy and were more familiar with the concepts of pollution and climate change (Sánchez-Emeterio & Figueira, 2019). Therefore, given the ignorance of key aspects of the circular economy, it is necessary to train circular literacy in students and stakeholders.

Moreover, in order to successfully implement a circular economy and transition to a sustainable future, this circular literacy process must extend to all sectors of society. This requires a specific kind of knowledge that allows individuals to understand the deep connections between humanity and nature, and to act intentionally to create positive change (Zwiers et al., 2020).

Frameworks such as the European Skills Agenda, Green Competence, CEDEFOP Green Skills and Innovation, OECD Skills Strategies, World Economic Forum's Future of Jobs Report, Global Cleantech Innovation Programme, Green Growth

Knowledge Platform, and UN Environment Programme Green Economy Initiative all provide indicators of knowledge and skills that can contribute to a conscious and faster green transition. In addition, initiatives such as the NewEuropean Bauhaus Initiative emphasise the importance of incorporating principles of sustainability into design, education, and practice, which is essential for developing the skills and competencies needed for a sustainable and circular economy.

The Circular Literacy Framework (Zwiers et al., 2020) is based on the concept that circular literacy is the knowledge-based ability to develop circular economy models. The framework consists of a set of dimensions and indicators that allow to evaluate the potential of circular economy thinking and practices towards sustainable development (see Table 1). It identifies three types of knowledge: (1) systems knowledge, (2) target knowledge, and (3) transformation knowledge. Systems knowledge is related to the structures, processes, and variabilities of complex systems and provides understanding of complexity through integrated research approaches. Target knowledge identifies targets of future development through the

Table 1 Circular literacy framework based on Zwiers et al. (2020)

Type of knowledge	Dimension	Indicator
System knowledge	Understanding of complexity	Consideration of the interconnections between diverse components in biosphere, technosphere, and sociosphere (or natural, social, and technological systems)
		Consideration of the self-organising properties of systems
Target knowledge	Establishing the circular economy	Consideration of de-materialisation and resource efficiency of economic and consumption practices
		Consideration of a closed-loop approach and feedback systems
		Consideration of resilience and regeneration of natural capital
	Preservation of natural livelihoods	Consideration of planetary boundaries
		Consideration of impacts throughout the lifecycle
	Intra- and intergenerational justice	Consideration of social justice
		Consideration of substantive, economic, and political inclusion
	Diversity and uniqueness	Consideration of diversity, local/cultural uniqueness
		Consideration of the creation of local capacities
Transformation knowledge	Epistemologies of complexity	Integration and adoption of different knowledge and various perspectives
		Addressing multiple means and actors
	Reflexivity and innovation	Relational and (self-)reflective knowledge
		Anticipation of unpredictable developments
		Creativity and innovativeness

Note: Adapted from 'Circular literacy. A knowledge-based approach to the circular economy', by J. Zwiers, M. Jaeger-Erben and F. Hofmann, 2020, *Culture and Organization 26,* Article 2 transdisciplinary creation of scenarios of just and sustainable futures, with the participation of decision-makers, stakeholders, and citizens. Transformation knowledge guides the transition from the current situation to a future one, involving cultural, social, and human resources in the development and stability of societal relations with nature (Zwiers et al., 2020).

Application of this framework for analysing a set of publications about circular economy allowed to identify that the economic vision is dominant, which demands the inclusion of the political and sociocultural visions in future research so to strengthen circular economy thinking (Zwiers et al., 2020). Moreover, sustainability competences such as systems thinking and futures literacy are useful when linked to sustainability values, as otherwise such competences could be used for unsustainable actions. Futures literacy empowers learners to create their visions for a sustainable future by providing them with the knowledge, skills, and attitudes to understand the future as a variety of alternatives (Bianchi et al., 2022).

Another relevant skill framework when it comes to the skills needed for the green transition is the Inner Development Goals framework (Growth that Matters AB, 2021). It encompasses a set of skills and qualities that indicate the kind of knowledge that is needed to deal with the complex societal issues involved in the green transition, particularly the Sustainable Development Goals, and the issues identified in UN's Agenda 2030 (see Table 2). It identifies five types of skills: (1) being—relationship to self; (2) thinking—cognitive skills; (3) relating—caring for others and the world; (4) collaborating—social skills; and (5) acting—driving change. Being—relationship to self—is related to establishing deep connections with our thoughts, feelings, and body as a way of being present, intentional, and calm in the face of complexity. Thinking—cognitive skills—include acceptance of different views, evaluation of information, and understanding of the wholeness of the world as a means to make wise decisions. Relating—caring for others and the world—is the ability to connect to others (including neighbours, future generations, or the biosphere) as a way of creating just and sustainable systems and societies. Collaborating—social skills—involves the capacity to embrace stakeholders with different values, skills, and competencies when working on shared concerns. Acting—driving change—includes qualities that help us act honestly, creatively, originally, and persistently.

Finally, in 2022, the Joint Research Centre (JRC), the European Commission's science and knowledge service created the GreenComp, identifying a set of sustainability competences to feed into education programmes to help learners develop knowledge, skills, and attitudes that promote ways to think, plan and act with empathy, responsibility, and care for our planet and for public health (Bianchi et al., 2022). The GreenComp is intended to be a living document, meaning it will be adjusted through an iterative process of stakeholder engagement. The GreenComp currently identifies four areas of skills for the green transition that bear some resemblance to circular literacy, namely, (1) embodying sustainability values, (2) embracing complexity in sustainability, (3) envisioning sustainable futures, and (4) acting for sustainability (see Table 2).

Table 2 Inner Development Goals and GreenComp frameworks based on Bianchi et al. (2022) and Growth that Matters AB (2021)

Framework	Dimension	Skills
Inner development goals	Being—relationship to self	Inner compass
		Integrity and authenticity
		Openness and learning mindset
		Self-awareness
		Presence
	Thinking—cognitive skills	Critical thinking
		Complexity awareness
		Perspective skills
		Sense-making
		Long-term orientation and visioning
	Relating—caring for others and the world	Appreciation
		Connectedness
		Humility
		Empathy and compassion
	Collaborating—social skills	Communication skills
		Co-creation skills
		Inclusive mindset and intercultural competence
		Trust
		Mobilisation skills
	Acting—driving change	Courage
		Creativity
		Optimism
		Perseverence
GreenComp	Embodying sustainability values	Valuing sustainability
		Supporting fairness
		Promoting nature
	Embracing complexity in sustainability	Systems thinking
		Critical thinking
		Problem framing
	Envisioning sustainable futures	Futures literacy
		Adaptability
		Exploratory thinking
	Acting for sustainability	Political agency
		Collective action
		Individual initiative

Note: Adapted from 'GreenComp—The European sustainability competence framework', by G. Bianchi, U. Pisiotis, and M. Cabrera, 2022, (https://doi.org/10.2760/13286) and 'Inner Development Goals: Background, method and the IDG framework', by Growth that Matters AB, 2021, (https://static1.squarespace.com/static/600d80b3387b98582a60354a/t/61aa2f96dfd3fb39c4fc4283/163854%203258249/211201_IDG_Report_Full.pdf, Accessed 30 March 2023)

In sum, there is an increasing recognition of the need to train transversal skills in addition to technical skills in order to achieve the green transition. In the past years, different frameworks on the transversal skills needed for the green transition have been emerging. This work, and specifically how it relates to training, education, and the work field, is still ongoing. Across frameworks, there are some common skills that are emerging, namely, (1) how we relate to ourselves (e.g. 'self-reflective knowledge' in the circular literacy framework, 'being' in the Inner Development Goals, and 'embodying sustainability' in the GreenComp), (2) how we relate to others (e.g. 'systems thinking' in the circular literacy framework, 'relating' and 'collaborating' in the Inner Development Goals, and 'embracing complexity' in the GreenComp) and how we move to action (e.g. 'acting' in the Inner Development Goals, 'acting for sustainability' in the GreenComp).

4 Dealing with Resistance

Across the different frameworks on skills for the green transition, how we relate to ourselves, how we relate to others, and being able to act is seen as key. However, there are several barriers that can prevent individuals from taking action in the face of climate change. Randall and Brown (2015) identify the following barriers: (1) there is a shared responsibility towards climate change and action is needed from many players, (2) harm done by climate change is not immediately visible, (3) the facts about climate change are difficult to deal with, (4) individuals have different ways of avoiding the truth about climate change, and (5) difficult emotions related to climate change (e.g. anxiety, guilt, or certain unsustainable behaviours are tied to their identity).

First, given that there is a shared responsibility towards climate change, it is unsurprising that the response to climate change is inherently moderated by social context (Clayton, 2020). Geiger and Swim (2016) carried out two studies that showed that individuals inaccurate perceptions about the opinion of others about climate change led them to self-silence. Individuals seem to expect to be perceived as less competent in a climate change conversation when others do not share one's opinion. Clayton (2020) points out that this self-silencing can ultimately lead to social isolation, while social support can be a great source of resilience. In addition, different social ties seem to have a different effects on behaviour. Colleagues seem to have an effect on cost-saving activities and altruistic behaviour, while neighbours have a more important influence on working on solving local community problems, volunteering, and recycling (Videras et al., 2012).

Second, when considering that individuals can deny the truth about climate change, there are several ways in which they can do this (Randall & Brown, 2015). Often avoiding the truth is thought of as avoiding the facts about climate change, but it can also display as denying the meaning of the facts (e.g. downplaying the seriousness of climate change), denying the implications of climate change (e.g. denying the effects climate change will have on people), denying the connection between climate change and individual lives (e.g. denying individual responsibility to undertake action), denying the emotional significance of climate change (e.g. denying difficult emotions related to climate change), denying the practical significance of climate change (e.g. denying any practical implications) or denying the irreversibility of climate change (e.g. denying that immediate action is needed now).

Given that climate change does lead to difficult emotions, it is unsurprising that individuals have these barriers to fully contemplate the impact of climate change. Clayton (2020) discusses climate anxiety as a response to both direct and indirect effects of climate change. Individuals can experience anxiety as a result of personal experiences with climate change, but anxiety can also be caused by individuals' perceptions about climate change even when they have not experienced any direct impacts of climate change. Climate change distress, a feeling consisting of both anxiety and hopelessness related to climate change, shows a small but significant positive associations with general depression, anxiety, and stress (Searle & Gow, 2010). Since climate change is a real and significant threat, it is rational to experience some worry. Moreover, given that it is a threat that is ongoing and still developing, adaptation to this threat is difficult (Clayton, 2020). Studies have found that climate change worries individuals around the world, In Europe, 20–40% of individuals stated being 'very worried' in 2016 (Steentjes et al., 2017), while 69% of Americans were at least 'somewhat worried' and 29% 'very worried' in 2018 (Ballew et al., 2019).

However, since climate change is a real threat, and anxiety can serve the adaptive function of signalling this threat and motivating individuals to take appropriate action (Barlow et al., 2019), Clayton (2020) points out the importance of not pathologising anxiety in response to climate change. Apart from creating the impression that an anxious response to climate change is inappropriate, it could direct attention away from social responses and responsibilities towards individuals. Randall and Brown (2015) point out that it is important to recognise that climate change needs an individual, social, and political response. As mentioned before, the need for multistakeholder responsibility in the green transition can be a breeding ground for resistance to undertaking action when this is not appropriately addressed. This was confirmed in a qualitative study that found that almost no participants believed individual behaviour change could make a meaningful contribution to climate mitigation (Kenis & Mathijs, 2012). Moreover, the reality is that climate change is partly a societal issue. The same qualitative study showed that individuals are critical of awareness-raising campaigns aimed at changing individual behaviour, since they seem to want to shift the responsibility towards individuals instead of society or larger organisations (Kenis & Mathijs, 2012). In addition, Ojala (2015a) found that in youngsters, feelings of societal powerlessness and social influences by parents and peers contribute to climate scepticism. A qualitative study found that being skeptical about strategies for climate change was a more important barrier for behavioural engagement in climate mitigation than being skeptical about climate change (Kenis & Mathijs, 2012). Other research has confirmed that low socioeconomic class contributes to a feeling of powerlessness in the face of climate change, while higher-status participants believe that action against climate change is good and achievable (Kennedy & Givens, 2019). In line with this, personal beliefs about climate change are predictive of supporting climate policy in individuals with a higher socio-economic status (Eom et al., 2018).

Feelings of worry and powerlessness seem to be prevalent when it comes to climate change. In in-depth interviews with individuals concerned about climate change, powerlessness seemed to be experienced by all individuals no matter how behaviourally engaged they are in mitigating climate change (Kenis & Mathijs, 2012). Participants indicated feeling powerless with respect to different aspects of climate change, namely, (1) the scale and complexity of climate change, (2) difficulty identifying the main cause of climate change, (3) the lack of strategic vision with regard to climate mitigation, and (4) lacking a credible alternative.

Albrecht (2011) has coined the term 'eco-paralysis' signifying the inhibiting effect that anxiety about climate change could have on undertaking action in the green transition. Indeed, considering that feelings of worry, anxiety, and powerlessness are prevalent when it comes to climate change, the question arises whether this leads to inaction or whether it can be a motivating factor for climate mitigation. Research on the effect of anxiety on behaviour has led to conflicting results. One study shows that distress about climate change was positively related to behavioural engagement (Reser et al., 2012), while another study showed that individuals reporting experiencing eco-anxiety indicated being more likely to adjust their behaviour (American Psychological Association, 2020). Another study however found no relation between climate anxiety and behaviour in either direction (Clayton & Karazsia, 2020). This could possibly indicate the possible motivating and paralysing effects of climate anxiety (Clayton, 2020).

Different ways of coping with these negative feelings related to climate change have been identified. Ojala (2012, 2015b) found that meaning-focused coping strategies, in which individuals try to gain perspective on climate change issues by seeing them in a larger historical perspective, or attempting to find hope by trusting societal actors, led to the most positive results in terms of higher behavioural engagement with climate change, higher life satisfaction, higher positive affect, and lower negative effect. On the other hand, problem-focused coping (e.g. talking to others about the problem, or informing oneself about the issues) led to higher levels of behavioural engagement with climate change, but also higher negative affect. Emotion-focused coping led to low behavioural engagement with climate change and high negative affect. Another healthy coping mechanism could be to tap into the restorative potential of contact with nature (Howell & Passmore, 2013). Martin et al. (2020) found that different types of contact with nature, as well as individual differences in connectedness with nature contributed differently to aspects of health, well-being, and pro-environmental behaviour. While the amount of nature in the neighbourhood seemed unrelated to well-being or pro-environmental behaviour, being in nature at least weekly was positively associated with general health and sustainable behaviour in the household. Feeling connected to nature shows a positive association to both well-being achieved by personal development and growth, and sustainable behaviour. Apart from looking for a connection to the natural world, Randall and Brown (2015) identified a need for justice and equality, striving for enlightened self-interest, and the challenge and creativity that comes with looking for solutions to the green transition as possible motivators for behavioural engagement.

Moreover, research shows that behavioural engagement with climate change is positively related to happiness (e.g. Brown & Kasser, 2005; Corral-Verdugo et al., 2011; Jacob et al., 2009). Individuals who feel more positive emotions after posing climate-positive behaviour or having been in contact with positive climate messages, are more likely to be more behaviourally engaged (Doell et al., 2021). In addition to leading to more positive effects, engagement with climate change mitigation can buffer against the negative effects the perception of climate change can have on mental health (Bradley et al., 2014) and stimulate social ties, feelings of empowerment, and personal development (Bamberg et al., 2018).

Randall and Brown (2015) identified some actions that can be important when trying to change your behaviour, namely, to (1) talk about your feelings related to climate change with others, (2) take account of relationships and systems that could influence the behaviour you are trying to change, (3) understand what helped you achieve change in the past, and (4) plan ahead, assess what is realistic and allow for the appropriate time frame to achieve change. Given that a social response to climate change is necessary, and relationships and systems can influence the behaviour individuals are trying to change, talking with others is of importance. However, the reactions of others can often be negative when talking about behavioural engagement to mitigate climate change (Randall & Brown, 2015). First of all, we tend to think we can convince others to change their behaviour by stating facts, but people have certain biases when hearing information that causes them to screen out or split off certain information that they think does not apply to them, makes them feel uncomfortable or challenges their sense of identity and expectations of life. Second, when we are communicating from a place of anxiety, it makes us into poor communicators. Third, given that climate change can cause anxiety, listeners can tend to project their anxiety and make the communicator the scapegoat for their anxiety (Randall & Brown, 2015).

Alternatively, instead of trying to convince others with information, achieving empathy and making others feel understood ironically leads to them being more open to the message you are trying to deliver. Empathy can be achieved by imagining what it is like to be the other person, being accepting of their point of view, and avoiding blame, criticism, and judgement. It can help to show understanding for the real dilemmas people are facing, such as being part of an international family and having difficulty finding sustainable alternatives for travel. In addition to showing empathy, paying attention to the different levels in a conversation can be helpful. Apart from the specific message that is being delivered or the content of the conversation, both individuals have a certain mood, agenda, and perception of each other during the conversation. Trying to be mindful of the mood of the conversation, by showing empathy, will probably make the receiver more receptive to the message you are trying to deliver. Apart from showing empathy, promoting self-efficacy by believing in people's willingness to act, nurturing their potential, and praising them for what is already done well can also help engage them (Randall & Brown, 2015).

In sum, worry, anxiety, and powerlessness are common reactions to climate change. Apart from these negative feelings, individuals can experience other barriers that stop them from undertaking action in the face of climate change. Social influences and shared responsibility play an important part in tackling the green transition. Showing empathy and understanding for the dilemmas that people face can help people be more receptive to behavioural engagement in the green transition. Given that people play an important part in the green transition and are often overlooked, the last part of this chapter focuses on the consumer perspective on circular economy.

5 Consumer Perspectives on Circular Economy

The consumers' role in a circular economy is of high importance. A systemic circular economy transition can be accomplished only by adequately combining the material, strategic, and consumer perspectives (Schöggl et al., 2020). According to a study by Chen et al. (2021), 13 social criteria play a role in the sustainable consumption transition: (C1) health concerns, (C2) consumer's readiness to reuse products, (C3) the effect of social power from family, friends, and colleagues, (C4) the authority, (C5) consumer's wealth, (C6) consumer's knowledge, (C7) the knowledge about solutions to environmental problems, (C8) consumer's knowledge about the effects of environmental problems, (C9) consumer's awareness about environmentally friendly consumption, (C10) interest in sustainable issues, (C11) purchasing behaviour based on environmental impact, (C12) consumer's opinions based on their ratings, and (C13) consumer's opinions in the form of words of mouth or recommendations. However, several complexities hinder the transition to a circular economy; these include behavioural barriers such as resistance to change, lack of awareness, and consumers' unwillingness to pay for circular products and services.

Therefore, in Austria, the CE4ALL project (open4innovation, n.d.) was conducted with a specific focus on the affinity to circular economy strategies of the Austrian population. This is of importance, as especially consumer behaviour and inclusion aspects related to the circular economy have received little attention, as they need to be considered in a very differentiated way. Hence, suitable target group-specific measures are needed to specifically change consumption behaviour in the direction of circularity.

We conducted a literature research, focus group discussions, and a survey to capture various aspects of the circular economy, such as social inclusion and consumer behaviour. The literature analysis is focused on the following elements: circular economy in general, social inclusiveness within the circular economy, and structurally disadvantaged people within the scope of inclusiveness. Moreover, to better understand the end consumer's contribution to the circular economy, we selected the following five R-strategies for further discussion in focus groups: reduce, reuse, recycle, repair, and rethink. In addition, we implemented a survey featuring statistical representation by age, sex, and spatial distribution among

Austrian federal states where we excluded the R-strategy recycle because of its incomparability to the other four strategies.

Based on an in-depth analysis of the currently prevailing target group-specific consumption behaviour and taking into account inclusion aspects, both incentives and enablers for strengthening the circular economy need to be identified and inhibitors should be uncovered. For this purpose, several R-strategies of the circular economy (Kirchherr et al., 2017) should be integrated with relevant product groups and target groups based on the typology by Stelzer and Heyse (2017) in a multidimensional mapping in a novel framework, the Circular Economy Cube (CEC). The idea of the CEC can be seen in Fig. 1.

For each sub-cube, i.e. for an R-strategy, a product group and a milieu, the resulting circularity score represents the affinity to promoting the circular economy. After a pre-selection of consumer-relevant R-strategies and product groups based on findings from a systematic literature review potential circular economy behaviour was discussed in three consecutive focus groups with people of different milieus with a specific inclusion of underprivileged people. In a further step, the results from the focus group discussions were used for an Austrian-wide survey to get insights into the milieu-specific behaviour as well as acceptance and effectiveness of R-strategies for certain product groups. The survey was also used to test nudges to promote circular economy-affine behaviour. The effect of behaviour-changing triggers is quantified and validated with the help of statistical models.

The results from the survey combined with those from the focus groups were combined to elaborate milieu-specific measures and recommendations for action to

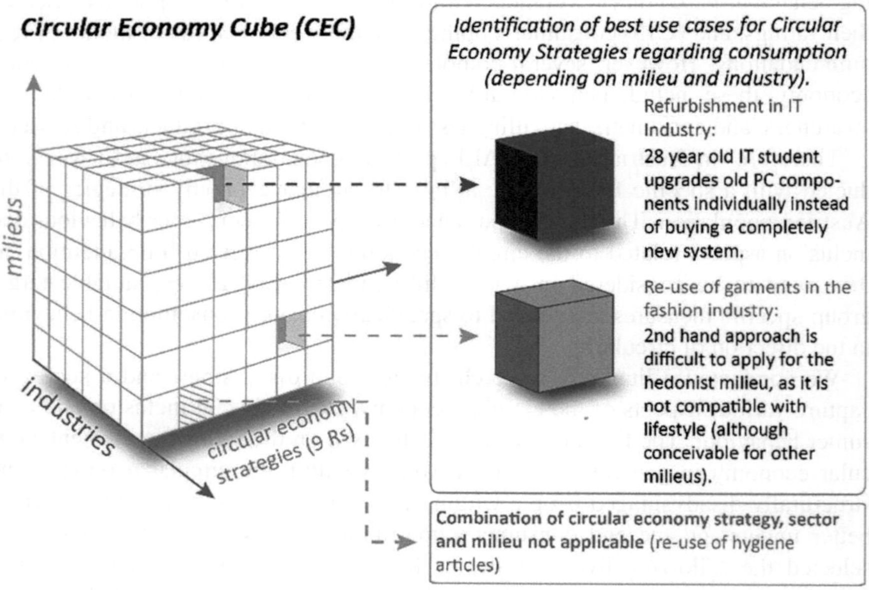

Fig. 1 The concept of the Circular Economy Cube (Anderluh et al., 2023)

change consumption behaviour towards strengthening the circular economy, as well as implications for consumption and RTI policy in order to make the concept of the circular economy the predominant and accepted system. For the sub-cubes in the CEC target-group-specific measures were elaborated for changing consumer behaviour and strengthening the circular economy affinity.

Key findings are the importance of both the known opportunities for and the existing capabilities to apply circular economy strategies from a consumer perspective. The influence of a person's affiliation to the milieus of society has turned out to be not very clear-cut, whereas the ability to make ends meet with the available income does have an influence on the affinity to the circular economy. The project revealed a lack of affinity especially when it comes to sharing things. On the opposite repairing things is already a widely accepted strategy above all for electronic and household devices (see Fig. 2). This goes in line with the tested nudges which

			Coping with income					
			With great difficulties	With difficulties	With some difficulties	Quite easily	Easily	Very easily
Reduce	Clothing	Agree	5.69	5.36	5.29	5.43	5.18	5.34
		Do	6.00	5.32	5.36	5.44	5.34	5.47
	Appliances	Agree	5.34	4.87	4.84	4.49	4.63	4.40
		Do	5.82	5.09	5.05	5.05	4.95	4.83
	Furniture	Agree	5.43	4.92	5.02	4.87	4.79	4.72
		Do	5.74	5.28	5.11	5.07	4.99	4.95
	Toys	Agree	6.10	5.72	5.80	5.66	5.85	6.12
		Do	5.59	5.40	5.53	5.48	5.58	5.35
Rethink (share)	Clothing	Agree	2.33	3.08	2.69	2.50	2.42	2.04
		Do	4.25	4.06	3.81	3.94	4.02	3.36
	Appliances	Agree	2.55	3.14	2.93	2.67	2.45	2.04
		Do	4.12	4.09	3.90	3.89	4.04	3.56
	Furniture	Agree	2.20	2.57	2.07	1.92	2.14	1.42
		Do	3.91	4.11	3.75	3.69	3.92	3.36
	Toys	Agree	3.03	3.41	2.93	2.95	3.13	2.88
		Do	4.45	4.16	3.79	3.76	4.00	3.40
Reuse	Clothing	Agree	3.84	4.14	4.10	3.63	3.61	3.42
		Do	4.83	4.60	4.47	4.22	4.48	4.26
	Appliances	Agree	4.31	4.22	4.18	3.54	3.73	3.30
		Do	4.82	4.49	4.37	4.12	4.46	4.02
	Furniture	Agree	4.40	4.74	4.53	3.71	4.11	3.72
		Do	5.17	4.72	4.40	4.26	4.48	4.22
	Toys	Agree	4.73	4.57	4.68	4.52	4.49	4.51
		Do	5.04	4.65	4.40	4.29	4.50	4.26
Repair	Clothing	Agree	4.30	4.35	4.24	4.31	4.45	4.43
		Do	4.96	4.95	4.67	4.84	4.74	4.83
	Appliances	Agree	5.39	5.45	5.49	5.59	5.78	5.56
		Do	5.45	5.28	5.12	5.26	5.18	5.12
	Furniture	Agree	5.01	5.01	4.88	5.06	5.05	4.84
		Do	5.24	4.89	4.79	4.81	4.82	4.83
	Toys	Agree	4.86	4.55	4.38	4.40	4.41	4.48
		Do	5.11	4.75	4.44	4.32	4.28	4.34

Fig. 2 Affinity for the circular economy according to coping with income per R-strategy and product group; scale for approval (agree) and implementation (do): 1 (not at all/light grey)—7 (fully/dark grey). *Note:* Adapted from 'CE4ALL—Kreislaufwirtschaft für alle: Incentives, Enabler und Inhibitoren für einen nachhaltigen Konsum', by A. Anderluh, A. et al., 2023, *Wien: BMK*. Available at: https://fti-ressourcenwende.at/resources/pdf/schriftenreihe-2023-2_ce4all.pdf (Accessed 24 Dec 2023)

have in part produced good results in encouraging appropriate behaviour, such as the repair bonus, which is already used in practice.

6 Conclusion

Strategic documents on the circular economy increasingly mention the importance of social aspects, including multistakeholder engagement. Social circular economy emerges as a framework to more explicitly include the social aspect in the circular economy. Social circular economy combines the concept of circular economy with social enterprises to overcome the potential pitfalls of circular economy, in which societal challenges can be overlooked by prioritising economic gain, and social enterprises, in which negative impact on the environment could be overlooked by prioritising social issues. Even though several taxonomies on the social aspect in circular economy are emerging, they propose different approaches and translate them differently into practice. Social aspects of the circular economy can include multiple actors, such as employees, organisations, consumers, communities, society, and multistakeholder engagement. There is a broad range of social aspects identified across different taxonomies, including features like well-being, social justice, labour practices, and infrastructure availability. This broad range and lack of a clear definition of the social aspect within social circular economy poses a challenge for measuring social impacts accurately. As such, there is still a need for an overarching framework with an explicit definition of social aspects in the circular economy in order to accurately assess social impact.

Similarly, different frameworks are emerging related to the transversal skills needed for the transition to a circular economy. Currently, the work on these frameworks is still ongoing and how they translate to training, education, and the work field is still unclear. Across frameworks, there does seem to be a common focus on (1) relating to oneself (e.g. self-reflective knowledge, authenticity, presence, openness, valuing sustainability, …), (2) relating to others (e.g. knowledge of intra- and intergenerational justice, empathy, trust, co-creation skills, systems thinking, …), and (3) moving to action for the green transition (e.g. establishing the circular economy, courage, creativity, collective action, individual action, …).

In order to move to action, individuals need to overcome several barriers, such as (1) the shared responsibility of many actors towards climate changes, (2), dealing with the difficult facts of climate change, (3) lack of immediate visibility of the consequences of climate change, (4) avoiding the truth about climate change, and (5) negative emotions related to climate change. As such, it is key to recognise social influences and shared responsibility in the green transition when dealing with the resistance individuals face in making the necessary changes for the green transition. Being empathic towards individuals' struggles can make them more receptive to behavioural engagement in the green transition.

Research on consumer perspectives on the circular economy confirms the importance of social influences when it comes to socio-economic status and engagement in the circular economy. In addition, people are differently engaged in different circular economy strategies (e.g. repairing things was widely accepted, while the sharing economy was not). This points out the importance of looking at specific behaviour and circular economy models when researching the circular economy.

Acknowledgements Part of this work was conducted within the CE4ALL project (No. 889847) which received funding by the Austrian Research Promotion Agency (FFG) and within the E³UDRES² Erasmus+ project (No. 101004069) which received funding from the European Commission.

References

Albrecht, G. (2011). Chronic environmental change: Emerging 'psychoterratic' syndromes. In I. Weissbecker (Ed.), *Climate change and human well-being: Global challenges and opportunities* (pp. 43–56). Springer. https://doi.org/10.1007/978-1-4419-9742-5_3

American Psychological Association. (2020). *Majority of US adults believe climate change is most important issue today*. Accessed March 30, 2023, from https://www.apa.org. https://www.apa.org/news/press/releases/2020/02/climate-change

Anderluh, A, Moser, M., Pellegrini, T., Richter, L., Zahra, M., Hackl, R., Raffler, C., & Sempoch, C. (2023). *'CE4ALL - Kreislaufwirtschaft für alle: Incentives, Enabler und Inhibitoren für einen nachhaltigen Konsum'*. Berichte aus Energie- und Umweltforschung. Wien: BMK. https://fti-ressourcenwende.at/resources/pdf/schriftenreihe-2023-2_ce4all.pdf

Ballew, M. T., Leiserowitz, A., Roser-Renouf, C., Rosenthal, S. A., Kotcher, J. E., Marlon, J. R., et al. (2019). Climate change in the American mind: Data, tools, and trends. *Environment: Science and Policy for Sustainable Development, 61*(3), 4–18. https://doi.org/10.1080/00139157.2019.1589300

Bamberg, S., Rees, J. H., & Schulte, M. (2018). 8 - Environmental protection through societal change: What psychology knows about collective climate action—And what it needs to find out. In S. Clayton & C. Manning (Eds.), *Psychology and climate change* (pp. 185–213). Academic Press. https://doi.org/10.1016/B978-0-12-813130-5.00008-4

Barlow, D. H., Durand, V. M., & Hofmann, S. G. (2019). *Abnormal psychology: An integrative approach*. Cengage Learning.

Bianchi, G., Pisiotis, U., & Cabrera, M. (2022). *GreenComp - The European sustainability competence framework (no. EUR 30955 EN)*. European Commission. https://doi.org/10.2760/13286

Bradley, G. L., Reser, J. P., Glendon, I., & Ellul, M. (2014). Distress and coping in response to climate change. In K. Kaniasty, K. A. Moore, S. Howard, & P. Buchwald (Eds.), *Stress and anxiety: Applications to social and environmental threats, psychological well-being, occupational challenges, and developmental psychology* (pp. 33–42). Logos Verlag Berlin GmbH.

Brown, K. W., & Kasser, T. (2005). Are psychological and ecological well-being compatible? The role of values, mindfulness, and lifestyle. *Social Indicators Research, 74*(2), 349–368.

Chen, C.-C., Sujanto, R. Y., Tseng, M.-L., Fujii, M., & Lim, M. K. (2021). Sustainable consumption transition model: Social concerns and waste minimization under willingness-to-pay in Indonesian food industry. *Resources, Conservation and Recycling, 170*, 105590. https://doi.org/10.1016/j.resconrec.2021.105590

Clayton, S. (2020). Climate anxiety: Psychological responses to climate change. *Journal of Anxiety Disorders, 74*, 102263. https://doi.org/10.1016/j.janxdis.2020.102263

Clayton, S., & Karazsia, B. T. (2020). Development and validation of a measure of climate change anxiety. *Journal of Environmental Psychology, 69*, 101434. https://doi.org/10.1016/j.jenvp.2020.101434

Corral-Verdugo, V., Mireles-Acosta, J. F., Tapia-Fonllem, C., & Fraijo-Sing, B. (2011). Happiness as correlate of sustainable behavior: A study of pro-ecological, frugal, equitable and altruistic actions that promote subjective wellbeing. *Human Ecology Review, 18*(2), 95–104.

Doell, K. C., Conte, B., & Brosch, T. (2021). Interindividual differences in environmentally relevant positive trait affect impacts sustainable behavior in everyday life. *Scientific Reports, 11*(1), 20423. https://doi.org/10.1038/s41598-021-99438-y

Doughnut Economics Action Lab. (2023). *About Doughnut Economics | DEAL*. Accessed March 28, 2023, from https://doughnuteconomics.org/about-doughnut-economics

Drèze, K., Scarpa, R., & Simonyan, G. (2022, June 12). *The doughnut economics & social economy: An economy that serves both people and the planet*. Social Economy Europe. Accessed March 28, 2023, from https://www.socialeconomy.eu.org/2022/06/12/the-doughnut-economics-an-economy-that-serves-both-people-and-the-planet/

Ellen MacArthur Foundation. (2021). *The circular economy glossary*. Accessed November 23, 2022, from https://ellenmacarthurfoundation.org/topics/circular-economy-introduction/glossary

Eom, K., Kim, H. S., & Sherman, D. K. (2018). Social class, control, and action: Socioeconomic status differences in antecedents of support for pro-environmental action. *Journal of Experimental Social Psychology, 77*, 60–75. https://doi.org/10.1016/j.jesp.2018.03.009

European Commission. (2019). *The European green deal*. Accessed March 30, 2023, from https://eur-lex.europa.eu/legal-content/EN/TXT/?qid=1576150542719&uri=COM%3A2019%3A640%3AFIN

European Commission. (2020). *A new circular economy action plan for a cleaner and more competitive Europe*. Accessed July 25, 2022, from https://eur-lex.europa.eu/legal-content/DE/TXT/?uri=COM:2020:98:FIN

European Commission. (2023). *Social economy in the EU. Internal market, industry, entrepreneurship and SMEs*. Accessed March 28, 2023, from https://single-market-economy.ec.europa.eu/sectors/proximity-and-social-economy/social-economy-eu_en

Everett, E. (2022). Combining the circular economy, doughnut economy, and permaculture to create a holistic economic model for future generations. *Environmental Sciences Proceedings, 15*(1), 19. https://doi.org/10.3390/environsciproc2022015019

Fonteneau, B., Neamtan, N., Wanyama, F., Morais, L. P., & de Poorter, M. (2010). *Social and solidarity economy: Building a common understanding*. Publications International Training Centre of the ILO.

Geiger, N., & Swim, J. K. (2016). Climate of silence: Pluralistic ignorance as a barrier to climate change discussion. *Journal of Environmental Psychology, 47*, 79–90. https://doi.org/10.1016/j.jenvp.2016.05.002

Giannoccaro, I., Ceccarelli, G., & Fraccascia, L. (2021). Features of the higher education for the circular economy: The case of Italy. *Sustainability, 13*(20), 11338. https://doi.org/10.3390/su132011338

Growth that Matters AB. (2021). *Inner development goals: Background, method and the IDG framework*. Accessed March 30, 2023, from https://static1.squarespace.com/static/600d80b3387b98582a60354a/t/61aa2f96dfd3fb39c4fc4283/163854%203258249/211201_IDG_Report_Full.pdf

Howell, A. J., & Passmore, H.-A. (2013). The nature of happiness: Nature affiliation and mental well-being. In C. L. M. Keyes (Ed.), *Mental well-being: International contributions to the study of positive mental health* (pp. 231–257). Springer Netherlands. https://doi.org/10.1007/978-94-007-5195-8_11

Hug, D. (2022, February 3). *Social economy, sustainability and the consumer-citizen protection*. Accessed March 30, 2023, from https://law.kuleuven.be/ccm/blog/?p=294

Jacob, J., Jovic, E., & Brinkerhoff, M. B. (2009). Personal and planetary well-being: Mindfulness meditation, pro-environmental behavior and personal quality of life in a survey from the social justice and ecological sustainability movement. *Social Indicators Research, 93*(2), 275–294. https://doi.org/10.1007/s11205-008-9308-6

Kenis, A., & Mathijs, E. (2012). Beyond individual behaviour change: The role of power, knowledge and strategy in tackling climate change. *Environmental Education Research, 18*(1), 45–65. https://doi.org/10.1080/13504622.2011.576315

Kennedy, E. H., & Givens, J. E. (2019). Eco-habitus or eco-powerlessness? Examining environmental concern across social class. *Sociological Perspectives, 62*(5), 646–667.

Kirchherr, J., & Piscicelli, L. (2019). Towards an education for the circular economy (ECE): Five teaching principles and a case study. *Resources, Conservation and Recycling, 150,* 104406. https://doi.org/10.1016/j.resconrec.2019.104406

Kirchherr, J., Reike, D., & Hekkert, M. (2017). Conceptualizing the circular economy: An analysis of 114 definitions. *Resources, Conservation and Recycling, 127,* 221–232. https://doi.org/10.1016/j.resconrec.2017.09.005

Lagun Mesquita, P. (2021). *The social dimension of sustainable product development.* Accessed February 23, 2023, from http://urn.kb.se/resolve?urn=urn:nbn:se:bth-21285

Leiting, T., Külschbach, A., & Stich, V. (2023). Development of a platform business model for co-creation ecosystems for sustainable furniture. *Journal of Innovation Economics & Management, 40*(1), 81–107. https://doi.org/10.3917/jie.pr1.0129

Luukkanen, J., Vehmas, J., & Kaivo-oja, J. (2021). Quantification of doughnut economy with the sustainability window method: Analysis of development in Thailand. *Sustainability, 13*(2), 847. https://doi.org/10.3390/su13020847

Malagón-Vélez, L. E. (2021). Social and solidarity economy conceptual contributions to the circular economy. *Cuadernos de Administración (Universidad del Valle), 37*(70) Accessed March 28, 2023, from https://www.redalyc.org/journal/2250/225069432005/html/

Martin, L., White, M. P., Hunt, A., Richardson, M., Pahl, S., & Burt, J. (2020). Nature contact, nature connectedness and associations with health, wellbeing and pro-environmental behaviours. *Journal of Environmental Psychology, 68,* 101389. https://doi.org/10.1016/j.jenvp.2020.101389

Mies, A., & Gold, S. (2021). Mapping the social dimension of the circular economy. *Journal of Cleaner Production, 321,* 128960. https://doi.org/10.1016/j.jclepro.2021.128960

Ojala, M. (2012). How do children cope with global climate change? Coping strategies, engagement, and well-being. *Journal of Environmental Psychology, 32*(3), 225–233. https://doi.org/10.1016/j.jenvp.2012.02.004

Ojala, M. (2015a). Climate change skepticism among adolescents. *Journal of Youth Studies, 18*(9), 1135–1153. https://doi.org/10.1080/13676261.2015.1020927

Ojala, M. (2015b). Hope in the face of climate change: Associations with environmental engagement and student perceptions of teachers' emotion communication style and future orientation. *The Journal of Environmental Education, 46*(3), 133–148. https://doi.org/10.1080/00958964.2015.1021662

Old, R., Rumpenhorst, I., Schmidt, I., Nicolau, M., & Slaughter, R. (2022). *Discussing the social impacts of circularity.* Collaborating Centre on Sustainable Consumption And Production (CSCP) gGmbH. Acccessed March 30, 2023, from https://www.cscp.org/wp-content/uploads/2022/06/CIAP_social-impact_report.pdf

open4innovation. (n.d.). *CE4ALL—Circular economy for all—Incentives, enablers and inhibitors for sustainable consumption.* Accessed February 27, 2023, from https://nachhaltigwirtschaften.at/en/projects/circular-economy/ce4all.php

Oslo Roundtable on Sustainable Production and Consumption. (n.d.). *Defining sustainable consumption.* Acccessed March 30, 2023, from https://enb.iisd.org/consume/oslo004.html

Padilla-Rivera, A., Russo-Garrido, S., & Merveille, N. (2020). Addressing the social aspects of a circular economy: A systematic literature review. *Sustainability, 12*(19), 7912. https://doi.org/10.3390/su12197912

Randall, R., & Brown, A. (2015). *In time for tomorrow?: The carbon conversations handbook.* Surefoot Effect Community Interest Company.

Raworth, K. (2018). *A healthy economy should be designed to thrive, not grow | TED Talk (video).* Accessed March 28, 2023, from https://www.ted.com/talks/kate_raworth_a_healthy_economy_should_be_designed_to_thrive_not_grow

Reser, J. P., Bradley, G. L., Glendon, A. I., Ellul, M. C., & Callaghan, R. (2012). *Public risk perceptions, understandings and responses to climate change and natural disasters in Australia, 2010 and 2011*. National Climate Change Adaptation Research Facility Gold Coast.

Robinson, S. (2018). *Social circular economy–opportunities for people, planet and profit*. Social Circular Economy. Accessed March 30, 2023, from https://circulareconomy.europa.eu/platform/sites/default/files/social_circular_economy_2017.pdf

Roh, T. H. (2016). The sharing economy: Business cases of social enterprises using collaborative networks. *Procedia Computer Science, 91*, 502–511. https://doi.org/10.1016/j.procs.2016.07.129

Rovanto, S., & Finne, M. (2022). What motivates entrepreneurs into circular economy action? Evidence from Japan and Finland. *Journal of Business Ethics*. https://doi.org/10.1007/s10551-022-05122-0

Sánchez-Emeterio, G., & Figueira, C. (2019). *Ecosocial literacy: Circular economy conceptions in initial teacher training*. Presented at the the Future of Education. Accessed March 30, 2023, from https://conference.pixel-online.net/FOE/files/foe/ed0009/FP/5551-SED4018-FP-FOE9.pdf

Scarpellini, S. (2022). Social impacts of a circular business model: An approach from a sustainability accounting and reporting perspective. *Corporate Social Responsibility and Environmental Management, 29*(3), 646–656. https://doi.org/10.1002/csr.2226

Schöggl, J.-P., Stumpf, L., & Baumgartner, R. J. (2020). The narrative of sustainability and circular economy - A longitudinal review of two decades of research. *Resources, Conservation and Recycling, 163*, 105073. https://doi.org/10.1016/j.resconrec.2020.105073

Searle, K., & Gow, K. (2010). Do concerns about climate change lead to distress? *International Journal of Climate Change Strategies and Management, 2*(4), 362–379. https://doi.org/10.1108/17568691011089891

Smith, G. (2005). Green citizenship and the social economy. *Environmental Politics, 14*(2), 273–289. https://doi.org/10.1080/09644010500055175

Steentjes, K., Pidgeon, N. F., Poortinga, W., Corner, A. J., Arnold, A., Böhm, G., et al. (2017). *European Perceptions of Climate Change (EPCC): Topline findings of a survey conducted in four European countries in 2016*. Accessed March 30, 2023, from https://orca.cardiff.ac.uk/id/eprint/98660/7/EPCC.pdf

Stelzer, M., & Heyse, M. (2017). *Die Lebensführungstypologie. Eine integrative Typologie der Lebensführungen in der Bundesrepublik Deutschland, 2017*.

United Nations. (2021). *The circular economy, cooperatives and the social and solidarity economy*. Accessed March 28, 2023, from https://www.un.org/development/desa/cooperatives/2021/08/02/the-circular-economy-cooperatives-and-the-social-and-solidarity-economy/

United Nations, Department of Economic and Social Affairs. (2020). *Frontier technology quarterly: Does the sharing economy share or concentrate?* Accessed February 24, 2023, from https://www.un.org/development/desa/dpad/publication/frontier-technology-quarterly-does-the-sharing-economy-share-or-concentrate/

Videras, J., Owen, A. L., Conover, E., & Wu, S. (2012). The influence of social relationships on pro-environment behaviors. *Journal of Environmental Economics and Management, 63*(1), 35–50. https://doi.org/10.1016/j.jeem.2011.07.006

Villalba-Eguiluz, U., Egia-Olaizola, A., & Pérez de Mendiguren, J. C. (2020). Convergences between the social and solidarity economy and sustainable development goals: Case study in the Basque Country. *Sustainability, 12*(13), 5435. https://doi.org/10.3390/su12135435

Zwiers, J., Jaeger-Erben, M., & Hofmann, F. (2020). Circular literacy. A knowledge-based approach to the circular economy. *Culture and Organization, 26*(2), 121–141. https://doi.org/10.1080/14759551.2019.1709065

"It's Only Waste If You're Not Innovative": Understanding Micro-Scale Circular Economy Practices as Social Innovations

Carla Nogueira, Hugo Pinto, and João Filipe Marques

Abstract Circular economy and social innovation are becoming increasingly important concepts in addressing the environmental and social challenges faced by modern society. This chapter presents an exploratory study that examines the relationship between circular economy, sustainability, and social innovation, using qualitative data from an Intentional Sustainable Community—PPAUW. The study analyzes how circular economy practices developed at the micro-scale by this community can be understood as social innovation and their importance for sustainable development. The findings provide insights into the role of social innovation in promoting circular economy practices and sustainable development. The practices analyzed have a strong DIY—Do It Yourself—component that comes from the creativity of the members of this community. Creativity is an essential dimension that is directly related to the representation of innovation that the interviewed members reveal to have. The study highlights the challenges faced by circular economy practices, such as producing aesthetically pleasing items and making the practice more widespread. The conclusions on the multifaceted nature of innovation and its relationship to social needs, creativity, and sustainability provide valuable insights for the field of circular economy. The study underlines relevant aspects for a valuable framework for understanding how circular economy practices can promote sustainability, creativity, and social well-being.

C. Nogueira (✉) · H. Pinto
CinTurs – Research Centre for Tourism, Sustainability and Well-Being, Faculty of Economics, University of Algarve, Faro, Portugal

CES – Centre for Social Studies, University of Coimbra, Coimbra, Portugal
e-mail: cfnogueira@ualg.pt; hpinto@ualg.pt

J. F. Marques
CinTurs – Research Centre for Tourism, Sustainability and Well-Being, Faculty of Economics, University of Algarve, Faro, Portugal
e-mail: jfmarq@ualg.pt

C. Rego et al. (eds.), *Transitioning to a Circular Economy*, Studies on Entrepreneurship, Structural Change and Industrial Dynamics,
https://doi.org/10.1007/978-3-031-77661-8_5

Keywords Circular economy · Sustainability · Social innovation · Intentional sustainable communities · Case study

1 Introduction

Circular economy, sustainability, and social innovation are key concepts that have gained increasing attention in recent years. While circular economy is a concept aimed at promoting resource efficiency, sustainability focuses on meeting the needs of the present without compromising the ability of future generations to meet their own needs. Social innovation, on the other hand, involves the development of new ideas, practices, and institutions that meet social needs more effectively than existing solutions. This chapter presents an exploratory study that aims to understand the relationship between circular economy, sustainability, and social innovation.

Circular economy is relevant today because of the urgent need to address the interconnected challenges of climate change, resource scarcity, and environmental degradation. By reducing waste and preserving resources, circular economy practices can help mitigate the environmental impacts of production and consumption while promoting economic growth and job creation. However, the circular economy must still fulfil its potential, and some tensions and challenges must be addressed. This chapter argues that the circular economy—as a practice rather than a public policy—is deeply connected to what social actors define as their needs and how they act to meet them.

This study starts from a constructivist perspective to analyze the circular economy as a way of meeting needs. It introduces the concept of social innovation to explain how circular economy practices can be categorized as innovative activities. In meeting needs, individuals attach different meanings to what they perceive as needs and what they consider social innovation. Therefore, the chapter focuses on the PPAUW—an Intentional Sustainable Community (ISC)—as a case study to analyze their representations of social innovation and circular economy practices. PPAUW is an ISC with no defined legal status in Wageningen, province of Gelderland, the Netherlands. Using qualitative the study examines how circular economy practices can be understood as social innovation and their importance for sustainable development.

Overall, this chapter contributes to the literature on circular economy, sustainability, and social innovation by providing insights into the role of social innovation in promoting circular economy practices and sustainable development. The case study presented in this chapter offers a perspective on how intentional sustainable communities can be a context facilitator of creativity, enabling the development of innovation practices that align with social needs and sustainability goals. The findings of this study have implications for policymakers, businesses, and communities interested in promoting circular economy practices and sustainable development through social innovation.

The chapter is divided into five parts. The first part provides a literature review on the intersection of circular economy, social innovation, and sustainability. The second part presents the methodology used in the study. The third subsection focuses on the results analysis, which includes a presentation of the innovation representations and circular economy practices observed in the community, as well as an analysis of these practices according to the dimensions of social innovation. Finally, some conclusions are highlighted as well as some limitations, and directions for future research. This chapter ends with policy recommendations based on the study's findings.

2 Exploring the Intersections of Circular Economy, Intentional Sustainable Communities, and Social Innovation

2.1 The Role of Circular Economy in Sustainable Development

The circular economy (CE) is a concept that has received growing academic interest in recent years, especially since 2012 (Kirchherr et al., 2017). As a concept, it has been strategically appropriated by public policies to respond to Sustainable Development Goals, being present in several reports of the European Commission (EC, 2020). The concept has become popular since policymakers introduced it in China and the European Union as a solution to enable countries, businesses, and consumers to reduce of the environmental consequences of production and consumption patterns, allowing the expansion of the product life cycles (ibid., 2020).

The circular economy emerges as a counterproposal to the linear economy model. It is based on transforming goods at the end of their useful life by originating new flows of use (Prieto-Sandoval et al., 2018). This means "reusing wherever possible, recycling what cannot be reused, repairing what does not work and reproducing what cannot be repaired" (Stahel, 2016:435). In this sense, the basic principles of the circular economy are the 5Rs—Rethink, Refuse, Reduce, Reuse, and Recycle—to close the life cycle of materials, ensuring greater resource efficiency and diminishing waste (Stahel, 2016).

The concept of circular economy has become increasingly important in the context of sustainable development in recent years. The broad utilization of the concept has limited its heuristic value, and currently, CE is considered one of the new buzzwords in sustainability studies and transitions (Murray et al., 2017). Despite this, many argue that the circular economy underpins a set of principles amenable to practical operationalization that can be useful for implementing sustainable development strategies (Ghisellini et al., 2016). This more practical component may therefore contribute to deconstructing these criticisms around the idea of sustainability—mainly for being too vague and not very operational or, as Naudé (2011:352) puts it, for being "a theoretical dream rather than an implementable reality".

Circular economy is an alternative to the traditional linear economic model of "take-make-dispose" (Geissdoerfer et al., 2017; Stahel, 2016; Ghisellini et al., 2016). It promotes a system that is based on minimizing waste and the use of resources. A review of the literature on circular economy and sustainable development shows a need to move towards a more sustainable economic model, given the current challenges related to climate change, resource depletion, and environmental degradation (EC, 2020; Ghisellini et al., 2016). It has been argued that the circular economy can contribute to a more sustainable future by promoting sustainable production and consumption patterns, reducing waste and pollution, and creating new opportunities for economic growth (Circle Economy and Ecofys, 2018).

The shift to a more circular model that prioritizes product life extension, reuse, and recycling requires a systemic change in how we design products, organize supply chains, and consume goods and services. Some studies have found that adopting circular economy practices can result in substantial environmental benefits, such as reduced greenhouse gas emissions, lower use of natural resources, and reduced waste generation (EC, 2020). Despite the benefits of the circular economy, its implementation has several challenges. The transition towards a circular economy requires significant investments in new technologies, innovative business models, and changes in consumer behaviour (Cooper, 2017). Additionally, there is a need to address potential trade-offs between sustainable development's environmental, economic, and social goals (Kirchherr et al., 2017).

Although the circular economy is a strategic goal of the EU and implies a set of changes in social structures, the role of the micro-scale[1] has been increasingly highlighted. Micro-scale communities, such as neighbourhoods, local communities, and households, can play an essential role in promoting circular economy practices by enabling and empowering individuals to implement sustainable solutions at the local level (Goyal et al., 2016). Several studies have highlighted the potential for micro-scale communities to contribute to circular economy practices (Dapp, 2018). For example, the household level is considered a critical site for promoting resource efficiency, such as reducing food waste or using renewable energy. Local communities, such as neighbourhoods or cities, can also promote circular economy practices by fostering collaborative consumption through community gardens or tool-sharing schemes. Moreover, micro-scale communities can help develop circular economy practices tailored to the community's specific needs and resources and promote a sense of ownership and accountability among community members.

However, some scholars have also identified challenges and limitations associated with promoting circular economy practices at the micro-scale. For example, individual actions may be limited by access to information, financial resources, and infrastructure, which may require broader systemic changes to address. Moreover,

[1] "Micro-scale" refers to small-scale, local or individual actions or initiatives.

promoting circular economy practices at the micro-scale may require a shift away from individualistic models of consumption and production and towards more collective and systemic models that prioritize the public good over individual interests.

Micro-scale communities can be valuable in promoting circular economy practices by enabling and empowering individuals to implement sustainable solutions locally and by fostering collaboration and innovation (Goyal et al., 2016). Policymakers, practitioners, and educators should therefore recognize and support the role of micro-scale communities in circular economy initiatives and work to address the challenges and limitations associated with promoting circular economy practices at this level.

Circular economy has been promoted as a promising approach for achieving sustainable development, but it has also faced criticism and challenges from various perspectives. One main criticism of the circular economy concept is its focus on technological solutions to environmental problems rather than addressing the underlying social and economic structures that drive unsustainable consumption and production (Kirchherr et al., 2017; Prieto-Sandoval et al., 2018). Critics argue that circular economy often overlooks the role of power relations and interests in shaping unsustainable practices and that more attention should be given to social justice and equity issues (Stahel, 2016).

Another criticism of circular economy is its potential to reinforce existing patterns of socioeconomic inequality by redistributing resource inequalities rather than addressing the root causes of inequality (Stahel, 2016; Cooper, 2017). For example, some scholars have suggested that circular economy initiatives benefit private sector stakeholders more than marginalized groups or small-scale producers (Cooper, 2017). Others have raised concerns about the potential impacts of circular economy on labour and employment, such as displacement or precarious working conditions (Murray et al., 2017).

Additionally, some critiques of circular economy have focused on the limits of technological solutions to environmental problems, such as the potential for rebound effects or unintended consequences. Some scholars have argued that circular economy should be holistic and incorporate a broader range of environmental, social, and economic indicators to ensure that sustainability objectives are met in multiple dimensions (Dapp, 2018). Others have advocated for a greater emphasis on demand reduction and lifestyle changes rather than just improving resource use efficiency (Pomponi & Moncaster, 2017).

Overall, the literature suggests that while circular economy has potential benefits for sustainable development, it also faces significant criticisms and limitations. The limitations of technological solutions and sustainable lifestyles are particularly important for this chapter. The case study presented is an example of how circular economy practices can emerge from the micro-scale and be focused on addressing those challenges. In order to prompt circular economy practices, the micro-scale is necessary to reflect on the importance of creativity and social innovation.

2.2 Circular Economy, Social Innovation, and Creativity

Social innovation has been defined as the process of developing new and creative solutions to social problems or needs, which involve new forms of collaboration, participation, and empowerment of citizens, communities, and stakeholders (BEPA, 2010). This chapter starts from Cajaiba-Santana's (2014) "structuration perspective", which emphasizes the importance of a holistic view of the social innovation phenomenon. In this proposal, the author argues that the actions of actors and social structures should be conceived as dualistic and interdependent. According to this view, social innovation is understood as interactively influenced by individuals and social structures. This means that social innovation is based on collective actions within a particular social system and that the institutions of that system determine these actions. Understanding what social innovation means also involves understanding the institutional environments in which actors are embedded and how these environments constrain or enable their actions.

Adopting Cajaiba-Santana's "structuration perspective" implies understanding innovation as a result of a continuous process subject to the mechanisms of action formation and the agency-structure duality (Giddens, 1984). Thus, innovation involves reciprocity mechanisms between individuals. It is necessary to consider aspects such as values, culture and objectives, people, knowledge and technology, and finally, how these elements are assimilated into social practices (Piteira, 2014). In this sense, social innovation can be defined as "(…) an idea that deliberately attempts to better satisfy explicit or latent social needs and problems, resulting in new or improved capabilities, and in the transformation of social and power relations, aiming at social change and the establishment of new social practices that positively affect the lives of individuals" (Pinto et al., 2021:65).

Social innovation is seen as a critical driver of sustainable development. It enables the creation of new forms of value that go beyond economic and financial considerations and address social, environmental, and cultural challenges (Koval et al., 2023). In the circular economy context, social innovation can play a crucial role in facilitating the transition to a more sustainable and equitable economy by creating new business models, social enterprises, networks, and platforms that foster circularity and social inclusion (Marchesi & Tweed, 2021).

The relationship between circular economy and social innovation has received increasing attention in the academic literature, with a growing number of studies exploring the potential synergies, tensions, and challenges of these concepts. Some studies have focused on the role of social innovation in promoting circular economy by identifying the drivers, barriers, and impacts of social innovation initiatives that aim to foster circularity and social inclusion (e.g. Boons & Lüdeke-Freund, 2013). The authors emphasize that social innovation and circular economy are closely intertwined, and both are crucial for achieving sustainability. It is argued that social innovation can enhance circular economy practices by promoting collaboration and knowledge sharing among stakeholders, co-creating solutions, and enabling experimentation and learning. At the same time, circular economy can facilitate social innovation by creating new opportunities for local communities, fostering social inclusion, and promoting social entrepreneurship. Boons and Lüdeke-Freund (2013) also highlight the importance of social innovation in driving circular economy transitions. They argue that social innovation can help overcome some of the barriers to circular economy, such as lack of awareness, institutional resistance, and inadequate policies. Moreover, social innovation can generate new business models and forms of governance that are more inclusive, participatory, and adaptive.

More specifically, social innovation can contribute to circular economy practices by addressing cultural and behavioural barriers that prevent adopting such practices. For instance, social innovation initiatives focused on education and awareness-raising can create a culture of reuse and recycling and encourage citizens to adopt sustainable lifestyles (Marchesi & Tweed, 2021). In this chapter, we argue that intentional sustainable communities can be a fertile dimension for the emergence of socially innovative circular economy practices.

Intentional sustainable communities are a type of community where individuals voluntarily choose to live according to a set of sustainability principles (Liftin, 2009; Kunze, 2012). It is a type of community where there is a sharing of a substantive good, which is the adoption of sustainable ways of life, environmentally, economically, and socially (Kirby, 2003). Moreover, they are also communities of practice, as they promote social learning, providing environments where different members can engage, learn, and reproduce new practices and skills (Russell, 2013). ISCs, like other types of contemporary intentional communities, are built from practices, symbols, values, and ideas. However, the specificity of these communities is that those dimensions are oriented towards what members believe to be a cohesive, fair, and sustainable society. In this sense, these communities stimulate scenarios for human action and for the production of new contexts (material, social, and imaginative) resulting from historical, environmental, and imaginative contingencies (Waerther, 2014; Kirby, 2003), where creativity and social innovation play a key role.

In fact, circular economy is a systemic approach that requires new ways of thinking about resource use and production processes. Creativity plays a critical role in this process by enabling individuals and organizations to identify innovative solutions to complex challenges, such as reducing waste, increasing resource efficiency, and promoting sustainable consumption and production practices (Borowski & Salzano, 2017). Moreover, creativity is critical for engaging consumers and stakeholders in circular economy practices and is necessary for addressing the social dimensions of circular economy. The circular economy cannot be achieved solely through technological solutions; it also requires changes in social norms, behaviours, and governance structures. Social innovation and creativity are essential for developing new models of governance, community engagement, and stakeholder participation that enable more inclusive and equitable forms of circular economy (Cooper, 2017).

The circular economy concept is not limited to industrial processes and products. However, it extends to the world of arts and crafts, where do-it-yourself (DIY) artefacts play a crucial role in promoting a circular economy. In a circular economy, the value of products and materials is preserved by keeping them in use for as long as possible. Do-It-Yourself (DIY) artefacts have the potential to contribute to a circular economy by providing a platform for individuals to repair, repurpose, and upcycle existing products and materials (Barbero, 2016). This can reduce waste, increase resource efficiency, and enhance creativity. DIY artefacts also promote a culture of sharing and collaboration, which can help to build resilient communities and reduce environmental impact (Chen, 2012).

While DIY artefacts and maker spaces have the potential to contribute to a circular economy, some challenges need to be addressed. One of the challenges is the need for more awareness and understanding of the circular economy concept (Chen, 2012). This can lead to a lack of interest in DIY artefacts and maker spaces. Another challenge is the availability of resources, such as tools, equipment, and materials which can limit the ability of individuals to engage in DIY activities (ibid., 2012). However, there are also opportunities, such as the potential to create new business models based on circular principles and engage in sustainable practices that can lead to a more resilient and equitable society. The case study that this chapter focuses on demonstrates the importance of micro-scale practices for sustainable development, particularly in the field of circular economy, where creativity and DIY play a central role.

3 Methodology

This chapter aims to understand the importance of circular economy practices emerging from the micro-scale, identifying potentials and barriers. Furthermore, it also aims to analyze these practices as examples of social innovation. To this end, the study uses a qualitative methodological strategy based on the case study of the sustainable intentional community PPAUW, using semi-structured interviews as a data collection technique.

A case study is a research strategy based on an in-depth empirical investigation of one or a few phenomena, intending to explore the configuration of each case and develop and evaluate theoretical explanations (Yin, 2014). This means that the case is not just a unit of analysis but a theoretical category. The case is not spatially delimited a priori, and this delimitation, spatial or otherwise, results from a theoretical conceptualization used by the researcher (Rueschemeyer, 2003). PPAUW was chosen as a case study because it is an off-the-grid community with good environmental performance indices (Nogueira et al., 2022) that has as its central mission the reduction of its environmental impact through the reuse of discarded material.

Data was collected through semi-structured interviews with PPAUW members. These interviews aimed to understand how these individuals construct the meanings of the notion of social innovation and which practices developed by the community are considered the most innovative. The semi-structured interview is an instrument whereby the researcher has a set of questions oriented in a script or guide, but where there is the possibility of changing the sequence of questions and where the interviewer is free to ask additional questions if he/she considers it necessary to explore the interviewee's answers (Bryman, 2012).

Due to the limitations imposed by the pandemic of COVID-19, the interviews were conducted online. There are a number of disadvantages to conducting interviews online which relate to three main dimensions (Bryman, 2012): access to the Internet, the loss of trust associated with information exchanges made in an online context, and the lack of physical presence which hinders the ability to pick up visual cues to understand how the interviewee is feeling during the interview. While the disadvantages associated with the first dimension are more difficult to overcome, the research team sought to develop strategies to minimize the others. In relation to the second dimension, the Zoom platform was used through an institutional account. This allows the association of an access password to the Zoom meeting, which was shared only with the interviewee, to avoid potential fraud. The invitations to the interviews were made by email, which allowed to engage in some previous conversations with the interviewees to create mutual trust and a sense of commitment. The questions were not sent out in advance, but some information was provided on the dimensions and topics to be addressed during the interview. The lack of physical contact was partially compensated by the use of video, in addition to audio, which allowed the interviewer to check possible facial expressions that were important for the interview.

The recruitment process began with a general contact where the study's objectives were presented and information about the data collection process was provided. At this stage, a "contact person" was identified who mediated the contacts. This person was asked to contact possible interviewees who should meet the following criteria: have a high degree of commitment to the community, have lived for more than 3 years, and be directly involved in the innovative activities developed by the community. An interview with a founding member was also requested. These criteria are due to the specific theme to be addressed. On the one hand, it implies that participants know the innovative dimensions of the community. On the other hand, participants should have a high level of commitment to the community and a certain residence length.

The interviewer exchanged emails with the possible participants in order to ascertain and confirm whether they met the established criteria. It is important to stress that there is always a particular bias introduced by the person who selects the individuals who seem to fulfil the criteria. This is one of the limitations of this research, but it was a possible strategy given the context. Determining the final number of interviews to be conducted considers the principle of saturation (Guerra, 2006). In this sense, the total number of interviews was appropriate for the theoretical and conceptual problematization.

A total of five interviews were conducted: with the founder and four members who are part of a group that recently joined this ISC, with an average duration of 1h14m. Three men and two women were interviewed, all of Dutch nationality. Except for the founder, the other interviewees have lived in PPAUW for about 3 years. All the interviewees, except the founder, have higher education. Two hold Master's degrees in areas related to agriculture and rural development and innovation, and three have a degree in arts. All interviewees have a history of occupying spaces and activism against private property. Closely associated with the Dutch *kraker* movement, these individuals see the occupation of property as a form of protest against the capitalist system, a way of speaking out against current property speculation, and also as a form of expression of their individuality. The experience of occupying spaces was relevant for these interviewees to have chosen PPAUW as their place of residence. In addition, they also see PPAUW as a space where they can live without spending much money in communion with nature, which allows them to have time to be creative and express their values, namely, regarding material detachment and the freedom associated with it.

The focus of the interviews was on three main dimensions: (a) the characteristics of the community, socially, economically, and environmentally; (b) the interviewees' representations of their understanding of social innovation; and (c) the description of the practices that they considered to be the most socially innovative. Interviews were then transcribed using NVivo qualitative data analysis software (QSR International Pty, Ltd., version 11.1, 2015) and analyzed using content analysis techniques.

4 Circular Economy Practices at PPAUW Community

4.1 Case Description

The intentional sustainable community of PPAUW is an informal group with no defined legal status, located in Wageningen, province of Gelderland, the Netherlands. At the moment, 13 people are living in PPAUW, a number that has been fluctuating since the beginning of the project in 2014. It is a community in a rural context but relatively close to the centre of Wageningen. PPAUW is on private land of about 4 hectares, which has been under occupation by the community members once it was in a state of disrepair and had no use. Initially, the founder had problems with the local government and the police because the occupation took place without the owner's knowledge. What followed was an effort by the founder to demonstrate his plans to rehabilitate the land. Currently, the members' residences and the activities developed in PPAUW have the landowner's consent and are tolerated by the local government.

The founder and the members of PPAUW have a history of squatting on empty land and buildings. They are used to living with a certain insecurity and volatility regarding their residence place. This ISC is commonly defined as an "itinerant eco-circus" because of the imminent possibility of moving elsewhere. For this reason, its members do not build permanently, and all their dwellings are caravans or tiny houses that can be moved. Changes are predicted for this ISC in the near future as the municipality is preparing a zoning plan for the area where PPAUW is located. This community is expected to disappear at the end of 2023, and a new phase will begin in a new area. PPAUW rejects the idea of private property and operates on a low-income logic. Each member is responsible for the upkeep of his or her mobile home and pays around €100 a month for common expenses such as food and infrastructure repairs.

At the social and relational levels, the community has faced some difficulties. It underwent moments of restructuring with the departure of several members who were in PPAUW since the beginning and with the entry of some young people who are part of the current nucleus. Challenges still persist despite the new ideas that the newer members have brought to the community. Most are related to divergences in the members' different views, values, and ways of life. There is a set of individuals with anarchist ideologies who reject all forms of control. Another group, more focused on ecology values, has a vision more focused on the community's economic, environmental, and social sustainability. The entry of the new members allowed the introduction of some principles of sociocracy in the decision-making and conflict management processes that have had some success in balancing the different visions and consolidating the commitment to the community's core values: creativity, freedom, and sustainability.

The community has benefited from the proximity to Wageningen University. Wageningen is a small municipality known for its university, mainly responsible for its number of inhabitants. One of the scientific areas in which the university excels is agriculture, and PPAUW has attracted many students to research its territory. In addition, the community organizes "action camps" lasting approximately 3 days, during which university students can learn and develop permaculture techniques. Currently, it is an open community that receives between 50 and 500 visitors per event, depending on the activities to be developed.

This community is off the grid since it is not connected to the public energy and water networks, which represents an added challenge, especially concerning managing waste from human activity. To address this problem, members have developed composting and greywater purification techniques. They use dry toilets, separating solid and liquid waste. Both the dry toilets at PPAUW were made from discarded materials. Dry toilets, also known as composting toilets, are a type of toilet that treats human waste in a dry environment, without using any water for flushing. This process allows the waste to decompose into a nutrient-rich fertilizer that can be used in agriculture or gardening. Dry toilets are important to sustainability and circular economy due to several factors, such as:

- Save water by eliminating the need for flushing, which can save thousands of gallons of water per year.
- Reduce water pollution by preventing the discharge of untreated sewage into waterways.
- Eliminate the need for expensive sewage treatment plants, reducing energy and resource consumption.
- Produce valuable compost, reducing the need for synthetic fertilizers, which can have negative environmental impacts.

- Can be used in off-grid or remote locations where traditional sewage infrastructure is not available.
- Can be used to promote sustainable agriculture by providing a local source of organic fertilizer.
- Can be used in disaster relief situations where traditional sewage infrastructure is unavailable or destroyed.
- Can promote community resilience by reducing dependence on centralized sewage infrastructure.
- Can reduce greenhouse gas emissions by preventing the release of methane, a potent greenhouse gas, from traditional sewage treatment processes.
- Can help to promote a more circular economy by converting waste into a valuable resource.

Regarding energy production, several vehicles have solar panels. Vehicles with solar panels, also known as solar-powered vehicles, are becoming increasingly popular in ecovillages and other sustainable communities. These vehicles are equipped with solar panels that generate electricity from the sun's rays, which is then used to power the vehicle's motor and other electrical components. Solar-powered vehicles offer several benefits, including reducing dependence on fossil fuels, lowering carbon footprint, promoting local energy production, increasing community resilience, encouraging sustainable transportation, and promoting a circular economy by upcycling solar energy to power vehicles that would otherwise use non-renewable energy sources. This practice is an important part of circular economy and sustainability efforts as they reduce dependence on fossil fuels, promote local energy production, increase community resilience, and provide a sustainable transportation option that reduces the environmental impact of transportation.

In addition, the community won a public tender for lighting a road near the land where it is located. Energy generated by the community powers the lighting, and local trees are used as lampposts. When the road is not in use, the lights automatically switch off, ensuring energy efficiency. However, energy management is challenging, especially in the winter when there are few hours of sunshine. At this stage, members must adapt their needs and reduce energy consumption to the minimum necessary.

Still, regarding transportation PPAUW has been developing several other experiences. At the moment, the community has a gas vehicle (LPG), an electric tricycle, a caravan with an exhaust gas neutralizer, and several bicycles, two of which are electric. Transportation plays a significant role in the circular economy, and it is important to consider its impact on the environment and society (Stahel, 2016). In this regard, the use of various eco-friendly transportation modes, such as electric tricycles, bicycles, and gas vehicles with an exhaust gas neutralizer, is critical in promoting sustainable transportation and circular economy. These transportation modes contribute to reducing greenhouse gas emissions, increasing energy efficiency, and promoting local energy production, which in turn helps to reduce the environmental impact of transportation. Additionally, the use of different transportation modes can also provide unique micro-experiences for individuals, enabling them to engage more closely with their surroundings, interact with their community, and promote sustainable living practices (Dapp, 2018). By utilizing eco-friendly transportation modes, communities can create a more sustainable and resilient transportation system, reducing their dependence on non-renewable resources, and promoting a circular economy that seeks to reuse and recycle resources while minimizing waste. Therefore, the importance of eco-friendly transportation and micro-experiences cannot be overstated in building a more sustainable and circular economy.

It is also possible to highlight several energy experiences regarding cooking developed in the community. Only local energy is used for cooking, and there are four main pieces of equipment: a solar dryer for dehydrating food, a solar light lens that allows cooking at the focal point, and two solar ovens. Figure 4 shows the solar dryer on the right and the solar light lens on the left. A solar dryer is a device that uses solar energy to dry food, herbs, fruits, vegetables, and other agricultural products. It consists of a box-like structure with a clear plastic or glass top that allows sunlight to enter and heat the contents inside. The solar dryer also has vents or fans to circulate air, removing moisture from the food and drying it out. A solar light lens is a device that uses a convex lens to concentrate sunlight onto a specific point, creating a focused beam of light that can be used for cooking. The solar light lens works by directing sunlight onto a pot or cooking vessel, heating it up, and allowing food to be cooked. The lens is usually mounted on a stand or tripod, allowing it to be adjusted to capture the maximum amount of sunlight.

Sustainable cooking artefacts play a crucial role in promoting circular economy and sustainability. The use of eco-friendly cooking equipment, such as solar dryers, solar light lenses, and solar ovens, enables individuals and communities to reduce their dependence on non-renewable resources and minimize their environmental impact. By utilizing only local energy for cooking, communities can promote the development of local renewable energy infrastructure and reduce the carbon footprint of their cooking practices. Additionally, the use of sustainable cooking artefacts can also promote a circular economy by encouraging the reuse and recycling of resources. For example, solar dryers and ovens utilize solar energy, a renewable resource that can be used repeatedly without creating waste, while the solar light lens concentrates sunlight to heat food, reducing the need for non-renewable energy sources. The promotion of sustainable cooking artefacts can also foster a sense of community and collaboration, with individuals coming together to share knowledge, skills, and resources for sustainable cooking.

Despite facing challenges such as the possibility of relocation, diverging views among members, and being off the grid, the community has successfully developed innovative solutions to manage waste, generate energy, and promote local consumption. The use of composting and greywater purification techniques, solar panels, and local energy for cooking and transport are examples of the community's commitment to sustainability. Additionally, the community has benefited from the collaboration with Wageningen University, which has allowed for research and learning opportunities for both community members and students.

4.2 Social Innovation and Creativity

One of the objectives of the case study is to understand the representations that the members of PPAUW have of the innovation practices they develop. In this research, innovation is understood from a constructivist perspective (Cajaiba-Santana, 2014) that emerges from agency and interactions between actors (Piteira, 2014). This means that each actor can attribute different meanings to the innovation process. Initially, an exploratory analysis was conducted on the information provided by the interviewees to understand which dimensions of analysis they attach more importance to (Namey et al., 2008), and consequently determine the trend of their viewpoints regarding representations of innovation.

In general, members of PPAUW perceive the innovation process as the creation of something new that seeks to respond to social needs creatively. For the PPAUW interviewees, the relationship between innovation and needs is more profound than the mere way of meeting them. In the speeches analyzed, it is implied that innovation processes and practices involve individual initiative, focused on restructuring and redefining the needs of each person. This re-evaluation is essential to the congruence between the skills and resources available for the development of innovation and the social needs that this innovation seeks to satisfy:

> Innovation is about creating new things that serve our needs. But in today's world new needs are created for profits. If we try to meet all those needs innovation becomes destructive. Innovation often is a tool of consumerism and capitalism. I think innovation is necessary but aligned with the boundaries and needs of all life on earth". (Interviewee 4, founding member, 8 years' residence)

One of the distinctive characteristics of this ISC is the use of abandoned materials and their re-use and reconfiguration into new artefacts that are useful to the community. There is a focus on DIY, i.e. the development of artefacts using discarded materials. As already mentioned, DIY artefacts can be valuable in promoting a circular economy beyond industrial processes and products. DIY artefacts allow individuals to repair, repurpose, and upcycle existing products and materials, preserving their value for as long as possible, and increasing resource efficiency (Barbera et al., 2016). The promotion of sharing and collaboration culture through DIY artefacts can help to build resilient communities and reduce environmental impact. Therefore, the emphasis on creativity was expected in the discourses of these interviewees, also due to their artistic background. There is a direct association between innovation and creativity, and participants often confuse the two notions.

> For me, innovation is, above all, creativity. Thinking of solutions to problems and then putting your hands to work and developing them. Reproducing what others do just because it works may be useful but it's not innovative. (Interviewee 3, events and education producer, 3 years' residence)

The link between innovation and creativity is relevant. Chen (2012) states that the two notions are often not distinguished and should be understood as parts of knowledge production and its transformation into value creation (social or commercial). Creativity (the generation of ideas) precedes innovation (the implementation of those ideas), or in the words of Gurteen (1998:6), "creativity is divergent thinking and innovation is convergent thinking". This is the logic behind the interviewee's excerpt: the ability to create ideas and "get down to business" to develop and apply them.

In fact, this data highlight the importance of creativity in the circular economy transition as it enables individuals and organizations to identify innovative solutions to complex challenges such as reducing waste, promoting sustainable consumption and production practices, and increasing resource efficiency. Technological solutions alone cannot achieve a circular economy; it also requires changes in social norms, behaviours, and governance structures (Borowski & Salzano, 2017; Cooper, 2017), and this focus on creativity in PPAUW is an example of that.

From a sociological perspective, creativity, like innovation, is a social phenomenon because the creative and innovative process involves the manipulation of symbols, rules, technologies, and materials and because individual motivations and capabilities derive from the cultural and institutional frameworks in which individuals are situated (Burns et al., 2015). The interviewees corroborate this idea, and when asked about what they understand by innovation, they highlight the importance of the community as an enabler of the process:

> Here at PPAUW I can be creative and develop things because I have time to think and free materials to test ideas. And that, for me, is very important for innovation because if I had a 9 to 5 job and lived in a "concrete jungle", my head would be busy with other things, I would be watching TV and I wouldn't be innovative. So, maybe that's it, innovation is the freedom to have time and space to think and to be creative. Or at least it's part of it. A big part of it". (Interviewee 5, multipurpose member, 3 years' residence)

In this sense, innovation and creativity are related to the embeddedness of actors in particular environments (Granovetter, 1985). Creative activities are socially contextualized, i.e. the creative capacity of individuals depends on the characteristics of contexts and vice versa. In this sense, the context can serve as a facilitator or blocker of creative action and subsequently welcome or reject innovation. PPAUW functions, for its members, as a context facilitator of creativity, which enables the development of innovation practices. As the most mentioned representations by the interviewees are related to the dimensions of needs and creativity, the practice that will be analyzed next is a set of practices that can be seen from a circular economy perspective: the production of discarded material and its transformation into artefacts that meet community needs.

4.3 *Circular Economy at PPAUW*

The mission of PPAUW is to use materials discarded by other individuals and institutions, categorized as waste (or of no use), and give them a new life and new uses. In the view of the interviewees, this is a potentiality because they believe that an excess of waste and obsolescence should be combated. However, it is also a difficulty because the artefacts they produce look unappealing by the standards of contemporary societies:

> Therefore, it is usually quite capital intensive if you want to make things really like look super neat. If you buy new wood and you have precisely the parts that you need, then, yeah, it will look fancier than if you use recycled wood. But it's also very sad that we're throwing away so much as society (Interviewee 5, multipurpose member, 3 years' residence).

In PPAUW, this is one of the primary purposes: to use end-of-life materials, repair them or transform them into new utensils or goods. The images presented in the case description show some examples of this practice. There are two main reasons why circular economy practices at PPAUW can be analyzed as a social innovation. On the one hand, although the idea of reuse, recycling, and reduction is not recent, it has only just started to emerge as a strategic objective in Europe and formalized as a scientific concept, which shows its centrality in the discussion on transitions to sustainability (Prieto-Sandoval et al., 2018). On the other hand, PPAUW members identify their needs and seek to meet them through the reuse of materials, using creativity and art to give them new uses, thus, being in line with what is understood as social innovation (Mulgan et al., 2012). The interviewees themselves consider these community-developed practices as the most innovative. Although they serve to meet their own needs, they end up showing those who visit them how the circular economy works in practice:

> We use a lot of, like, scrap materials from the society around us and we are doing, we are now creating these of great facilities with a low budget, as compared to maybe a very innovative project in a fancy sustainable neighbourhood, or something. Because, this is only waste if you are not innovative. So, we show society that closing cycles or living in an off-grid way is not necessarily super high tech and very expensive, so to say, because there's a lot of waste in our society that can be used. (Interviewee 2, multipurpose member, 3 years' residence).

Furthermore, everything is managed collectively, from the identification of the need to the development of the idea to the search for materials and the realization of the solution. The members of PPAUW have a dynamic, non-formal, and non-hierarchical organizational structure due to the anarchist ideology of some of its members. Nevertheless, they have weekly meetings in which they discuss structural issues for the community, the collection of materials, and the development of artefacts being one of them.

> This also has a bit of social innovation because we are all invested in the process. And sometimes, I think that my needs are a priority and someone else may have another idea and we have to reach a consensus because we don't have enough material for everything. That part of the process is also important. It shapes the way we relate and the way we adapt to the different needs that exist here. (Interviewee 1, responsible for composting and communication, 3 years' residence)

Table 1 describes the main social changes that occur in the development of these circular economy practices and identifies their main potentials and limitations.

Table 1 Circular economy at PPAUW

Impact stage	Actor interaction	Social relations	Power relations	Social practices
Identification of the need	All PPAUW members are involved in the discussion on priority needs. There is no interaction at this stage with actors outside the community.	Social relationships embody a set of management of the different needs of the group.	At this stage, there are no individuals with greater power in the process as the community has a horizontal structure.	The needs are discussed in weekly group meetings where some techniques of sociocracy and non-violent communication (even if informally) are used.
Development of ideas and prototypes	Prototype development is done collectively but managed by members who have more skills to develop certain artefacts.	Social relationships are shaped through the collection of end-of-life materials free of charge from other institutions or individuals.	Power relations changed due to the non-use of money to acquire the materials necessary for the development of artefacts.	Social practices that contradict dominant consumption models and therefore do not feed the production of materials.
Scale up and dissemination	They try to involve other actors, namely their visitors to show the possibility of developing artefacts using recycled and reused materials.	They try to make visitors want to invest in DIY and offer practical courses on how to give new life and use materials considered as "waste".	There is difficulty in shaping power relations at the stage of scaling up and diffusing the practice as the appearance of the artefacts developed is not appealing for investment.	The social practices that change are at the level of the community and its visitors. However, there are an increasing number of neighbours who seek to understand whether PPAUW members are interested in keeping some material before discarding it.
Social change	There are no elements to verify how actors interact at this stage of social change because practice is still located at the micro level of the community.	Not applicable.	The practice has the potential to alter power relations by taking power away from industry and transferring it to individuals, as it implies that they develop the artefacts they consider necessary for their needs without resorting to buying new products.	There is a potential to change social practices, strengthening the idea of the 4R's of the circular economy which would imply a greater commitment from individuals to reduce, reuse, recycle and restore the goods they use. At this stage, it is only a potential idea as social practices are still very much at the level of recycling only.

Source: Author's elaboration based on Mulgan et al. (2012) and Cajaiba-Santana (2014)

Circular economy practices are still at an incipient stage, occurring only in specific niches, as is the case of PPAUW. There is still a long way to go before these innovations can be scaled and disseminated to other contexts, thus introducing social change. As already mentioned, there remains an insistence on individual efforts to recycle to the detriment of reducing consumption, reusing goods, and repairing them to develop new functions. This is a trend that should be attempted to counter in the coming decades due to the political commitment to implement the circular economy in businesses and organizations (EC, 2020). Nevertheless, like any other transition process, this requires a long adaptation period for social change towards sustainability. Specifically, in the case of PPAUW, the main difficulty encountered lies in the aesthetic attractiveness of the artefacts produced. In this sense, it is crucial to find investment to improve the aesthetics and make these artefacts attractive so they can eventually be disseminated.

5 Conclusion

The case study presented in this chapter sheds light on the multifaceted nature of innovation and its relationship to social needs, creativity, and sustainability. The members of PPAUW view innovation as a response to social needs, and they emphasize the importance of aligning it with available resources and skills. However, there is a concern that innovation driven solely by profit and consumerism can have destructive effects on society and the environment. Instead, innovation should be aligned with the needs of all life on earth to promote sustainability and social well-being. In this process, the role of creativity is of great importance because creativity and innovation are closely related once they should be understood as distinct parts of knowledge production and transformation into value creation. The ability to generate and apply ideas is essential for innovation, and it is socially contextualized. PPAUW serves as a context facilitator of creativity, enabling the development of innovation practices.

The circular economy practices of PPAUW, which involve using discarded materials to create new utensils and goods, are a prime example of how innovation can be aligned with sustainability goals. By reusing and recycling materials, PPAUW is promoting sustainable living and combatting excess waste and obsolescence. However, the challenge of producing aesthetically pleasing items can be a hindrance to widespread adoption of circular economy practices.

PPAUW's collective decision-making process fosters social relationships and shapes how the community relates to one another. By developing artefacts collectively and distributing power relations horizontally, PPAUW is challenging traditional power dynamics and promoting individual empowerment. These community practices have the potential to alter power relations by taking power away from industry and transferring it to individuals and demonstrate how circular economy practices can promote social innovation, foster community engagement, and lead to sustainable living. While there are challenges to making the practice more widespread, its potential impact on society is significant. One of the main challenges faced by circular economy practices developed at PPAUW is producing aesthetically pleasing items and making the practice more widespread. By understanding these challenges, researchers and practitioners can develop strategies to overcome them and promote more sustainable living.

However, this study is not without limitations. One of them is related with small sample size. The study may have limited generalizability due to the small sample size. PPAUW is a specific community, and the findings may not be applicable to other communities or contexts. Besides that, there is a potential self-report bias once it is relied on self-report data from PPAUW members. Participants may have provided socially desirable responses or overestimated their level of creativity or innovation. The study focused primarily on PPAUW's circular economy practices and their relationship to social innovation, creativity, and sustainability. Other aspects of innovation, such as technological or scientific innovation, were not explored. In order to overcome some of these limitations future lines of research are suggested. For example, future research could examine how PPAUW's circular economy practices and community decision-making processes compare to those of other sustainable communities. This could provide a broader perspective on the effectiveness of PPAUW's practices and identify areas for improvement. Other possible paths include examining the impact of PPAUW's practices on the broader community.

Based on these findings, there are several policy implications that can be considered. First, policymakers should prioritize the alignment of innovation with social needs and sustainability goals, rather than solely focusing on profit and consumerism. This can be achieved through creating incentives and regulations that promote sustainable innovation practices, such as circular economy practices. Second, policymakers should prioritize creativity as a crucial component of innovation. By investing in education and training programs that develop creativity and innovative thinking, individuals can be equipped with the skills necessary for developing sustainable and socially responsible innovations. Third, policymakers should prioritize community-based innovation practices that empower individuals and foster social relationships. This can be achieved through supporting the development of community-based organizations and initiatives that encourage collective decision-making and horizontal power dynamics. Lastly, policymakers should consider the potential impact of circular economy practices on waste management and sustainability. They should develop policies that encourage the adoption of circular economy practices by individuals and organizations, as well as supporting research and development of new technologies that promote sustainable waste management practices.

Acknowledgements This work was supported by National Funds through FCT—Fundação para a Ciência e Tecnologia within the scope of the PhD Grant SFRH/BD/117398/2016 (Carla Nogueira), of the Scientific Employment Stimulus Program INST/00052/2021/CP2792/CT0001 (Hugo Pinto) and Project UIDB/04020/2020 (João Filipe Marques and Carla Nogueira). The authors deeply acknowledge the support of all those interviewed in the Los Portales community.

References

Barbero, S. (2016). Subjects of change: Micro & small enterprises. In L. Bistagnino (Ed.), *micro-MACRO*. ed. Ambiente.

BEPA. (2010). *Empowering people, driving change. Social innovation in the European Union*. Bureau of European policy advisors. European Commission. https://doi.org/10.2796/13155

Boons, F., & Lüdeke-Freund, F. (2013). Business models for sustainable innovation: State-of-the-art and steps towards a research agenda. *Journal of Cleaner Production, 45*, 9–19. https://doi.org/10.1016/j.jclepro.2012.07.007

Borowski, I., & Salzano, M. (2017). Re-inventing business models through circular economy concepts: An exploratory research. *Journal of Cleaner Production, 168*, 1531–1542.

Bryman, A. (2012). *Social research methods* (4th ed.). Oxford University Press.

Burns, T. R., Machado, N., & Cortea, U. (2015). The sociology of creativity: Part I: Theory: The social mechanisms of innovation and creative developments in selectivity environments. *Human Systems Management, 34*(3), 179–199. https://doi.org/10.3233/HSM-150839

Cajaiba-Santana, G. (2014). Social innovation: Moving the field forward. A conceptual framework. *Technological Forecasting and Social Change, 82*, 42–51.

Chen, K. (2012). Organizing creativity: Enabling creative output, process, and organizing practices. *Sociology Compass, 6*(8), 624–643. https://doi.org/10.1111/j.1751-9020.2012.00480.x

Circle Economy and Ecofys. (2018). *Circular economy: A powerful force for climate mitigation*. Retrieved from https://www.circle-economy.com/publications/circular-economy-a-powerful-force-for-climate-mitigation/

Cooper, T. (2017). Which way to turn? Product longevity and business dilemmas in the circular economy. In J. Chapman (Ed.), *Routledge handbook of sustainable product design, chapter 27* (pp. 405–423). Routledge.

Dapp, M. (2018). Toward a sustainable circular economy powered by community-based incentive systems. In H. Treiblmaier & R. Beck (Eds.), *Business transformation through Blockchain* (pp. 153–181). Springer.

European Commission. (2020). *Closing the loop—An EU action plan for the circular economy*. EC.

Geissdoerfer, M., Savaget, P., Bocken, N. M. P., & Hultink, E. J. (2017). The circular economy—A new sustainability paradigm*? Journal of Cleaner Production, 143*, 757–768.

Ghisellini, P., Cialani, C., & Ulgiati, S. (2016). A review on circular economy: The expected transition to a balanced interplay of environmental and economic systems. *Journal of Cleaner Production, 114*, 11–32.

Giddens, A. (1984). *The constitution of society*. Polity.

Goyal, S., Esposito, M., & Kapoor, A. (2016). Circular economy business models in developing economies: Lessons from India on reduce, recycle, and reuse paradigms. *Thunderbird International Business Review, 60*, 729–740.

Granovetter, M. (1985). Economic action and social structure: The problem of embeddedness. *American Journal of Sociology, 91*(3), 481–510. http://www.jstor.org/stable/2780199

Guerra, I. C. (2006). *Pesquisa qualitativa e análise de conteúdo - Sentido e formas de uso*. Principia Editora.

Gurteen, D. (1998). Knowledge, creativity and innovation. *Journal of Knowledge Management, 2*(1), 5–13. https://doi.org/10.1108/13673279810800744

Kirby, A. (2003). Redefining social and environmental relations at the ecovillage at Ithaca: A case study. *Journal of Environmental Psychology, 23*, 323–332. https://doi.org/10.1016/S0272-4944(03)00025-2

Kirchherr, J., Reike, D., & Hekkert, M. (2017). Conceptualizing the circular economy: An analysis of 114 definitions. *Resources, Conservation and Recycling, 127*, 221–232.

Koval, V., Arsawan, I. W. E., Suryantini, N. P. S., Kovbasenko, S., Fisunenko, N., & Aloshyna, T. (2023). Circular economy and sustainability-oriented innovation: Conceptual framework and energy future avenue. *Energies, 16*, 243. https://doi.org/10.3390/en16010243

Kunze, I. (2012). Social innovations for communal and ecological living: Lessons from sustainability research and observations. *Journal of the Communal Studies Association, 32*(1), 50–67.
Liftin, K. (2009). Reinventing the future: The global ecovillage movement as a holistic knowledge community. In G. Kütting & R. Lipschutz (Eds.), *Environmental governance: Knowledge and power in a local-global world*. Routledge.
Marchesi, M., & Tweed, C. (2021). Social innovation for a circular economy in social housing. *Sustainable Cities and Society, 71*, 102925. https://doi.org/10.1016/j.scs.2021.102925
Mulgan, G., Tucker, S., Ali, R., & Sanders, B. (2012). *Social innovation - What it is, Why it matters and How it can be accelerated* (pp. 1–52). Oxford Said Business School Skoll Centre for Social Entrepreneurship - Young Foundation.
Murray, A., Skene, K., & Haynes, K. (2017). The circular economy: An interdisciplinary exploration of the concept and application in a global context. *Journal of Business Ethics, 140*(3), 369–380. https://doi.org/10.1007/s10551-015-2693-2
Namey, E., Guest, G., Thairu, L., & Johnson, L. (2008). Data reduction techniques for large qualitative data sets. In G. Guest & K. M. MacQueen (Eds.), *Handbook for team-based qualitative research* (pp. 137–162). AltaMira Press.
Naudé, W. (2011). *Entrepreneurship and economic development: Theory, evidence and policy*. IZA Discussion Paper, 5711.
Nogueira, C., Marques, J. F., & Pinto, H. (2022). Intentional sustainable communities and sustainable development goals: From micro-scale implementation to scalability of innovative practices. *Journal of Environmental Planning and Management*. https://doi.org/10.1080/09640568.2022.2106553
Pinto, H., Nogueira, C., Guerreiro, J. A., & Sampaio, F. (2021). Social innovation and the role of the state: Learning from the Portuguese experience on multi-level interactions. *World, 2*(1), 62–80. https://doi.org/10.3390/world2010005
Piteira, M. (2014). *Construção Social da Inovação*. Clássica Editora.
Pomponi, F., & Moncaster, A. (2017). Circular economy for the built environment: A research framework. *Journal of Cleaner Production, 143*, 710–718.
Prieto-Sandoval, V., Jaca, C., Ormazabal, M., & Santos, J. (2018). Towards a circular economy: A multi-level analysis of business models in the mobile industry. *Journal of Cleaner Production, 172*, 3454–3465. https://doi.org/10.1016/j.jclepro.2017.10.071
Rueschemeyer, D. (2003). Can one or a few cases yield theoretical gains? In M. James & D. Rueschemeyer (Eds.), *Comparative historical analysis in the social sciences* (pp. 305–336). Cambridge University Press.
Russell, H. (2013). Sharing conventions: Communities of practice and thermal comfort. In N. Spurling & E. Shove (Eds.), *Sustainable practices: Social theory and climate change* (pp. 103–114). Routledge.
Stahel, W. R. (2016). The circular economy. *Nature, 531*, 435–438.
Waerther, S. (2014). Sustainability in ecovillages—A reconceptualization. *International Journal of Management and Applied Research, 1*(1), 1–16. https://doi.org/10.18646/2056.11.14-001
Yin, R. K. (2014). *Case study research: Designs and methods* (5th ed.). Sage.

Universal Design Applied to Inclusive and Sustainable Economy for All

María-Rosa Oria-Segura and Miquel-Àngel Muñoz-Medina

Abstract The territorial sustainability of the economy and access to quality employment, considering the real possibilities of each territory, are closely and indissolubly related. Often, the design of training strategies for employment and entrepreneurship is presented as a series of dissociated actions: sometimes focusing on the economic sphere (in the promotion of a sector *product*, its sustainability and profitability), other times only focused on the social sphere (with equity measures towards a specific interest group, such as immigrants, gender groups and ethnic minorities) or in the design of training programs that do not take into account the needs and strategies of the productive sectors and the new trends and occupational niches (place of destination of the people trained). Statistical data impose the reality on us. Despite all the efforts made, the employment of PwID is much further away than that of people without disabilities.

This chapter establishes that support for the group of People with Intellectual Disabilities (PwID) can and must be an element of social cohesion and economic development of the territories. We describe the experience we have developed and demonstrate that we have been able to promote employment with training that has put PwID in contact with the real world of work.

Keywords Social inclusion · Inclusive education · Social economy · New business model · Social and comparative education

---

M.-R. Oria-Segura
Department of Sciences of Education, Faculty of Education and Psychology,
University of Extremadura, Badajoz, Spain
e-mail: mros@unex.es

M.-À. Muñoz-Medina (✉)
UCM, UAB, UEx, Madrid, Spain

C. Rego et al. (eds.), *Transitioning to a Circular Economy*, Studies on Entrepreneurship, Structural Change and Industrial Dynamics,
https://doi.org/10.1007/978-3-031-77661-8_6

1 Introduction

The territorial sustainability of the economy and access to quality employment, considering the real possibilities of each territory, are closely and inextricably linked. Often, the design of training strategies for employment and entrepreneurship is approached as a series of dissociated actions: sometimes focused on the economic sphere (in the promotion of a *product* of the sector, its sustainability and profitability), others only focused on the social sphere (with equity measures towards a certain interest group, such as immigrants, gender groups and ethnic minorities) or in the design of training programmes that do not take into account the needs and strategies of the productive sectors and the new trends and occupational niches (place of destination of the people trained).

This chapter establishes that support for the group of People with Intellectual Disabilities (hereinafter, PwID[1]) can and must be an element of social cohesion and economic development of the territories. They can generate employment opportunities not only for this group per se but also in compliance with the Universal Design Principles that aspire to inclusive strategies valid for multiple situations (it is also necessary to consider the auxiliary positions that can be generated and that can be filled by non-disabled people). We understand that the dissemination of this knowledge should not be addressed in separate publications but can and should be really integrated in society and be present in general reflections. Based on our experience, we believe that entrepreneurship for employment in PwID is not possible if it is not approached simultaneously and in a coordinated way from three areas: economic-business, training-academic and social. In other words, in a systemic way and considering the strategies deployed by all stakeholders, according to their legitimate interests.

Statistical data impose reality on us. Despite all the efforts made, the employment of PwID is much further away than that of non-disabled people: The Spanish Institute of Statistics (INE, 2022a) estimates that the employment rate of people with disabilities in Spain is 26.7% (not only PwID, but any type of disability). This data contrasts sharply with the 64.3% employment rate of non-disabled people, 89.6% of whom are employees. This means that for disabled people—more specifically PwID—it is advisable to train not only the individual person (PwID) in entrepreneurship, but the inclusive entrepreneurial attitude must be transversal: from the person who starts up a business to the society that welcomes him/her.

Of course, national and international policies that provide opportunities for the three spheres to work together are essential as a kick-start. We analyse why they should only be a kick-start and push factor because, in the economic sphere, if the activity itself is not profitable and becomes eternally dependent on aid and

[1] In this text we consider, as subject of interest, only PwID. Sometimes, if necessary, we compare with other more accepted disabilities, such as sensory (blind or deaf) or motor disabilities, but highlighting the difference.

subsidies, it is doomed to fail. It will not be sustainable and will not be able to develop on its own.

If it is already difficult for Universal Design approaches to gain ground in industry and, in general, in society—it is a matter of making everything that is produced accessible to all people—, in the field of educational, PwID training for inclusive employment (and entrepreneurship when possible), there is still a lot to be done to train teachers, ordinary students and their families. Much remains to be done to ensure that Universal Design for Learning (UDL) is present in schools and in all types of training centres and not only in specialised ones.

PwID are often segregated from the educational system either directly (in special centres in the most extreme cases) or covertly through *diversification* programmes and alternative itineraries that lead to lower qualifications and fewer job training opportunities, as shown by educational statistics (from the State School Council, those collected by the IOÉ Collective in its numerous studies, CERMI collection, etc.).

For decades PwID reached the same levels of schooling, with no major differences: at most, they reach a level of Compulsory Secondary Education (ISCED Level 2) or, at most, a Baccalaureate level or equivalent in a Vocational Training Cycle (ISCED Level 3), but they are prohibited from further training, as legislation has always reserved access to the "higher" level (ISCED levels 5–8) on the basis of *ability*, without taking into account other post-secondary training (ISCED level 4) that can also be offered to PwID. In this way, PwID becomes a forgotten group, with barriers that are not always understood and inflexible glass ceilings. For example, we have observed that jobs are offered for profiles with a minimum level of training and maximum age requirements, which in this group is difficult to reach.

Finally, regarding the third area, it will be analysed whether the fragmentation into social entities specialised in different disabilities may have contributed to divergent rather than convergent forces—in terms of the main objective of providing inclusive employment opportunities and revealing points of tension.

Once the three areas indicated have been analysed, a proposal will be made that includes training, advice, and consultancy for these areas. This is not a utopian proposition, but rather this chapter presents real examples of the experiences promoted by the University of Extremadura and which, since 2017, have culminated in a "success case" carried out in the Community of Extremadura until the 2021–22 academic year. This experience shows how an institutional call offered by the European Social Fund, articulated through Fundación ONCE, has branched out and has generated new real employment opportunities for the PwID group. Furthermore, it will be compared with other initiatives developed in other Autonomous Communities of Spain (in addition to Catalonia), in order to transcend the anecdote and extract useful principles to similar contexts.

2 Contextual Framework

In this section, using everything that is already firmly established in Social Economy or Entrepreneurship, we want to build a framework that provides us, together with the economic parameters, with distinctive pedagogical features to shape the experience we are going to talk about in Sect. 3.

Thus, the first thing would be to understand that the Social Entrepreneurship Model for disability that we make in this proposal, results from the belief that when we talk about Social Inclusion—and above all related to the issue of employment, we talk in terms quoted by Álvarez-Ramírez (2023, p. 32) of *ableism* (European version) or *disableism* (US version), which is something to overcome, although both terms were common throughout the twentieth century and still persist in today's world. In fact, this is a problem because it leads to a negative concept of disability in general and of PwID in particular, because intellectual matter becomes a glass ceiling in an universe—that of disability—overcrowded with ceilings. This brings us to the first question of our contextual framework.

2.1 *Correcting Disabled People or Helping to Ensure that their Disability Is Not a Social Disadvantage?*

This question may seem extemporaneous, but traditionally, disability has been approached from the clinical paradigm of *deficit* and the very term ableism represents a clear example of this. Although we struggle, from educational spheres, from associations, etc., to achieve a much more social and interactive paradigm, previous centuries have given rise to attitudes and ways that have remained engraved in the collective subconscious and are really difficult to eradicate.

We are not going to describe here some widely known practices, since Antiquity, such as the infanticide of people with obvious disabilities (in Greek or Roman culture) or the conception of every human being as a divine creature, but at the same time considering any difference as something to be feared, a kind of supernatural punishment (in the theocratic Middle Ages). Later, we find the first educational experiences of sensorial disability in the Renaissance period.[2] Nowadays, in the Western society, we can establish the beginning of the systematic study of intellectual disability in the second half of the nineteenth century (i.e. Itard, Pinel, or a little bit later, at the beginning of the twentieth century, Binet), associated with the

[2] Understanding the origins of the ideals and principles that move us, or against which we struggle, is important to understand the context of the experience we are about to show. However, if we found similar circumstances (warrior or theocratic societies, or societies that focus on the human being as the *centre* of society) we can expect to find similar attitudes of infanticide, condemnation, or interest in trying to help these persons in their progress (although with natural adaptation to their specific contexts).

progress of scientism in many epistemological and health fields. The consequent tendency was the characterisation of individuals based on symptoms, syndromes, or typologies over and above anomalies according to statistics. This explains why psychological, medical, and diagnostic tests became indispensable to achieve objective classification and, consequently, to label individuals. In the deficit paradigm, it's important to correct the deficiencies observed in these individuals. But in case of failure, these people cannot avoid being labelled. They will suffer a lifelong stigma as carriers of anomalies inherent to their nature. Depending on the level of economic, social, and civic development of the society in which the individual lives, and as long as they are considered by society as *people worthy of attention*, they may be provided educational and supportive care, but they are usually segregated in specific institutions. Usually, the reasons for this are apparently good: "they will be well cared for, with specialised staff in their care…". But the perverse side effect is that the rest of the community is not involved in the problem. In individualistic societies, the focus is on the individuals directly affected, their families, and specialised staff. In welfare states, on the other hand, public administrations are also seen as responsible for solving this problem.

After a century of existence, the deficit paradigm has been replaced by the paradigm of social diversity. The UNESCO Salamanca Statement (1995) can be considered as the foundational document of this fresh paradigm, focusing on the needs of human beings. In 2001, another medically relevant support came from the International Classification of Functioning, Disability and Health (ICF), established by the World Health Organization (WHO, 2023). The ICF presented a very different picture. From now on, what is relevant is not only the physical impairment or the intellectual impairment suffered by the person but the relationship with a set of opportunities offered by his or her environment. What is really important is not the damage itself, but whether this type of person finds his or her possibilities for participation in the social and economic environment altered. In other words, the focus is not on the individual, but on all the possibilities in the environment to prevent them from becoming second-class citizens. This marks a difference between the two models. The old model only focused on the subject. In the new model, there are at least two points of action: The person and his or her environment. In 2006, The United Nations Convention on the Rights of Persons with Disabilities (CRPD) made the social and legal areas of the paradigm even wider. Subsequently, when this convention experienced a massive dissemination, The Office of the High Commissioner of Human Rights (OHCHR) developed an Advocacy Toolkit on how to present persons with disabilities socially. The toolkit contained some comments such as "disability is not a person's identity" so "omit mention of an individual's disability unless it is pertinent to the story" (OHCHR, 2008, p. 30). And other recommendations such as this one: mass-media should "show employees/employers with disabilities working together" (*ibídem*) to promote inclusive socio-occupational models in the collective worldview. Finally, the United Nations initiated in April 2018 the process leading to the Disability Inclusion Strategy, "to strengthen system-wide accessibility for persons with disabilities and the mainstreaming of their rights" (UN, 2019, p. 1). The World Bank continued these considerations "to integrate disability into development through its analytical work, data, and good-practice policies"(WB, 2023), publishing achievements in different emerging countries, considering disability in development, and linking it to equitable opportunities and a way to build sustainable communities.

As we can see, important instruments have been made to achieve a change, but unfortunately, after two decades, these recommendations are still necessary in the world of disability in general and almost unknown in the field of PwID. The main problem is attitudes, as a result of fear, ignorance, misinformation, or pure apathy, and that the deficit model, as it has been quoted by Álvarez-Ramírez (2023, p. 34), is highly compatible with capitalistic systems and other oppressive models such as patriarchy or colonialism. But there are more reasons for resistance to change. First, this society has become accustomed to *someone or something* taking care of PwID, without reflecting that all of us are called to create opportunities for PWD to *convivir*[3] with the rest of society, not only in a limited familiar or associative circle, or together with other people with disabilities, but also we must make PwID become active persons in our society.[4] Third, in Western society, economic models are not conducive to inclusive features. Competitive systems demand returns and, for this reason, PwID may be at a disadvantage. But this does not mean that PwID cannot be considered as *productive people*. This society values the more productive people but too often forgets there is no single standard of productivity. Unfortunately, the dominant economic models related to the care of PwID focus mainly on the welfare aspect and forget about cost-effectiveness with the effects mentioned in Sect. 3.1. Fourth, the theoretical developments related to declarations have not been rapidly translated into national legislation with the desired agility. Even in those countries where declarations have been adopted, none of them have gone beyond the level of mere stopgap, bureaucratic changes, or declaration of intent. Finally, it is not easy to identify a single collective target, and this is for two reasons:

- There is profusion of terms, from several spheres and points of views: educational, social, legal, clinical, economical, statistical… As a result, we can find subgroups and categories in several fields that hide functional realities (as we will see in Sect. 2.4). The issue of legal recognition of disability is particularly sensitive. Sometimes, the legal terms become outdated and synonymous with clinical categories. Consequently, the clinical paradigm does not disappear and even extends to other areas.[5]
- People themselves join together in associations that refer to their classic clinic category (association of *this* or *that* syndrome, disability, etc., for example,

[3] This Spanish word literally means "live *together* with".

[4] It is a problem not only for PwID, but also for elderly people, immigrants, women… groups that are also vulnerable to the economic system.

[5] For example: from 1995, as product of this "invasion", in the Spanish education system, after defining what special educational needs were (in the second Additional Regulation of LOPEGCE) it was necessary to clarify that "having special education needs" and "being disabled" where different conditions, and a person could have one without having the other, o could have both.

autism, Down…), thus losing the sense of *needs model* proposed by international instruments mentioned above. Sometimes, even the groups themselves reject certain options because they consider them to be too inclusive and are afraid of being diluted as a group of strength, in order to raise their demands. In this sense, for example, Spanish intellectual disability groups refuse terms such as *functional diversity* and continue to defend the term disability, because each person is *diverse* and they consider that the term *functional diversity* hides their real situation.

Finally, to answer the question that heads this Sect. 2.1, and in accordance with the model established by the aforementioned international instruments of UNESCO, UN, WHO, WB, etc., in the experience we show in Sect. 3, we have not tried to turn PwID into persons whose functional diversity goes unnoticed. This is not about camouflaging or constructing a parallel universe in which PwID can feel safe or forget their disability. We have paid attention to the following idea: disability puts them at a disadvantage. Our approach is based on the concept of Social Capital of these people, related to Social Justice (both aspects developed by Liamputtong, 2022) and linked to critical pedagogy theories, social and economic movements, and a collective model of society. From this angle, we think it is positive to promote non-segregated environments, with different strategies in order to ensure that their condition, ever-present, does not provoke a situation of inferiority, not having the opportunity for everyone, disabled and non-disabled people, to coexist and understand each other. In this way, PwID can feel that they are capable people because they really are, and society must see them like this and, therefore, accept that there are opportunities for all of them.

Educating to eliminate disadvantages implies that educators should not only have pedagogical or didactical knowledge focused on disability (these are necessary, but not the only ones). The fact that teachers and non-formal trainers can offer real opportunities involving knowledge of new productive realities, as they have a long-term vision of the skills to be developed in the training phase. And this also implies that training should not be organised around the needs caused by disability, classified into specific types (organic, psycho-emotional, intellectual, evolutional…). On the contrary, it could be an organisation of training according to the issues that they should face at work, which could become an obstacle to integration into the labour market (or any other area of their lives). To be more specific, we find the following items:

- Comprehension difficulties in different languages (not only the spoken language), in different situations (not only an academic situation).
- Motor development could be an obstacle for work actions.
- Self-care skills.
- Assertiveness, dependence or passivity, self-esteem, or social skills.
- Skills that nowadays are basic in any curriculum, for example, training in Information and Communication Technologies (ICT).

To sum up, we believe that the basic philosophy in training for employment should be linked to enhancing skills and competences in a normalised context, and not *correcting* to PwID.

2.2 Interdisciplinarity in Training

The interdisciplinarity[6] that inspires the experience of university training for the PwID's employment introduced in the Sect. 3 has several senses. Number one, the most obvious and traditional one, is to *bring together areas of knowledge*, but with a special nuance: these areas interact with each other. The concept of interdisciplinarity has a specific effect, when it comes to understanding the modular organisation of educational contents. In the last quarter of the twentieth century, *module* meant a complete set that can be taught even separately (López García-Cano, 1988), in a progressive sense, linked to the theory of microtasks and supported by operational classifications such as Bloom's Taxonomies. However, this view has evolved substantially. In our approach, a *module system* would be, following the definition of Padilla Arias (2012), implemented for university learning: a system that "helps to rationalise the offer […] to respond to specific social demands" (p. 73), built from "critical thinking and creative action" (p. 74), and proposing a "critical and interdisciplinary learning, where the essential is that the student becomes a protagonist actor" (p. 74).

Interdisciplinarity is also intended from the very beginning of the program. We are talking about the need to combine, from the planning stage, initial university training with the needs of the labour market, which can be called *interdisciplinarity of professional fields*. Usually, training for the employment of PwID is carried out by an isolated institution, such as the educational system, the public employment system, or private entities for the care of PwID. But these are separated actions, in particular fields, without lasting goals. This is a phenomenon recently described by researchers such as Queiruga-Santamaría et al. (2002), who have argued that "there should be communication between the educational sphere and the labour sphere, so that the offers matches demands of the business structure, because the target is that both, sharing common interests, can move together in the same direction" (Queiruga-Santamaría et al., 2002, p. 91).

The third sense of interdisciplinarity has been applied to the skills and competences promoted through training, so that multidisciplinary competencies (the so-called *soft skills* in the labour market) have been worked on (Alex, 2009; Calero López & Rodríguez-López, 2020). Usually, training for employment (in general, not only for PwID) can be very generic or very specific, or too academic (indeed,

[6] Note that we use *interdisciplinarity* and not *multidisciplinarity*. This is an important detail, because *inter*, according to the dictionary of *Real Academia de la Lengua Española*, is a prefix meaning contact, mixture or relationship of mutual cooperation, whereas *multi* has a connotation of coexistence, but does not imply that there is a relationship.

hard skills) or only focused on the emotional or personal part. In today's world, soft skills training has a high added value for companies and businesses and is in high demand, as jobseekers show a high level of concern, high levels of proactivity, and a genuine interest in lifelong learning. But these skills should be focused on real jobs in their environment; otherwise, they will fail and will not achieve their goal.[7] In our experience, the development of soft skills takes place in Sect. 3.2.3.

The fourth sense of interdisciplinarity, added by us, is working together with *many typologies of PwID in the same group*. Usually, when it is spoken about inclusion, it means that people of a specific collective (such as deaf people, blind people, down people…) are introduced into an ordinary majority group without disabilities. This type of inclusion requires *specific support*, but the so-called *ordinary group* is actually made up of people with different characteristics and with their own emotional and health needs who also need individualised support. Among PwID there is also a wide range of characteristics, but they forget them when they act within the circle of people with the same type of disability. It is true that each typology of PwID (Down's, borderline, autism spectrum disorder, social phobia, mental development delay, etc.) has some common characteristics, and if there are specific associations in groups, this fact makes them stronger and more visible. But, as stated in the UNESCO and WHO documents mentioned above and as associations say, disability does not define a person, just as our hair colour, height, or any other psychobiological characteristic does not define us in the eyes of other ordinary people.

2.3 *Dialogical Character of Planning and Implementation of Training*

The model underlying our experience was to explore possibilities of applying the principles of "Learning communities" (LC) in two-fold levels. First, in the planning and implementation of training. Normally LC are applied as an instrument for the emancipation of vulnerable people, due to problems of illiteracy, socio-economic status, and have become drivers of change in the educational and social reality of the environment. The conditions for the egalitarian dialogue that contribute to overcoming disadvantages have been widely described in the last two decades. It is said that:

> We are moving towards a new approach to education. Education for development wants to support and transmit sustainable development. This sustainable development must raise a

[7] In other spheres this is obvious: it is not possible to train future workers ignoring productive needs or scientific advances, or without knowing the labour insertion niche. However, this example is not perfect for to two reasons: first, because professional skills are rather predefined and concrete and PwID needs specific job profiles that are not so predefined, but, at the same time, they need flexibility to adapt that job according to their needs, not to adapt PwID to profile. Second, if this situation only joins trainers, pupils, and workers on the same area of knowledge, interdisciplinarity disappears in its wide sense, because in addition to trainers (one job) and workers (another job), they are inserted in the same area of knowledge, generating the same background, with common rules.

> fresh biological, economic, political and cultural development. To develop this new approach in any educational centre, it is necessary to involve all sectors of the community by joining forces to create real learning communities, where everyone can participate, contribute and learn. (Barrio de la Puente, 2005, p. 131)

The liberating potential of LC revealed over the two last decades should not be used with PwID at the end of their environment, but it makes sense to put it into practice, as a theoretical framework, together with the parameters mentioned in Sect. 2.1.Thus, egalitarian dialogue is also a *conditio* sine qua non for successful and productive interdisciplinarity in every possible sense (Sect. 2.2). Thus, "epistemological and methodological pluralism" (Moya Otero, 2003) cannot be separated from a critical approach to education. Pedagogical theories with a critical approach from their origins (Carr & Kemmis, 1998, to the present day, such as Moya Otero, 2003; Becerra Hernández & Moya Romero, 2010; or Morales Zúñiga, 2014), question traditional patterns of learning. It is true that learning through dialogue has existed since classical Greco-Latin origins (Plato, Socrates, etc.), but traditional PwID training is mostly based on clinical models of skills training, emphasising epistemological and psychological curricular contents to the detriment of social and pedagogical sources. Ironically, these critical approaches are present in other fields (economy, social action), but they are rarely connected with the formal educational system, creating parallel worlds (academic world, working world, associations, etc.) that do not converge in the dialogical meetings we stand for here.

In the egalitarian dialogue, everyone should speak without fear or prejudice, without arrogating to themselves any kind of supremacy over others. The legitimacy of everyone's opinions does not depend on the position of the people who speak. Dialogue generates significant and high-quality learnings because it helps any person or entity to acquire a deeper understanding of the reality in which everyone interacts. This is so because all the varied points of view of all the interlocutors are taken into account, all are valid because they are all present in reality. A successful, productive, and indeed egalitarian dialogue requires that all interlocutors recognise that they need each other, without a feeling of self-sufficiency. On the other hand, the next requirement implies that they cannot, once they have learned the model, replicate it without the help of the rest of the group. Mutual respect is also necessary, and if someone coordinates the group, he or she should do as primus inter pares. The principle of egalitarian dialogue is not unique. It is one of the six principles of so-called "dialogical learning" (Barrio de la Puente, 2005), and all of them are recognised in our proposal. Cultural intelligence, which means that collective creation is different from the point of view of quantity and quality, and instrumental dimension of learning, which means that one learns a concrete thing and, at the same time, provides the community with new tools, through learning, and this generates a positive transformation of people and the environment. An interactive solidarity is developed, respecting all differences, understanding that difference enriches the process and the final result.

2.4 *Normalised Training from a Social and Educational Point of View*

Basic approaches in the training of PwID should be based on the principles of normalisation. For most Western countries, it is appropriate to invoke the principle of normalisation as defined by Nirje (1999) some decades ago (specifically, in 1969), and to recall that PwID should have living conditions similar, as far as possible, to those of the average Western country. Instead of efforts, first *integrative* and then *inclusive*, the desired results have not been achieved.

However, there is one European country, Italy, where invoking normalisation did not make sense. Since the 1970s, Italy, in a very strong commitment to inclusion (which was the result of a social pressure[8]), built a structure that led health staff, teachers, families, and local entities to work together on short-term plans (6 months) and medium-term plans (2 years). In this way, real empowerment was achieved, trying to enable disabled people, supported by families, to develop a personal and professional career. According to Eurostat, this de-facto normalisation gave disabled people qualifications. For this reason, it would be good to analyse the relationship between qualifications and employment opportunities.[9]

In the PwID employment training, it is especially obvious the lack of normalisation. At scholar level, (ISCED levels 0–2, and may be ISCED level 3 if provided in same educational institutions) compulsory education does not qualify for work. This is the case from the ISCED level 4 onwards, where disabled people in general and PwID in particular, feel that doors are closed to them. As a consequence, they have to turn to specialised institutions, losing the possible previous normalisation. In the case of access to higher levels (from ISCED level 5 to 8), we can find critical problems such as those cited by Pavone and Bellacicco (2016): "difficulties in applying for services, problems in accessing information, negative or indifferent behaviour on the part of the academic community" or "limited availability of the teaching staff in elaborating flexible didactic proposal" (p. 105).

[8] In Italy, it was a *total integration* model, meaning a radical inclusive option: in the 1970s, parents of disabled people started a strong social movement to get their children regular schooling, not in segregated institutions, but going to the same centres and levels, with necessary aids. That movement led to *Legge* 517/1977, which put the basis for further actions: *Legge* 17/1999; *Legge* 104/1992; *Legge* 17/1999; *Legge* 328/2000, and *Legge* 134/2015, with their correspondents' legislative extensions.

[9] This aspect has already been analysed in a previous research (Oria-Segura, 2006). Indeed, two decades ago, there was a striking phenomenon: the difference between rates of training and access to the labour market achieved by people with disabilities (severe or moderate) in Italy, compared to the same population in Spain.

2.5 Comparative-Functional Educational Planning

We stand for the presence of Comparative Education in theoretical fundamentals of our proposal, because it is a subject that offers a research methodology (García-Garrido, 1996; Ferrer, 2002; Vega Gil, 2011; Caballero et al., 2016), where it is essential to understand and analyse how to adopt successful experiences and their mechanisms. The second reason is that Comparative Education offers ways of categorising the units of study (Alexander et al., 1999; Bray et al., 2014) so that, even if the proposed experience belongs to one unit, the presentation of the structure allows further comparisons with other regions or even with other countries.

Furthermore, Kazamias (2009) Comparative-Functional Education raises the question of interdisciplinarity and the humanistic approach in the analysis of the role of the educational process in society, which implemented in our experience, involves analysing and correlating variables and functions that the training of PwID must develop in a specific society (Extremadura, in this case). It is possible that we find ourselves with two different social, economic and geographical realities and, consequently, with different structures. But we can pay attention to the variables of educational phenomena, how they interact with each other, and how they interact with functions of other social entities. In this way, we come to understand how the educational phenomenon should work and assess whether successful experiences can be transferred to other contexts.

With this initial approach to the definition of jobs in an economical activity, we conclude that the first step would be to reflect on tasks. Then the tasks are "organised and gathered according to standards and the set of tasks is assigned to a job" (Bilbao, 2000, p. 75). And the last step is to find people who fit the mould. It is logical that employers want to have the most versatile person possible to face future changes in their needs. However, if the needs are task-related, we can think of new ways of organising tasks that can accommodate new realities related to PwID. This involves a paradigm shift in work activity and training for this specific job. On the other hand, all training models for PwID tend to focus on handling tasks, with no possibilities for readjustment when the needs of the labour market change. It is important to clarify that there can be many functions to be performed from various polyvalent units. It is not a matter of belonging to the entity in question. In the end, what should be relevant is how the tasks are organised, choosing those who are able to perform them efficiently. This is the origin of the model developed in three phases of action in Sect. 3.

Another issue would be the poor statistical definition of disability. Care for people with disabilities is absolutely fragmented and segregated. On the other hand, the entities that study the problem change their methodologies, as the website of INE (2022b) recognises, and also offer inhomogeneous data, based on different aspects of the reality of people with disabilities. The main consequence is that statistics are hardly comparable when it's time to gather information from different sources. Some researchers consider that it is not possible to establish universal categories on disability and think that this universalism is as modern, colonial, and western construction (Mara-Danel et al., 2021, p. 37). Beyond the justice social debate, developed in the Sect. 2.1, these disagreements seriously affect the very conceptualisation of disability, being a real obstacle to understanding its magnitudes and the sectors that may be involved, which means that it is almost impossible to compare statistics on people with disabilities (comparisons between countries and within the same country, from private or public agency sources). This issue is highly relevant because experts consider that the poor definition of the phenomenon determines measurement problems that can sometimes be really serious (Huete-García & Quezada-García, 2012, p. 10). These authors put on the table large number of associated problems: fragmented data, outdated sources, and taking into account that each organism collects information for different purposes, the prevalence of data coming from the medical field that are not useful to answer crucial questions related to realities of discrimination, support needs, inclusion, social participation, empowerment… (Ibídem, p. 11). Problems that affect organisms such as Eurostat, which has changed its forms of classification, following the International Classification of Functioning, Disability and Health (ICF) 2001 and UN-Convention of rights of Persons with Disabilities (CRPD) 2006. However, if one consults online databases of Eurostat (2023), it is easy to see that all disability issues (115 in total) are listed in its Data Browser under the epigraph of Health. And there they include items related to health, but also to educational attainment level, employment status, incomes, home care, barriers for social integration… all of them. Moreover, Eurostat does not classify people who are already in the severe or moderate category. At present, Eurostat classifies them according to the concept of *dependency* in one or more domains.[10] This is a new obstacle to overcome, as it is difficult to compare past figures with current ones because the percentages do not include the same groups, which distorts the reading of the data significantly.

From the experience described in this chapter, we also miss another model of scientific documentation consistent with the social-critical approach mentioned in the previous sections. There is a lack of qualitative studies, comparative cases, or life stories that illustrate, beyond statistics, the reality of disability. In this regard, it is worth highlighting the work carried out for decades by Ebersold at the Organisation for Economic Cooperation and Development (OECD), who has also valued the influence of globalisation on inclusive education policies. This author tries to "identify how to decentre the comparative framework to develop a possible design and research structure for comparing inclusive education policies at the European level" (Ebersold, 2021, p. 142) and assumes that "approaches to disability echo the way a society mirrors and legitimates itself through norms and standards produced by institutional bodies" (*Ibídem*). We know that even qualitative studies have many problems. For example, and precisely related to university access for people with disabilities, some authors claim that "some small sample groups of the survey and/

---

[10] Assessments of severe or moderate disability were linked to the clinical approach. In the social approach of disability, physical injuries are one of the four domains. But this damage is related to these: daily living skills, difficulties to participate and environmental barriers. Therefore, this new classification is not comparable with the previous ones, as it includes more factors.

or particular types of disability are privileged, to the detriment, for example, of quantity/quality data triangulations" Pavone and Bellacicco (2016, p. 104). As a result, it is not easy to understand the different problems faced by people with disabilities at these educational levels.

2.6 Employment Social *Support*

It is extremely difficult, due to the privacy of the data and the way it is collected, to get an accurate picture of the employment/unemployment situation of PwID. Statistics show situations in ranges,[11] mixing different vital realities, with various possible situations. The non-specialised society does not distinguish between people with intellectual disabilities and people with mental disease which can also cause disability, but has a different bias.

Once again, we are faced with fragmented information, elaborated by different entities from incomparable data. Normally, these studies are developed in successive waves, producing outdated information. In 2003, Decisión 2001/903/CE launched The European Year of Disability, promoting accessibility of infrastructures, access to culture, the so-called Knowledge Society, and equal opportunities in education and training for people with disabilities. When Member States talk about educational guarantee, they are referring to the compulsory level. The main reason is clear. There is no government that recognises that it is attacking such a fundamental right as the right to education (quote from the UN Rapporteur on the Right to Education Tomasevski (2005). In Spain, complexity is added by the fact that educational competences are transferred to 19 autonomous governments (17 Autonomous Communities and 2 Autonomous Cities—Ceuta and Melilla). In 1997, the Colectivo IOÉ & CIMOP pointed out that, in 1986, five and a half million people (then, 15% of population) had some kind of disability problem. Furthermore, 40% of these people presented severe limitations. We can also find that, within the aforementioned group, there are 2.6 million working-age people, but only 553,000 people were actually working. This implies that more than 1.5 million people (about 5.04% of the Spanish population) were not working because of their disability, although they could actually do so. Since then, the now called "Survey on disability, personal autonomy and dependency" (EDAD by the Spanish initials) has been carried out in 1999, 2008, and 2020 (the latter published in April 2022 and updated in September). According to the INE's EDAD 2020, there were in Spain, about 1,583,600 people with disability, of whom 818,200 were women.

Here are some of the arguments related to labour insertion in 1997, which, in today's world can be considered current statements:

---

[11] For example, categorising by percentages, but including all kinds of disabilities. This may hide the fact that some disabilities, even if severe (sensory disabilities such as blindness and deafness, and motor disabilities when they affect lower limbs) are more employable than others with a similar percentage of severity.

> Taking into account that, except in special cases, functional limitations [...] do not invalidate the person for all types of jobs, but only for a range of jobs and that contributory pensions are compatible with simultaneous economical activities, it is obvious that people with disabilities are mostly excluded from the salaried labour market and this situation tends to get more serious in countries like Spain, with high rates of unemployment. (Colectivo IOÉ & CIMOP, 1997, p. 8)

The same text adds more significant reasons:[12] socio-economic levels, labour qualifications, economic recession due to the territory or the historical moment. Boldly, they dared to emphasise that positive discrimination measurements were necessary, but to a certain extent, because there was a real danger of generating new segregation areas, or the danger of promoting a culture of dependency (linked to collective or individual subsidy) instead of promoting genuine insertion.

2.7 Integral and Inclusive **Support** *from a Point of View of Social Business World*

In the world of social business entrepreneurship is an inherent feature, because it has to create new conditions and models that, as said (Ashraf et al., 2019), it is necessary to focus on disadvantaged beneficiaries, apply or create new hybrid theories, and develop collaborative entrepreneurship within the community. This is why in our experience we speak of *integral support*, because we consider support to PwID as integrating all perspectives of intervention: academic, social, and business/entrepreneurial.

Unity of action is indispensable from all possible angles, which is why it is necessary to have different social partners. In addition, a sustainable business model is essential. Geissdoerfer et al. (2018) provide one of the most recent, comprehensive, complete, and enlightening reviews on both topics; so it is a good resource to understand the theoretical background of our experience. But too often, when these topics are analysed, the disability side is forgotten or adopts the same protective patterns that we have mentioned in Sects. 2.1, 2.4, or 2.6, when they do not themselves become *the object* or *the target* (economic target) of that concrete business under the appearance of "social model".

In addition, an extended notion of *sustainability* refers to a way of management, to achieve a continuous customer's flow, to environmental issues, or to social issues. And it is in this last aspect that we consider our experience fits, because disability is still an unresolved issue for society, which is torn between pity and respect. And if we respect, we are, as a society, obliged to allow them to find a really productive job niche, not only as occupational therapy, or only working in outdated or obsolete

[12] Basically, any economic activity, if not competitive, needs to be at least profitable in the sense that it allows for economic viability in order to be truly sustainable.

productive sectors, destined to disappear.[13] Other specific items that also fit (al least, a little) in this approach are: circular economy, technology-based approach, entrepreneurship, innovation and "corporate sustainability, circular economy, business model thinking, strategic management, innovation management, and social entrepreneurship" (Dentchev et al., 2018, p. 696). But, very often, specific needs of PwID are hidden from these studies, even though there are topics that could be very appropriate for these persons: collective action, design thinking, hybrid organisation, and entrepreneurial ecosystems. From the behavioural approach, as defined by Dentchev et al. (2018, p. 700), we could extract the idea of importance about how government support is so important to promote a leadership in for more social initiatives for PwID, but this support cannot become the *only support*, or the *eternal support*. Moreover, in the Kleine and von Hauff (2009) triangle, it seems that the social approach is one of the three sides (besides economic and ecological), but if you care about social matters you distance yourself from the other two. This is a problem: it is true that a social business model cannot have dividends', but this does not mean that it is not profitable in an equity sense, without producing economic losses but allowing to make ends meet comfortably, not only as a subsistence or subsidiary economy.

Thus, the *integral and inclusive support* that heads this section is given by the consideration of work with PwID not as a mission related to corporate social responsibility, but as an integral part of the hard core of the model business. It is so sad to see, in relevant studies as carried out by Lüdeke-Freund et al. that inclusive patterns are deleted from their Table 1 when they try to rank the answers on "expectations of value-creating effects that may result from applying the [business model] patterns" (Lüdeke-Freund et al., 2018, p. 154).

2.8 Universal Design for All

Educationally, the way to make the ordinary curriculum accessible has been through curricular adaptations. The misunderstood adaptations consisted of reducing the curriculum, instead of understanding that the changes implied that the removal of elements went hand in hand with the introduction of new unique aspects that enriched the training giving an added value to people with disabilities.

Universal Design seeks to generate a physical product or service that is useful for everyone, without the need for adaptation. This is the utopian goal, but in training, it clashes with a basic principle, that of Piagetian *cognitive conflict*. According to this principle, when a learning objective is set, it is necessary to achieve an optimal adjustment between what a person knows how to do and the level of difficulty that is posed: if there is too great a mismatch, it discourages the person, and if it is too

[13] We can remember jobs in sectors such as gardening, stationery, printing, leather crafts, laundry, etc. These have been traditional occupational models for PwID in Spain, and continue to be so in the Autonomous Community of Extremadura, where we are developing our experience.

small, it does not provoke interest in the learner, who considers it too close (easy) to what he or she already knows or knows how to do. Therefore, in education, Universal Design for PwID could be understood as the principle of individualisation of teaching, so that, starting from a multi-purpose content, learning situations can be generated that are valid for all students. In this individualisation, the figure of the support teacher inside the classroom, described in Sect. 3, is essential. Only in this way, and in line with the Italian model mentioned in Sect. 2.4, can continuity in the work between teachers, perceived by pupils, the relationship with families and the connection with the course management, be guaranteed.

3 Inclusive-Labour Project Developed in the University of Extremadura (2017–2022)

The entrepreneurial model that we propose had several phases, being the formative part the most stable. On the other hand, the working part was where evolution experienced can be better appreciated. That evolution was due to contextual characteristics detected, as well as the socio-economic and family contexts. We are aware that this is a concrete experience, but we want to offer a *case study* as an example of an insertion path, offering social, economic, and educational keys that have made it possible.

This experience would not have been possible without the support of two entities that have been thanked in the acknowledgements ONCE Foundation (FONCE) and European Social Fund (ESF). In addition, SAPIEM and SEXPE were also essential to get funds for an educational experience extended to other areas (managers of special employment centres, and Labour Trainers). This shows that we are in favour of unity of action from all possible angles. First, as part of conditions of the FONCE and ESF calls for proposals, we did it by counting on several associative entities representative of the social sector[14] for their planning and execution. Second, as another condition from FONCE-ESF, we counted on the participation of a knowledge centre, such as University of Extremadura, to be involved. With these two types of partners, we could make a *good course* and we would have done a course similar to what is usually done with other universities participating in these calls. But we wanted to add a third distinctive seal: in the beginning, we wanted to try a *pure* entrepreneurial model, and in the third year, we saw that it was necessary to change the model, but for that, it was necessary to have experts on Special Employment Centres.[15]

[14] As Full Inclusion (Plena Inclusión, a Spanish federation of entities) and "Brothers' Foundation for Equality and Social Inclusion" (FUNDHEX, Fundación de Hermanos para la Igualdad y la Inclusión Social in Spanish).

[15] As a Social Business Factory SBF) and ZoomIn Social and Sustainable Impact. This second organisation was fundamental for the final successful model we proposed in the 2021–22 edition.

Thus, as a result of all the principles explained in Sect. 2 of this chapter, and adding the idea of *action on the ground*, our educational experience had two parts over the three editions: it started as an entrepreneurial action linked to the students admitted to the courses themselves. Subsequently, as a result of the analysis of the data that we collected over 2 years, we analysed the effects not only on the PwID themselves but also on their families and their associative environment, moving towards the triple formative model that we will talk about in Sect. 3.3.

3.1 Background of the Project

The European Union has made a major effort to build a more equal society related to disability. Training was identified as a strategic area in the European Disability Strategy 2010–2020 and its subsequent development (European Commission, 2010, 2017), just the area towards which our experience is directed.

Extremadura is considered a preferential region to receive support from European Funds, such as Horizon or Next Generation, with especially advantageous conditions due to its economic situation. Extremadura receives ERDF and ESF funds but hardly receives aid such as PERTE (and only with the theme of digital and sustainable agri-food cooperatives, not in areas related to the social economy or PwID).

In Spain, since 1982,[16] there has been a social objective to achieve access to the labour market for one of the most vulnerable groups: people with disabilities. Successive Ministries of Labour have signed agreements with entities in the sector, such as the Spanish Committee of Representants of People with Disabilities (Comité Español de Representantes de Personas con Discapacidad, CERMI, by its Spanish acronym), to promote improvements in this access; and then, the autonomous communities developed or implement similar initiatives. In 2013, for the third time, a legislative reform was approved, trying to reinforce institutional commitments, but at the same time, caused a perverse effect, as since 2014 the statistical data can no longer be compared with previous years, which means that the historical perspective of the evolution has been lost (something which has been outlined in Sect. 2.6, and which is pointed out by INE, 2022b).

Anyway, only with data from the last decade can we see clear trends. Taking into consideration the official statistics (INE, 2022c) for the period 2017–2021 (the period in which our experience took place) we offer an overview of the national and

[16] In Spain, three specific laws on disability have been enacted which, in addition, regulate access to the labour market for these people, with tax advantages for medium and large companies. They are: Ley 13/1982, the 7th April, about Social Integration for Disabled people (LISMI, by its Spanish acronym; currently abolished); Ley 51/2003, the 2nd December, about equal opportunities, non-discrimination and universal accessibility for disabled people (LIONDAU by its Spanish acronym; currently abolished); and Real Decreto-Legislativo 1/2013, 29th Novembre approving a revised text of the General Act of rights of disabled people and their social inclusion (currently in force).

regional employment situation, comparing the overall data for people with and without disabilities in Graphs 1, 2, 3 and 4 below.

At first sight, it seems that the statistical trends are clear and that the data are as expected: there are more non-disabled people active than not active, both in Spain as a whole (Graph 1) and in Extremadura (Graph 3), while for the same two areas but referring to disabled people (Graphs 2 and 4, respectively) the opposite phenomenon occurs. However, Extremadura is always about five points higher than the national average, i.e. the ratio between non-disabled and active non-disabled people in Spain is around 28% for the whole series of years, while in Extremadura it does not fall below 33%. Consequently, people with disabilities are also more unemployed in Extremadura, with the same differential of about five points. But the nuance is that the proportion of non-active disabled people is three times higher than the proportion of active disabled people, while at the national level, this proportion is only twice as high.

Extremadura has other characteristics that are important for understanding the experience: its business fabric is made up of a large majority of small and medium-sized enterprises (SMEs, PYMES by its Spanish acronyms). It is an eminently rural community, with a very dispersed population. It is made up of two provinces: Badajoz and Cáceres. Badajoz is an area that is located in the south in mountainous foothills and has 11 socio-sanitary zones[17] that group together the smaller municipalities, the most important of which are the city of Badajoz, Mérida (regional capital), and Don Benito-Villanueva de la Serena area (an association of two cities), together with other strategic cities due to their geographical location or their economy, such as Zafra or Almendralejo. In the province of Cáceres, the orography is even more complicated due to the presence of mountainous areas and a Natural

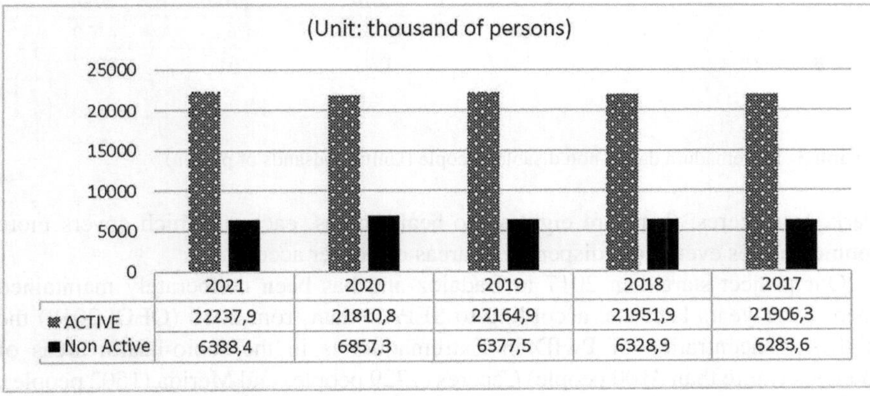

	2021	2020	2019	2018	2017
ACTIVE	22237,9	21810,8	22164,9	21951,9	21906,3
Non Active	6388,4	6857,3	6377,5	6328,9	6283,6

Graph 1 National data—non disabled people (Unit: thousands of person)

[17] The social and health areas are the administrative demarcations established by the Territorial Directorate of the Extremadura Service for the Promotion of Autonomy and Care for Dependency (SEPAD, by its Spanish acronym of *Servicio Extremeño de Promoción de la Autonomía personal y atención a la Dependenci*a) in order to organise its work.

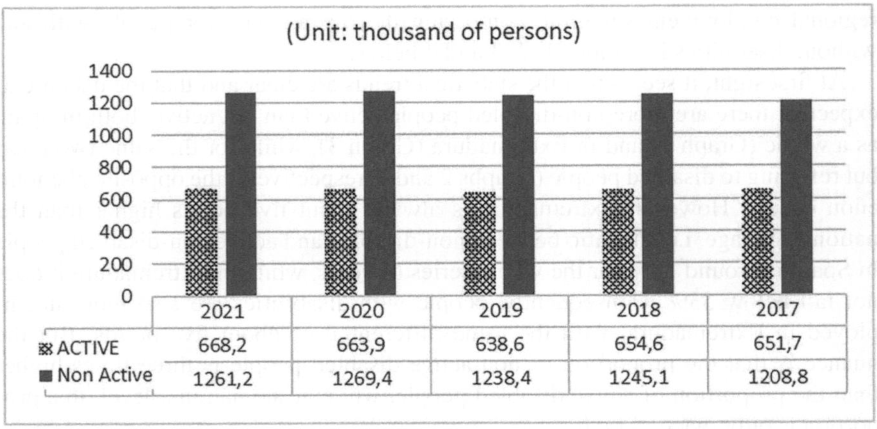

Graph 2 National data—disabled people (Unit: thousands of person). Source: Own elaboration, with data from INE (2022c)

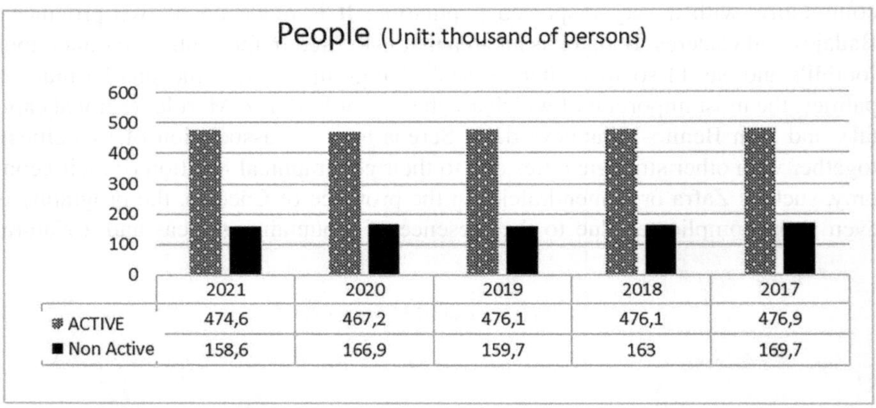

Graph 3 Extremadura data—non disabled people (Unit: thousands of person)

Park. In Cáceres, there are eight socio-health areas, each of which covers more municipalities even more dispersed in areas of poorer access.

Our project started in 2017 in Badajoz and has been deliberately maintained there for 3 years because, according to SEPAD data from 2016 (OED, 2018) the highest concentration of PwID in Extremadura is in the socio-health areas of Badajoz (more than 3100 people) Cáceres (1729 people) and Mérida (1502 people). Other areas that also have a considerable number of people are Plasencia Sur (1074 people), Don Benito-Villanueva (1015), Plasencia Norte (865), Zafra (851), and Coria (795). In 2017 Extremadura ranked as the second Spanish Community with the second highest rate (11%) of population with disabilities.

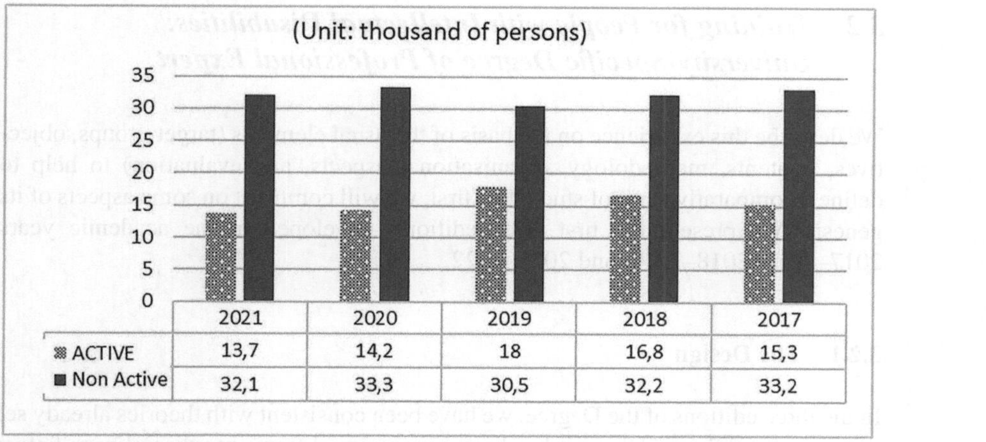

Graph 4 Extremadura data—disabled people (Unit: thousands of person)

We have received suggestions to extend the experience to the Mérida/Don Benito area and even to Cáceres, due to the difficulties of the population in general and of PwID in particular in terms of accommodation or public transport infrastructure. The University of Extremadura has sites in Badajoz, Mérida, Cáceres, and Plasencia, but since the model was in its initial stages, it was decided not to extend it until the model was well-defined. This has led to some recruitment difficulties and the students on the courses have come only from the same province, Badajoz, where the course was given. In fact, the maximum distance from which they travelled was over 80 km, the distance between Mérida and Badajoz, but distances within the region can be much greater than this.

On the other hand, the University of Extremadura (UEx), where the training experience took place, as recognised in its mission, "is a public institution committed to society, whose mission is the comprehensive training of people through the creation and dissemination of knowledge and culture by means of quality teaching and research, with a vocation for intellectual and cultural leadership" (UEx, 2014, p. 7). It should be noted that UEx has always sought to pay special attention to diversity in all its degrees, and for this reason, in 2004 UEx created the Student Care Unit (UAE, by its acronym in Spanish or *Unidad de Atención al Estudiante*) to attend to the needs of students with disabilities. The course had the character of its University Specific Degree because it was one of the conditions of the ONCE-FSE Foundation call, but it also became an opportunity to dignify the training that PwID usually receive, usually relegated to the non-formal sphere, outside the ISCED education levels, such as *occupational workshops'*.

3.2 *Training for People with Intellectual Disabilities: University-Specific Degree of Professional Expert*

We describe this experience on the basis of the usual elements (target groups, objectives, contents, methodology, organisational aspects, and evaluation) to help to define a comparative unit of study. But first, we will comment on some aspects of its genesis. We present the first three editions, developed in the academic years 2017–2018, 2018–2019, and 2021–2022.

3.2.1 The Design

In the three editions of the Degree, we have been consistent with theories already set out in Sects. 2.2 and 2.3, so it has been programmed from interdisciplinary fields: from the university field (with academic staff and university services experts in educational attention to the PwID and in Employment Guidance), from the professional field with business consultants (to match the needs of the PwID with the demands of the labour market) and with members of associations working with the PwID in the field (who know the needs of the PwID and the socio-family environment that surrounds them). In this way, pedagogues, economists, psychologists, educators, and members of associations in the field of PwID joined together in an egalitarian dialogue, where truly innovative and ground-breaking ideas arose, due to the fusion of different ways of doing things, which had not been done in Extremadura until that moment.

The coordination fell to the female professor who signs this article, as in the UEx only university professors can direct this kind of degree. The requirement to have a Specific University Degree[18] was one of the conditions of the call that financed the courses. Coordination was exercised with common premises previously defined and accepted by all, to avoid each area/person tending to approach the course from their own unique vision; the aim was to generate a true shared and transformative knowledge of reality. To ensure transparency, all the material generated was shared digitally in real time among the members of the management core team, first through email chains, and then through a closed virtual space, to which only the management team had access. Only the personal data of students and teachers, protected by current privacy legislation, was reserved for the director. This created a perception, on the part of the other partners, that there was real trust within the team, and helped contacts to become increasingly open.

[18] "University Specific Degrees" (*Título Propio* in Spanish language), as indicated in the preamble of the Regulations contained in the Resolución de 9 Noviembre 2022, are not official degrees, but specialisation studies. With them, each university shows its commitment to various social issues. As they are not official, there is more flexibility in offer, subject matter, implementation, and duration of these degrees but, at the same time, they must comply with certain established standards of structure and accountability in order to be approved by the Governing Board of the University.

The fact that it was organised from a Faculty of Education, and with a vision of critical and Social Pedagogy, facilitated the interdisciplinary nature of academic fields. University staff participated in its initial design (the coordinating professor, a psychologist from UAE and two workers from the Employment Guidance Office), as well as the FUNDHEx association (representing PwID) and the Social Business Factory (SBF) as a partner of the social economy (then participated by the other signatory of this article, who has been present in all editions, even when this entity ceased to participate). The initial design of the first edition (2017–2018) was a theoretical training with an internship which, for the PwID students, did not involve working directly, but rather outlining a business project and, through this delimitation of the project, achieving instrumental learning, which led them to analyse the reality of their province, and to come into contact, through interviews and in-situ visits, with the business reality. In the second edition (2018–2019), the same actors participated and the same methodology was applied.

After the second edition, several events occurred: the insertion results themselves, which did not reach the expected results; the indication, by FONCE, to introduce novelties; and the slowdown that the COVID pandemic forced us all to endure, which also revealed the problems of working with entities outside the reality of Extremadura. All these facts led to a change in the program: for the 2021–2022 edition, we counted more intensively on Plena Inclusión Extremadura, as it is a federative entity with greater presence and scope in Extremadura than Fundhex, and on ZoomIn as a partner specialised in social economy (instead of SBF), of which the second signatory of the article was also a member. In other words, the university scope has remained stable, but throughout the editions consultants and associations have been changed, with the intention that the work experiences should be as broad as possible, that the PwID association should cover most of the Extremadura territory, and that the results of access to the labour market should be greater.

Following the principle of normalisation[19] already defined, the scenario we presented to the students was of great importance. Since Piaget, we know that educators provide learning situations and scenarios, where people's minds can interact and internally generate their own learning. We sought to offer the most normalised environment possible: the ordinary classrooms of the Faculty of Education, sharing common spaces, such as the Library, the Bar, and the various university services in situ, spread around the campus, as would also be the case in other work environments.

And that is one of the big differences we raise: students attend the same university campus and the same faculties as other boys and girls of their same age. The pre-vocational or Vocational Training that PwID usually receive can sometimes prepare them for a job, but they receive it either in High Schools, together with younger children (under 18 years of age), or in specific, segregated centres, in a more closed

[19] Normalisation makes perfect sense in the context of our proposal for two reasons. First, because where, as in Spain, segregation has previously existed, it must be normalised and, second, because it is still present in the current legislation in Spain, specifically in Article 21 of the Real Decreto Legislativo. 1/2013.

structure, with more surveillance and with ways of doing things that end up generating artificial environments. On the other hand, our course puts them in the situation of having to go to an *unsupervised* and *non-special* environment, and in this way, the personal development experience changes.

Particularly relevant was the fact that there was no school bus to pick them up from their homes: they had to arrive at the Faculty by their own means, or by public transport. The fact that there was no bus route meant that they had to wait to get off at their bus stop. There have been cases of students who, on the first day, got lost on campus, but in such cases, they had the phone number of the support technician, whom they could call.[20] For many of them, having all kinds of normalised experiences for the first time was very important to achieve personal empowerment. Some students have revealed to us in the final satisfaction surveys that for the first time they felt they were treated as adults, not as children. On the other hand, families and associations confirmed to us that they all noticed in their sons and daughters the great change in the way they relate to each other, the way they speak, their level of self-initiative, and so on.

3.2.2 Student Characteristics

Our students in the three editions have had specific characteristics, because the FONCE-FSE call for applications already established a predetermined profile:

- To be between 18 and 30 years old
- Have a minimum of 33% of intellectual disability officially recognised by the competent public body in Spain.
- Be an active job seeker at the beginning of the course, through the Spanish National Youth Guarantee System.
- Not to suffer from mental illness (the field of mental health is left for other actions).

The selection was made by interview, upon request of the PwID or their relatives. In the interviews, it was checked whether they were people with a minimum of acquired skills in reading, writing, and arithmetic, that they could correspond to an academic level of primary school,[21] as well as that they had some personal autonomy in self-care and hygiene. In other words, we are talking about boys and girls with intellectual disability but potentially capable of carrying out paid work.

Another key factor is the presence of people with different disabilities, for the reasons already discussed in Sects. 2.1 and 2.2. What we have tried to do is to

[20] This has not been the common trend: it only happened in the first 2 weeks, and with concrete students, particularly those who had been most overlooked up to that moment. In the end, it turned out to be a powerful orientation exercise that they extrapolated to their usual movements, not only to get Faculty.

[21] They were not required to have a diploma, but to demonstrate their skills during the interview, in small tasks that were not perceived as an *exam*.

articulate a personalised training model that is empowering. As opposed to the clinical model, ours tries to generate opportunities so that, once the person is placed in the learning scenario, with the necessary elements of security and accompaniment, they are given the opportunity to demonstrate how far they are capable of going, and to relate to people who have *other* difficulties. And this greatly empowers these boys and girls, who in many cases feel for the first time responsible for their own action and see themselves capable of helping their peers and of being helped by them, without feeling inferior, but feeling that there are no limits, and this fact makes them advance more than with individualised clinical sessions. Other characteristics of our students can be seen in Sect. 4.1, when we explain the results.

3.2.3 Targets

The general target was set out in the FONCE-FSE call for proposals and has therefore been maintained throughout the three editions: "To bring students closer to, raise awareness of and motivate them towards employment as a viable and possible option for their personal -and professional—future. The approach is always based on the knowledge and development of the abilities and skills that are brought into play in the employment process".

The 12 specific objectives also responded to the approaches of the FONCE-FSE call, but the way of carrying them out has changed every course, involving different services in each call, in order to check which were more beneficial for the students. It is beyond the limits of this text to detail them all, but we can say that they were related not only to the contribution of skills or personal and work opportunities for the PwID but also to the change of attitudes in UEx itself and on the part of future employers or co-workers. Moreover, being in a Faculty of Education, where future of Primary and Secondary teachers are trained, the exchange between ordinary students and PwID students was an opportunity for future teachers who, in their classrooms, can keep in mind that *today's children* will be *tomorrow's adults*, also when they have intellectual disabilities, that they are *not eternal children*.

3.2.4 Curricular Contents

The contents of the programme have evolved along with its duration. Table 1 summarises the contents of each course, together with its duration expressed in hours.[22] In order to clarify who teaches it we have used the following acronyms: university teachers (D), trainers from university services (S), trainers from the world of associations (A), trainers from the world of work (L), and an artist (F) from Spanish Yehudi Menuhin Foundation (FYME by its Spanish acronym). Sometimes for the

[22] The conditions of the call for applications and the regulations of the University's own degrees made it necessary for the final equivalence to be 1 credit equals 10 h.

Table 1 Contents and duration by academic year

Academic year → Module ↓	2017–18 (300 h)	2018–19 (380 h)	2021–22 (300 h)
Social and working skills	60 h D + F		
Social, labour, and emotional skills		60 h D + F	
Social, cognitive, and emotional skills			50 h D
Use of community services	18 h D	24 h D	20 h D
Mathematical skills, logic, and accounting	30 h D	50 h D	30 h D
Expres-Arte (combination of art and proxemic communication, with MUSE methodology)			30 h F
Safety and health at work	12 h D + S	16 h D	
Job skills		40 h S	20 h A
Work environment			40 h S + L
New technologies and communication	50 h L	50 h L	
Technological skills: Communication and labour use of the Internet			40 h L
Entrepreneurial skills	70 h L	60 h L	
Internships	60 h A + L	80 h A + L	70 h A + L + D

Source: Own elaboration

same content there can be several acronyms, because each module was subdivided into *subjects*.

The change of name in some of the modules has been a response to the needs detected and a sign of the evolution of these modules as different actors have been incorporated. If they had been made only from the field of psychology, pedagogy, or economics, they would not have included all these aspects. These aspects condition participation in social and professional life.

The social skills module is the one that changed the most, because it is where the inclusive contents were introduced, which were worked on together with ordinary students from the university degrees: Plastic arts education, English and, in the third edition. Robotic classes and visits to exhibitions carried out by pupils from other schools in the city were added.

There has been one aspect present from the beginning, and it is related to the contents and methodology provided by one of the professors: specifically, the artist who came from FYME. This decision was taken because, from university institutions, a type of academic work is more common, and from the working world, a form of work adapted to the economic world, but in both spheres, work is done with adults and less attention is paid to the *human* aspect. However, there are other areas in which soft skills are also developed, which are not normally present either in the workplace[23] or at school: soft skills are also developed in the arts, hence the idea, taken in agreement with FYME representatives, to incorporate one of their professionals. This person worked with PwID students not as an extracurricular or playful activity, but by applying their in-depth knowledge of the arts (dance, in this case) to generate knowledge and positive attitudes towards certain proxemic communication issues.

3.2.5 Some Methodological and Organisational Issues

The main basic methodological principles present in this training have already been described in the previous sections: normalisation, inclusion, active and participatory methodology, seeking the integration of different experiences, taking into account the different paces and abilities in a personalised manner, and encouraging teamwork and the ability to learn by themselves.

For this reason, the contents have been worked on in different scenarios: in the classroom, integrated into ordinary subject seminars, museum spaces, municipal theatre, offices of municipal or regional bodies and even in non-formal spaces (the university library) or directly in informal spaces (cafeteria or reprographic service), as well as in work practice spaces. Because we are plastic beings, and different environments help us to develop. In informal spaces, it is essential to warn and train staff beforehand, so that they do not overprotect these students, but neither do they treat them without taking their realities into account; in this way, the PwID interacts spontaneously, but the environment has the capacity to be educational.

As for the teaching staff, a University of Extremadura rule for the University Specific Degree is that at least one-third of the training hours must be taught by university lecturers. The planning from a Faculty of Education made it possible to have teachers from areas such as educational sciences, psychology, mathematics, language didactics, plastic arts, or educational robotics. In addition, throughout the three editions, other faculties have been involved (such as the Faculty of Economics and Business Studies, the Faculty of Agricultural Engineering, Faculty of Sciences, and the Faculty of Documentation and Communication, all on the Badajoz Campus) and other University services such as the aforementioned UAE, the Employment Guidance Office (OOL by Spanish acronym of *Oficina de Orientación Laboral*), the Guidance and Teacher Training Service (SOFD by Spanish acronym of *Servicio de Orientación y Formación Docente,* which in the third edition provided its computer classrooms of the future so that students could practise with advanced technology),

[23] In the last two decades, ideas such as having staff play percussion instruments together, for example, to understand the need to be coordinated, have been incorporated as a technique in business management courses, but it has a much more incidental meaning.

the services of the Central Library, Onda Campus, etc. But, as already mentioned in Sect. 2.2, if there had only been university teachers from "different academic areas of knowledge", the training would have had a single academic imprint, very similar to which the PwID have already experienced throughout their school education. That is why in these editions we have always opted to incorporate as teachers experts in business, who work in the social economy sector, and experts in job preparation for PwID, coming from the partner *Plena Inclusión Extremadura*, who have less academic but much more experiential ways of communicating, and who bring the reality of work into the classroom.

Finally, with regard to schedules, it is worth commenting on one issue: although the structure of the contents was modular, it was modular because of the *sense* of the contents. However, when it came to teaching it, we wanted to create a cyclical time structure, both within the day and within the week and throughout the course. The conditions of the FONCE-FSE call imposed that it should take place in the morning and during university class hours (from Monday to Friday). This led to a design of 4 h of work per day, with a half-hour break in the middle of the day (the only exception to our ordinary university lectures). The classes had a duration of 1 or 2 h of continuous work, so each day they had two or three different teachers (as a way of preparing them for the continuous dealing with different people in a job). In addition, the contents were taught in a cyclical way, because it is much more effective for the PwID to present the contents regularly, little by little over many weeks, than to work on them intensively in a few days and never come back to them again. This cyclical nature was particularly important for the contents taught by the FYME artist to have the desired effect: this teacher only came for 1 h a week, but with a methodology so different from everything else that it became the *weekly event*. In order to work on the transition from academics to work, it was also ensured that the beginning had more presence of university teachers, and the end of the period of university classes, before the internships, was filled with more work-related content, so that by the time of facing the internships, both knowledge and skills would be well established.

The figure of the support teacher in the classroom has proved to be essential and indispensable as we have planned it. It does not mean a classic *Technical Educational Assistant* to accompany students in their self-care. It is a professional with a specialised qualification (special education teacher) who assumed teaching skills to help teaching and personalisation of contents; skills in conflict management in the classroom (or personal, because sometimes our students brought "problems from home"), continuity among the teaching staff and liaison with the course management.

Another human resource was essential for the good development of the internships: the Job Coach (PL by Spanish acronym of *Preparador Laboral*). In the first two editions, there were no internships in companies, but there were entrepreneurship internships that involved direct contact between these students and the Public Administration, entrepreneurs, suppliers... Therefore, it was necessary to have an expert, provided by the partner associations of our project, who guided them in the process of preparing interviews and professional visits to companies, and subsequently helped them in the analysis of the material collected. The interviews and visits were carried out, whenever possible, at the interviewee's place of work, for which it was necessary to have professionals from the city of Badajoz, with whom this PL contacted or helped the students to contact, using the models and letters that he had previously worked on. In the third edition, this expert figure became a "pure" PL provided by the federation of associations collaborating in that edition. His tasks consisted in analysing the students' skills after several months of attending the course, in order to establish the best match for an internship in real companies. He was also responsible, once the student's job profile was determined, for contacting and helping to establish the internship agreement, in direct contact with the course management.

Families have always played a very important role: since the first edition, at least three meetings have been held during the course (initial, middle, and final) so that they could express their concerns, share the changes they could appreciate in the students in environments that were beyond our reach, etc. They were also able to contact, upon request, the management team or the support technician if necessary. In the results, we will comment on some of the issues that arose from these interactions.

Finally, the course included a social part, precisely to normalise this type of experience. In the first two editions, and in line with the more entrepreneurial approach given to the course, an entrepreneurship day was held in the penultimate month of each course, in which students presented their findings and business model to an auditorium with free admission, and had the opportunity to share their ideas with entrepreneurs from the province who came to explain their origins.

As any other training activity, all three editions had their graduation ceremony in university style: with public and academic authorities, with representatives of FONCE, and with the students' families, with media (press, television, and radio ones) covering the event, and a subsequent feast.

3.3 Business Project in the Social Sector

Is there a competitive social initiative business sector in Extremadura? In other words, with the capacity to introduce value-added products or services to the market while serving its social purpose. The answer is difficult and complex. In addition to all the aspects pointed out in previous sections of this article—population dispersion, pre-eminent rural economy, and fabric, high unemployment in all sectors of the population—we can also find in the genesis of this business sector and in its legal structure some reason that prevents or limits the generation of more economically sustainable social businesses. Among some of the aspects detected—from the dialogue with the partner Plena Inclusión Extremadura and with the rest of the associations directly—we can consider.

Most of the Special Employment Centres in the region are Single Person Limited Companies (SLU, by Spanish acronym of *Sociedades Limitadas Unipersonales*) in which all or most of the capital is in the hands of the Associations or Foundations that promote them. Bearing in mind that the statutory purpose of these entities is fundamentally welfare, the promoted business is accessory. Thus, the main objective of the promoted businesses is to continue the assistance work, leaving in the background whether the activity, value proposition, and business model are in demand by the market and generate sustainable economic return. In other words, the purpose of the Special Employment Centres is to continue *occupying* PwID people.

Moreover, these businesses have no decision-making autonomy. Their corporate purpose is subordinated to the social action of the Association or Foundation that are *owners* of the business.

In a very particular affirmation or conclusion, by the authors of this article, it is very difficult to generate real and long-term employability opportunities for the PwID group if we start from productive units without a market focus and without value proposals appreciated and valued by their clients.

The social initiative business sector in Extremadura differs greatly from other models observed in Spanish territory, such as in Catalonia or the Basque Country, where most of the enterprises have autonomy and a clearly defined business model. In other words, a market approach. Where initiatives and businesses far removed from the traditional ones (handling, maintenance, gardening, laundry) are observed. These businesses are characterised by their high degree of professionalism and have therefore found their place in the market. Some of the entities studied include La Fageda d'en Jordà (the second largest yoghurts producer in Catalonia, behind only Danone), Taller Àuria in the province of Lleida (which, among other activities, produces cosmetics, perfumes, and catering for collectives) and *Grupo* Cooperativo TEB (with production lines dedicated to paints, insecticides and even a green division of agro-ecological products with high added value). These three entities alone employ around 2000 PwID, which is more than the objective active population of PwID in the Badajoz health region. The legal form of the model entities observed is that of a Social Initiative Associated Work Cooperative, which means that the *company* has management-decision autonomy with regard to the *social assistance project*. And, therefore, its main objective is the occupation and promotion of the autonomy of the PwID employed. Beyond the necessary welfare condition, but insufficient if we pursue a higher degree of autonomy and long-term employability that these other models have managed to achieve satisfactorily.

Taking all of the above into account, we worked in a coordinated and joint manner with Plena Inclusión Extremadura to define a roadmap to facilitate the professionalisation of the social initiative business sector and the design of market-focused business models.

This idea is guided by the objective of providing the Social Initiative Companies and Associations sector with tools that enable new activities, encourage entrepreneurship, intra-entrepreneurship, and cooperation among entities, increasing their competitiveness and capacity for innovation. We know, because it happens in other territories, that a strong social business sector implies a higher number of employed PwID. This is the main objective that guides our joint actions.

We proposed a flexible and long-term roadmap that included training, advice, and consultancy, impacting all the agents at the same time and with their necessary participation and involvement. This roadmap includes itineraries according to the degree of development of the initiatives or projects identified.

And we do so because we understand that increasing employability in the group of people with intellectual disabilities involves developing training and intervention programmes—that promote job creation—with the necessary participation of the PwID, the job coaches, and the people with management responsibility.

For this reason, the entrepreneurship model to which we refer and which we have been perfecting throughout the three courses also includes and takes into account the following objectives:

- Improve the competitive capacity of the existing business network—in the social sector—and its recruitment capacity—we are referring to Special Employment Centres.
- Facilitate the promotion of new business activities or new business lines through entrepreneurial and intra-entrepreneurial initiatives and cooperative initiatives. Those activities generate business models with greater added value. Beyond the traditional activities carried out by the Special Employment Centres—handling, gardening, laundry services, etc.
- Provide management analysis and management tools. Since many of the people who hold management positions in Special Employment Centres have previous training or experience in business management, as we could observe in the meetings with managers sponsored by *Plena Inclusion Extremadura*.
- Promote the labour competences of PwID, both within Special Employment Centres and in ordinary companies. And in this sense, in order to promote job opportunities in ordinary companies, it is necessary to know their inclusion strategy, the socio-labour conditions, and the detection of new employment niches. In this way, training can be adapted to the capabilities of the users, but also to the real and potential demands of the labour market.
- Structure and increase the figure of the job coach as an effective figure in PwID supporting itineraries and labour market insertion.

This work, carried out in parallel between the University of Extremadura and SAPIEM, jointly and in coordination with *Plena Inclusión Extremadura* allowed us to learn first-hand the needs of the sector (entrepreneurial strengthening) and its difficulties. And with this, we were able to adapt our intervention-training model based on the analysis and approach of the stakeholders. We deduced that in order to be successful, it was necessary to work jointly and not in isolation with: users and families, the associative sector, and the business sector (both protected and ordinary companies). And through a joint dialogue, to be able to find solutions and define a strategy that facilitates the main objective of our training: to improve the employability possibilities of the group of PwID (Fig. 1).

So that, through SAPIEM, we launched several initiatives aimed at strengthening the Social Sector, designing a pilot experience that would also allow us to learn about the needs first hand and expressed directly by the participants-protagonists of the initiative. Specifically, a course named *Challenges for Inclusive Employment* was designed—taught in November 2020—in which three itineraries were defined

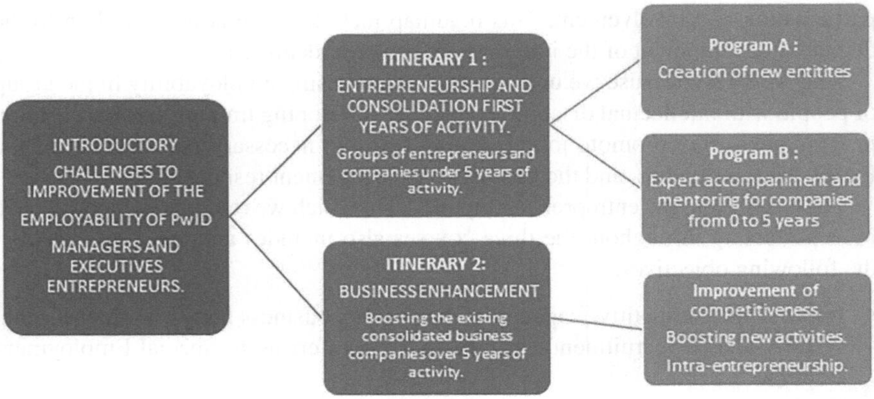

Fig. 1 Specific training for the business sector. Source: Own elaboration

for the three target groups: managers and directors of Special Employment Centres, job coaches, and PwID. In the first group, the future of the sector, challenges, and opportunities were addressed and a strategic diagnosis of the 12 participating social entities was carried out, also providing them with tools for strategic self-diagnosis and the formulation of value proposals and business models. In the second group, the essential figure of support staff, job coach, was addressed. And in the third group, with PwID people, the specific needs of the users, limits, barriers, but also opportunities and potential. Finally bring everyone together in the same space for reflection and obtain valuable qualitative information, from which to design training or support interventions *tailored* to the expressed needs.

4 Results

Some key findings have already been presented in this chapter, because they were necessary to explain the background to the experience. In this section, we summarise the data from the three editions in a comparative and juxtaposed way, interpreting some aspects in the light of the accumulated experience. All graphs in this section are self-made, so the source is omitted so as not to be repetitive.

4.1 Students Profile

We can start by considering gender (Graph 5), age (Graph 6), and disability characterisation (Graph 7) of each year group of students.

The first thing we see in Graph 5 is that the trend of gender bias is corroborated. Although we wanted to be equitable in this regard, we were unable to accommodate

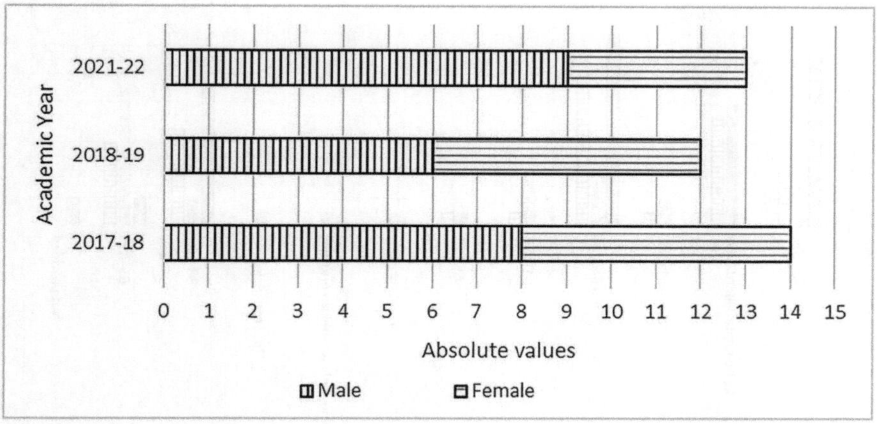

Graph 5 Gender of students

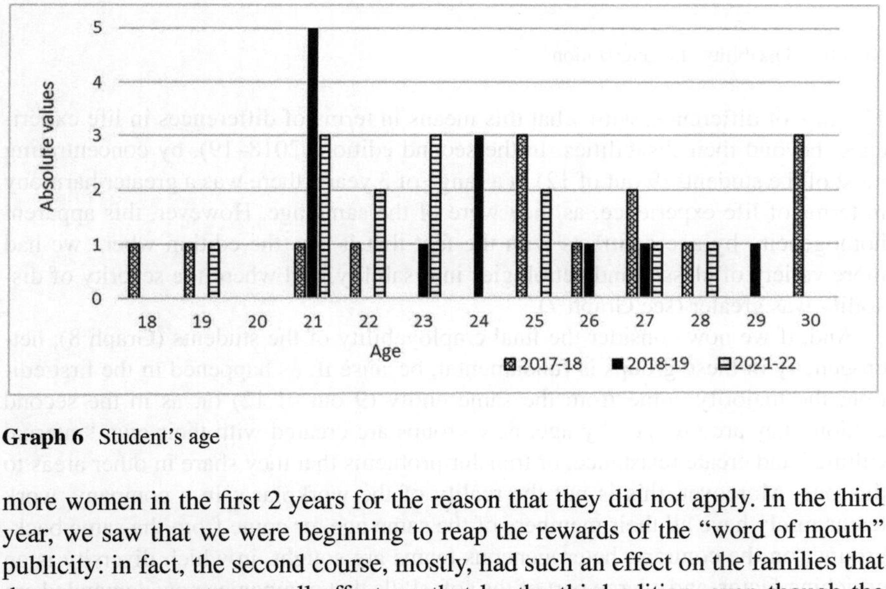

Graph 6 Student's age

more women in the first 2 years for the reason that they did not apply. In the third year, we saw that we were beginning to reap the rewards of the "word of mouth" publicity: in fact, the second course, mostly, had such an effect on the families that there was a very strong pull effect, so that by the third edition, even though the whole pandemic problem had happened, students were still remembering the course and its importance. The maximum number of people we could admit to the course was 15.

On the other hand, the age distribution (Graph 6) makes sense, since up to the age of 21 in Spain there are opportunities for training within the educational system: certainly, in segregated institutions, but families prefer this to the uncertainty of having them without schooling, or in special centres for the rest of their lives. Therefore, it is from the age of 21 onwards that there are more possibilities for enrolment. In the first year, there was a greater dispersion of ages, with up to

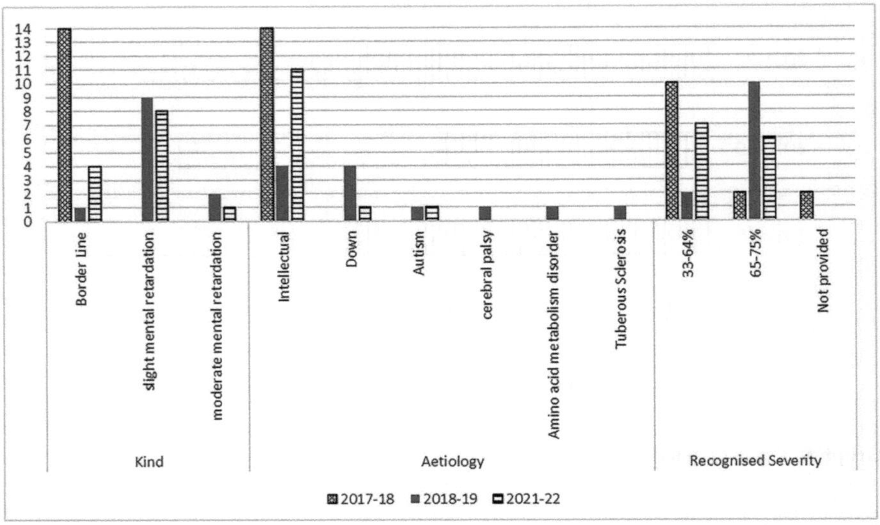

Graph 7 Disability characterization

12 years of difference, with what this means in terms of differences in life experience, beyond their disabilities. In the second edition (2018–19), by concentrating most of the students (9 out of 12) in a range of 3 years, there was a greater harmony in terms of life experience, as they were of the same age. However, this apparent homogeneity by age contrasts with the fact that it was the edition where we had more variety of classes and aetiologies in disability, and where the severity of disability was greater (see Graph 7).

And, if we now consider the final employability of the students (Graph 8), heterogeneity in these groups is fundamental, because if, as happened in the first edition, the majority come from the same entity (9 out of 12) or, as in the second edition, they are grouped by age, new groups are created with their own "internal culture" and create resistance, or transfer problems that they share in other areas to this one. Moreover, this is not the reality of the workplace: in a company, work teams rarely have all their members of the same age, or come from the same background; on the contrary, heterogeneous teams are sought, in which diversity is an enriching factor, and a large part of the soft skills that companies now demanded are related to this.

Thus, we believe that in the three editions, the composition of the group has been improving, to the point that the third edition is the one that has best-achieved balance in a group that is diverse in gender, age, type, aetiology, and severity. This, together with the change of orientation given to the course, as a consequence of the maturation of the business project presented in Sect. 3.3, has meant that the employment rate of our graduates, as shown in Graph 8, is higher than in the previous edition.

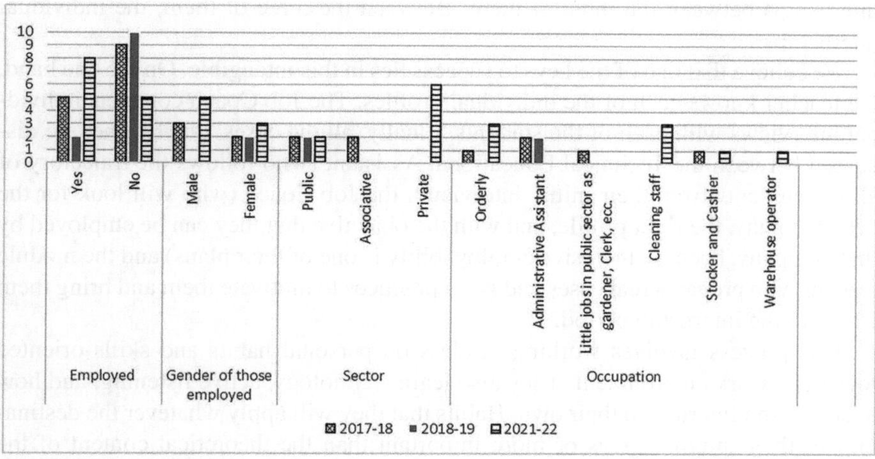

Graph 8 Occupation achieved after the courses

The employment data corroborate other aspects that we detected during the courses, which we are unable to detail here. But, at least with the third edition, after a gradual process of settling in and testing different versions, we see that the best training option is that of salaried employment, with real work experience in companies, which was the model finally implemented in the third edition. The challenge now is to diversify the employment possibilities. But we believe that we are on the right track, as the employment rate has shot up to more than 60% in the last edition, in a period when, economically, the society around them is not at its best.

One last important, non-quantitative result is that we took advantage of the pandemic period to redesign our model for the third edition. We spent more than a year calibrating the improvements. Taking into account not only academic models but also the associative experience of *Plena Inclusion Extremadura*, as well as the experience of expert consultants in the employability market for people with disabilities. In our regular meetings—weekly meetings for 1 year—we try to solve questions such as: Are there specific labour market niches for PwID? Which companies in our environment are implementing inclusivity objectives? Which could be the adequate itineraries—in each case—depending on the person? Ordinary or protected employment market? As well as debating about what kind of soft skills our students should develop in order to be prepared before starting their internships. Based on this qualitative analysis, we were able not only to design the internship but also to prototype for certain profiles what could be the best or most successful career opportunities.

One of the participating expert consultants in the field of social entrepreneurship is also a lecturer on the course. Specifically, the communication and technology module. He redesigned this module to serve as a springboard to internships in ordinary companies. For this purpose, the teacher had to coordinate with both the Technical Educational Assistant and the Job Coach in order to fit the individualised internships between the three of them. Between the three of them, the individual internships.

We believe that one of the keys to success lies in this intangible. On the one hand, the teacher knows each of the individual profiles. The Job Coach conducts individual interviews with each of the students. Finally, all the cases, one by one, are discussed between the Technical Educational Assistant (who follows the trajectory of all the students from their initial interview), the Job Coach (who will look for the best company, the right profile, and with the objective that they can be employed by that company, because inclusive employability is one of their plans) and the module teacher who prepares real cases and good practices to motivate them and bring them closer to the internship period.

This process involves working in class on personal habits and skills oriented towards a work environment. They also learn technology, active listening, and how to search the Internet on their own. Habits that they will apply whatever the destination of their internships as or more important than the theoretical content of the module is this other work.

It is, therefore, a multidisciplinary approach where we try to satisfy the desires and concerns of each student and their real chances of employability.

4.2 Future Lines of Action: The Training of Support Professionals for this Specific Model

The figure of the Technical Educational Assistant and the Job Coach that we have used in our course does not have its own training itinerary. Normally, due to the weight of the diagnosis, they usually come from psychology, and what they do is to establish a more or less detailed profile, looking for existing jobs or jobs that can be adapted to the specific needs of each person. Once the job position has been adapted, the professional disappears from the scene so that, in the best of cases, the person with a disability remains for life in the same job, performing the same function.

But production needs change, and with it the profile of the required worker, so we have detected that this figure should be more professionalised, with a profile that would be more like that of a coach, who accompanies and links his or her career to that of the person he or she is working with. We are not talking about a leap in the dark: the model we are proposing exists, but it is fragmented. The Italian model to which we have already alluded proposes Individualised Educational Projects (PEI, by its Italian acronym of *Piano Educativo Individualizzato*) that are drawn up for 6 months and 2 years. They are then reviewed throughout schooling. On the other hand, in a company, a human resources manager can, depending on the forecasts of changes in production or staffing needs, see the potential of his employees and propose (sometimes impose) training itineraries for them to develop new tasks in the company (or it is the worker himself who decides to be trained to progress in that or another company). However, it is true that in many cases, due to his own profile, the worker we are dealing with in this chapter will not have the ability to undertake new training routes, so that the Job Coach, that companion who follows his career and helps him to progress, is the one who can become the engine of change, not only for the worker but even for the company itself with which he works. This is because the Job Coach should not only be a specialist *in the person*, but also *in the activity* that he/she develops and in the business sector in which he/she is, so that he/she can be a figure that helps new ventures with a universal design within the social enterprise.

That is why we believe that it should not be a specific degree training, nor limited to a single sector, but that any professional in any field could have access to a post-graduate training itinerary, drawing from three sources: socio-economic, pedagogical, and psychological. In other words, for all activities, whether in the scientific, social, legal, technological, humanistic, etc. sectors, there is room for PwID performing specific tasks. But it is not possible to know what these tasks are if there is no specialist in that productive sector. The problem is that this person may not have the pedagogical and psychological training to work with these people, or he or she may have such training, but may not know enough about human resource development in business or sustainable social economy. Therefore, their training would be complementary, postgraduate, including aspects of all these areas, and with internships in the sector and with PwID. This implies that it would be necessary to coordinate training in the three areas we have been discussing: business, these trainers, and the PwID, so that they are able to create synergies among themselves.

This opens up a very interesting field of professional development that fits in with those organisations or companies that design strategic Diversity and Inclusion policies and plans. An approach that will require specialised professional figures or multidisciplinary work. All of this with the aim of maximising the employability of people with disabilities, but also to value the richness of diversity. It is therefore a question of going a step further, with institutions that educate inclusively but also with companies that are a mirror of the society in which they sell their products and services.

5 Conclusions

With the implementation of these training initiatives, we have been able to see, first hand, the opportunities and threats of this type of actions. We will start with the opportunities, and then comment on some of the threats and what we have learned from them.

The fact that the initiative was supported by a grant from a national and European project, which required not only the initial commitment of its promoters but also the support of the University's rector's team, facilitated, at certain times, readjustments with the project already underway. Economic readjustments, for which, initially, no routines were foreseen in the university structures, but for which solutions were sought within the legal possibilities, in order to facilitate the inclusive project.

Nothing else could be expected from an institution which, like the UEx, has had a marked social vocation since its origins.

The important volume of people who constitute the PwID in particular and people with disabilities in general in Extremadura society cannot be ignored by higher education. It could provide qualification training that would help labour market insertion, and which would exert a triple combined action such as the one proposed in Sect. 3.3. Training opportunities that are truly linked to access to the labour market are one of the pillars for the autonomy and dignified life of all people; otherwise, families and the PwID themselves sometimes consider them as little courses that lead nowhere.

We are aware that this discourse may be perceived as a threat by associations already engaged in this type of training, as we have been told during the course of our experience. But we have detected that this training does not achieve the objective of employment if it is not accompanied by a strengthening of the business fabric as we have already described in Sect. 3.3. In fact, in the training experience that we launched, as a spin-off in November 2020 and in which we tried to systematically and simultaneously train social enterprise managers, Job Coaches, and PwID, we detected that entrepreneurship opportunities are being lost by those who have management responsibilities, sometimes due to lack of knowledge and others, due to the association structure itself, which is more focused on assistance than on promoting employment. We do not wish to denigrate the well done work by the associative sector, but we do wish to highlight the need for greater professionalisation and cooperation between the University, which has powerful tools for innovation and development, and the entities, without ever losing sight of the principles of egalitarian dialogue. All of this with a view to enabling the PwID to have a doubly inclusive training path: for the people with whom they interact during their studies and for a real access to the labour market. This would be the basis of our *universal design of a higher education for the PwID* group which, among all the disabled people, is who has the most vetoed the personal training progression with the excuse that they are *not able* to.

In addition to training for employment, other types of systematic and continuous actions could be implemented to meet the needs of these groups for access to culture in an inclusive way: in programmes that, being available to all, allow spaces for encounters between PwID and non-disabled people. In this way, PwID are allowed to interact in normalised situations not only with ordinary people (which, somehow, even unconsciously, puts them at a disadvantage) but also with groups in which there are different disabilities. This allows them to understand and appreciate that everyone has *values* and *difficulties*, helps them to demystify that there are people who can and those who cannot, and empowers them to be truly aware that they can also be present in the university institution. We have experienced first-hand the empowering value of going to university for PwID, with students with autism spectrum who in less than a month improved their communication skills, or with people who, after going through our course, have signed up for vocational training courses because they have realised that they could take them, and had not done so before because they did not trust in their own possibilities. One of the intangible conclusions of our project is that when the objective is the empowerment of the person and not their assistance, the results are different. Believing *that they can* is one of the main keys to improving their employability.

We have also found two other threats that have to do with the economic part and with the traditional dependence on public subsidies in the sector, and that is what has also contributed to generating this resistance towards the training initiatives that we have launched. On the one hand, being a course within the European Operational Programme for Youth Employment, which also has other lines of action, we saw that, unintentionally, we found ourselves *competing in the same fishing ground*. We repeat once again, because this has been the greatest difficulty we have faced (the distrust of the entities that usually attend to the PwID) that it is not a matter of replacing anyone, but of establishing cooperative synergies. However, one of the most powerful entities in the sector in the city of Badajoz, where we have developed our course, which is also highly dependent on subsidies for this sector, did not perceive this and, at certain points, tried to hinder our actions. The second threat is that the very system of protection for PwID, which is undoubtedly necessary, does not empower but encourages dependence: many PwID are afraid to embark on a project of their own—as we saw in the first two editions—either because they lack economic resources or because they are afraid of losing the small economic contribution they receive from the State's social welfare services. Logically, we are not saying that people should risk their way of life, but we believe that a serious and rigorous study should be undertaken to break the vicious circle of *poor training-unemployment-economic dependence* in which the PwID are immersed.

Integrating into an institution such as the academic one implies that the institution itself must review some of its approaches. Much has been said about the need for all educational institutions, from school to university, to be open up to their environment. The problem is that often the people who teach do not really share the problems of the neediest groups, nor do they know the needs and continuous evolution of the labour market, and that is why it is so difficult for them to find valid solutions for practical purposes for their students. That is why it is important to make it clear that this model is not intended to replace anyone. In fact, in the second year we introduced a change that we maintained in the third edition: instead of coming to the University 5 days a week, they came only four, from Tuesday to Friday, so that they would not be disconnected for months from their association—at the same time that the association was able to monitor progress on an ongoing basis-. The morning schedule and throughout the university school calendar—not the school calendar, which is different—had a dignifying effect for the students and their families, but it clashed with the interests of some entities. Given that the Universities also develop their teaching work in the afternoon, putting this course in that timetable may make it easier for more people to attend, although the course cannot be given to a larger group without more support because otherwise, individualisation is lost.

Families and PwID themselves need and ask for this type of action, which confronts them with other types of activities and environment. They are aware of when they go to an activity that has no route or when they carry out a daily work but do not receive a salary for it. The aim of this experience, recognising the contribution of each previous action (school, associations, occupational workshops, etc.), is to offer another alternative for personal development. For example, a specific association wants its associates to come to these courses because they understand that their speech develops, and it is not because we use speech therapy techniques, but because they are immersed in an inclusive, natural environment, which stimulates them to speak and interact, and awakens their interest in the environment (this has also happened to us with autistic students). In other words, we are very clear that by promoting a global action on all productive sectors (business and social fabric and entrepreneurial initiatives) and making the PwID aware of these connections between the parts of the productive sector, a mutual empowerment takes place in which the University, as a knowledge centre, can play a catalytic role (not a substitute for) in the implementation of projects in different territories. We try to offer one special path of personal development, in a real inclusive context, really committed to a final solution for their state of unemployment. And, this really works.

Employability success recorded at the third edition can be explained because on that occasion we could really put into practice the egalitarian dialogue described in Sect. 2.3 between university academic sphere and social economy world. So, when we were planning and developing those editions, partners from both spheres were working closely, interacting in full confidence when a part took or executed a decision, delegating to each one what was proper to it, and providing constructive criticism that made them revise the uncritically established assumptions in each area. Our proposal that year contains elements with strong components coming from Social Pedagogy and Social Economy, which have different ways of doing things than traditional academic institutions of Higher Education, and also differences with respect to types of actions in which the clinical (neurological or psychological) aspect predominates.

Of course, this experience does not end only with the achievement of a job for the PwID. The project promoted thanks to ONCE Foundation and the ESF could be continued in more university training actions or in non-formal training. In the academic world, as said inclusive actions of the same way as those university classrooms that already exist for elder people. We believe that we should start thinking about generalising the *universities of disability*, as FONCE—FSE calls are promoting with its Unidiversity program. In universities, it would be like another university extension service. In these actions, even people who have not finished their primary education are accepted, and people do not go for financial gain, but for the pleasure of learning and socialising with other people. Why not create a University for Disabled People or, better still, use Universal Design into already existing Open classrooms, with PwID-complaint contents made for all? This would be an action that would dignify the active ageing of the PwID, and contribute to the really inclusive society.

Besides, non-formal educational experiences could be generated in the same structures that exist for ordinary people, as Popular Universities (in Spain they have been operating for over a century. This institution comes from the Social Pedagogy area: provides training that, after, can be officially recognised (for example, by official exams), and has a great versatility, so they are perfect for a really personalised training because they are not as constrained by specific rules as school institutions.

We believe, and our experience has shown, that without underestimating the importance of taking into account the individual reality of each PwID (as in any other personalised training), training actions for employ and entrepreneurship should be provided as standardised as possible, generated from the same economic and educational budgets as for the rest of the people, whether or not they are valued as PwID. This is a way to bring together the specific world of disability with the mainstreaming of work, education, and society, so that, without losing sight of their needs, they can share not only leisure or educational spaces, but professional spaces too.

Acknowledgements In this chapter, we explain an experience that has been possible thanks to the funding obtained in the framework of the Unidiversity Programme, in its successive calls by the ONCE Foundation, for the development of University Training Programmes for Employment aimed at young people with intellectual disabilities (18–30 years old) job seekers through Youth Guarantee Spanish National System, co-financed by the Youth Employment Initiative 2014–2020, of the European Social Fund (Youth Employment Initiative YIE—The ESF invests in your future). In addition, and to work with the employment sector, we thank to the Support Service for Entrepreneurial Initiative of the University of Extremadura (SAPIEM, by its Spanish acronym of *Servicio de APoyo a la Iniciativa EMprendedora*) which receives funding from the Extremadura Public Employment Service (SEXPE, by its Spanish acronym of *Servicio EXtremeño Público de Empleo*), to implement a second coordinated experience for the three partners: employers, trainers and PwID. Thanks also to all the teaching and support staff who have made its development possible, as well as to the academic authorities (Rectorate Teams and Dean's Office of the Faculty of Education) who have supported it since its inception.

References

Alex, K. (2009). *Soft skills*. S. Chand Publishing.

Alexander, R., Broadfoot, P., & Phillips, D. (1999). *Learning from comparing, new directions in comparative educational research*. Symposium Books.

Álvarez-Ramírez, G. E. (2023). *El capacitismo, estructura mental de exclusión de las personas con discapacidad.* CERMI. Accessed from https://back.cermi.es/catalog/document/file/winyt-b63kc-el-capacitismo-estructura-mental-de-exclusion-de-las-personas-con-discapacidad%2D%2D-cermi.es-85.pdf

Ashraf, M., Razzaque, M. A., Liaw, S. T., Ray, P. K., & Hasan, R. (2019). Social business as an entrepreneurship model in emerging economy systematic review and case study. *Management Decision, 57*(5), 1145–1161. https://doi.org/10.1108/MD-04-2017-0343

Barrio de la Puente, J. L. (2005). La transformación educativa y social en las comunidades de aprendizaje. *Teoría Educativa, 17*, 129–156.

Becerra Hernández, R., & Moya Romero, A. (2010). Investigación-acción participativa, crítica y transformadora Un proceso de permanente deconstrucción. *Integra Educativa, 3*(2), 133–156. Accessed from http://www.scielo.org.bo/scielo.php?script=sci_arttext&pid=S1997-40432010000200005&lng=es&tlng=es

Bilbao, A. (2000). Trabajo, empleo y puesto de trabajo. *Política y Sociedad, 34*, 69–81. Accessed from https://revistas.ucm.es/index.php/POSO/article/download/POSO0000230069A/24543/0

Bray, M., Adamson, B., & Mason, M. (2014). *Comparative education research. Approaches and methods* (2nd ed.). Comparative Education Research Centre, University of Hong Kong.

Caballero, A., Manso, J., Matarranz, M., & Valle, J. M. (2016). Comparative education research: A way for new researchers. *Revista Latinoamericana de Educación Comparada, 9*, 39–56. Accessed from https://www.saece.com.ar/relec/revistas/9/art3.pdf

Calero López, I., & Rodríguez-López, B. (2020). The relevance of transversal competences in vocational education and training: A bibliometric analysis. *Empirical Research in Vocational Education and Training, 12*, 12. https://doi.org/10.1186/s40461-020-00100-0

Carr, W., & Kemmis, S. (1998). *Teoría Crítica de la Enseñanza. La Investigación-Acción en la formación del profesorado*. Martínez Roca.

Colectivo IOÉ & CIMOP. (1997). *Discapacidad y trabajo en España. Estudio de los procesos de inclusión y exclusión social de las personas con discapacidad*. IMSERSO. Accessed from http://www.colectivoioe.org/uploads/b2f211c63c5afa933b8ec283ed399205f08a0ddb.pdf

DECISION 2001/903/CE del Consejo de Europa, de 3 de diciembre de 2001, sobre el Año Europeo de las personas con discapacidad 2003. Accessed from https://eur-lex.europa.eu/legal-content/ES/TXT/?uri=celex%3A32001D0903

Dentchev, N., Rauter, R., Jóhannsdóttir, L., Snihur, Y., Rosano, M., Baumgartner, R., Nyberg, T., Tang, X., von Hoof, B., & Jonker, J. (2018). Embracing the variety of sustainable business models: A prolific field of research and a future research agenda. *Journal of Cleaner Production, 194*, 695–703. https://doi.org/10.1016/j.jclepro.2018.05.156

Ebersold, S. (2021). Which decentered methodological framework is best for comparing inclusive education policies? In O. Giruad & M. Lallement (Eds.), *Decentering comparative analysis in a globalizing world* (pp. 141–165). Brill. https://doi.org/10.1163/9789004466609_007

European Commission. (2010). *European disability strategy 2010–2020: A renewed commitment to a barrier-free Europe*. Accessed from https://eur-lex.europa.eu/LexUriServ/LexUriServ.do?uri=COM%3A2010%3A0636%3AFIN%3Aen%3APDF

European Commission. (2017). *Progress report on the implementation of the European Disability Strategy (2010–2020)*. Accessed from https://ec.europa.eu/social/BlobServlet?docId=16995&langId=en

Eurostat. (2023). *Data browser-health-disability*. Accessed from https://ec.europa.eu/eurostat/databrowser/explore/all/popul?lang=en&subtheme=hlth.hlth_dsb&display=list&sort=category&extractionId=HLTH_DSI015

Ferrer, F. (2002). *La Educación Comparada actual*. Ariel.

García-Garrido, J. L. (1996). *Fundamentos de Educación Comparada*. Dykinson.

Geissdoerfer, M., Vladimirova, D., & Evans, S. (2018). Sustainable business model innovation: A review. *Journal of Cleaner Production, 198*, 401–416. https://doi.org/10.1016/j.jclepro.2018.06.240

Huete-García, A. & Quezada-García, M. (2012). *La discapacidad en las fuentes estadísticas oficiales. Examen y propuestas de mejora*. CERMI. Accessed from https://cermi.es/colecciones/coleccion-cermies/166

INE. (2022a). *El Empleo de las Personas con Discapacidad [Data set, EPD-2020]*. Accessed February 24, 2022 from https://www.ine.es/prensa/epd_2020.pdf

INE. (2022b). *El empleo de las personas con discapacidad [Data set]*. Accessed March 8, 2023, from https://www.ine.es/dyngs/INEbase/es/operacion.htm?c=Estadistica_C&cid=1254736055502&menu=resultados&idp=1254735976595

INE. (2022c). *Series 2014–2021.Empleo de las personas con discapacidad. Resultados por Comunidades Autónomas [Data set]*. Accessed March 8, 2023, from https://www.ine.es/jaxi/Tabla.htm?tpx=29872&L=0

Kazamias, A. M. (2009). Reclaiming a lost legacy: The historical humanist vision in comparative education. In R. Cowen & A. M. Kazamias (Eds.), *International handbook of comparative education. Springer international handbooks of education* (Vol. 22). Springer. https://doi.org/10.1007/978-1-4020-6403-6_78

Kleine, A., & von Hauff, M. (2009). Sustainability-driven implementation of corporate social responsibility: Application of the integrative sustainability triangle. *Journal of Business Ethics, 85*(Suppl 3), 517–533. https://doi.org/10.1007/s10551-009-0212-z

Liamputtong, P. (Ed.). (2022). *Handbook of social inclusion. Research and practices in health and social sciences*. Springer. https://doi.org/10.1007/978-3-030-89594-5

LIONDAU-Ley 51/2003, *de 2 de diciembre, de igualdad de oportunidades, no discriminación y accesibilidad universal de las personas con discapacidad. BOE n. 289, de 3 de diciembre*. Accessed from https://www.boe.es/eli/es/l/2003/12/02/51/con

LISMI-Ley 13/1982, *de 7 de abril, de integración social de los minusválidos. BOE» núm.* 103, de 30 de abril. Accessed from https://www.boe.es/eli/es/l/1982/04/07/13

LOPEGCE-Ley Orgánica 9/1995, *de 20 de noviembre, de la Participación, Evaluación y Gobierno de los Centros Docentes. BOE 279, de 21 de noviembre*. Accessed from https://www.boe.es/eli/es/lo/1995/11/20/9

López García-Cano, J. M. (1988). *El Diseño modular como marco de la formación permanente del profesorado. Modelo de actuación propuesto*. Accessed from https://dialnet.unirioja.es/descarga/articulo/117552.pdf

Lüdeke-Freund, F., Carroux, S., Joyce, A., Massa, L., & Breuer, H. (2018). The sustainable business model pattern taxonomy—45 patterns to support sustainability-oriented business model innovation. *Sustainable Production and Consumption, 15*, 145–162. https://doi.org/10.1016/j.spc.2018.06.004

Mara-Danel, P., Pérez-Ramírez, B., & Yarza-de-los-Ríos, A. (Eds.). (2021). *¿Quién es el sujeto de la discapacidad? Exploraciones, configuraciones y potencialidades*. CLACSO. Accessed from https://www.clacso.org/wp-content/uploads/2021/05/Quien-es-el-sujeto-de-la-discapacidad.pdf

Morales Zúñiga, L. C. (2014). Critical thinking in contemporary educational theory. *Actualidades Investigativas en Educación, 14*(2), 4–24. Accessed from https://www.scielo.sa.cr/scielo.php?pid=S1409-47032014000200022&script=sci_arttext

Moya Otero, J. (2003). Una ciencia crítica de la educación ¿pluralismo metodológico y/o pluralismo epistemológico? *Ágroa Digital, 6*, 1–26. Accessed from https://rabida.uhu.es/dspace/bitstream/handle/10272/3519/b15761782.pdf?sequence=1

Nirje, B. (1999). How I came to formulate the normalization principle. In R. In Flynn & R. Lemay (Eds.), *A quarter-century of normalization and social role valorization* (pp. 17–50). University Ottawa Press. Accessed from https://books.openedition.org/uop/2474?lang=es

OED-Observatorio Estatal de la Discapacidad. (2018). *Cartografía de la Discapacidad en Extremadura*. Accessed from https://www.observatoriodeladiscapacidad.info/wp-content/uploads/2018/05/CARTOGRAFIA-DISCAPACIDAD-EXTREMADURA.pdf

OHCHR. (2008). *Convention on the rights of persons with disabilities. [Advocacy Toolkit. Professional Training Series. N° 15]*. Accessed from: https://www.ohchr.org/sites/default/files/Documents/Publications/AdvocacyTool_en.pdf

Oria-Segura, M. R. (2006). La realización del derecho a la educación en personas con necesidades educativas especiales desde la óptica inclusiva en Italia y España. In L. M. Naya & P. Dávila (Eds.), *El derecho a la educación en un mundo globalizado. [Conference Proceedings of X National Congress of Comparative Education]* (pp. 486–501). Erein.

Padilla Arias, A. (2012). The modular teaching system: A curricular alternative of university higher education in Mexico. *Revista de Docencia Universitaria, 10*(3), 71–98. Accessed from https://dialnet.unirioja.es/descarga/articulo/4132176.pdf

Pavone, M., & Bellacicco, R. (2016). University: A universe of study and independent living opportunities for students with disabilities. Goals and critical issues. *Education Sciences & Research., 1*, 101–120. Accessed from https://journals.francoangeli.it/index.php/ess/article/view/3497/92

Queiruga-Santamaría, O., García-Álvarez, J., & Santos-González, M. C. (2002). Analysis of the effectiveness of vocational training plans for employment from the point of view of the labor market. *Fuentes, 24*(1), 90–103. Accessed from https://revistascientificas.us.es/index.php/fuentes/article/view/15258/18043

Real Decreto-Legislativo 1/2013, *de 29 de noviembre, por el que se aprueba el Texto Refundido de la Ley General de derechos de las personas con discapacidad y de su inclusión social. BOE. 289, de 3 de diciembre*. Accessed from https://www.boe.es/eli/es/rdlg/2013/11/29/1

RESOLUCIÓN de 9 de noviembre de 2022, *del Rector, por la que se ejecuta el acuerdo adoptado por el Consejo de Gobierno por el que se aprueba la Normativa Reguladora de los Estudios Propios de la Universidad de Extremadura. DOE n. 249, de 30 de diciembre*. Accessed from https://www.unex.es/organizacion/servicios-universitarios/secretariados/postgrado/noticias/index__html/normativas/22064232.pdf

Tomasevski, K. (2005). El derecho a la educación: panorama internacional de un derecho irrenunciable. In L. M. Naya Garmendia (Ed.), *La Educación y los derechos humanos* (pp. 63–90). Erein, Espacio Universitario.

UEx. (2014). *Plan Estratégico 2014–2018*. Accessed February 8, 2023, from https://www.unex.es/archivos/ficheros/PlanEstrategicoUEx.pdf

UN. (2019). *United Nations disability inclusion strategy.* Accessed February 20, 2023, from https://www.un.org/en/content/disabilitystrategy/assets/documentation/UN_Disability_Inclusion_Strategy_english.pdf

UNESCO. (1995) *World conference on special needs education: Access and quality*. MEC/UNESCO. Accessed February 20, 2023 from https://unesdoc.unesco.org/ark:/48223/pf0000098427

Vega Gil, L. (2011). *La educación comparada e internacional*. Octaedro.

WB. (2023). *Disability inclusion* [web site]. Accessed February 20, 2023. from https://www.worldbank.org/en/topic/disability#2

WHO. (2023). *International classification of functioning, disability and health [Browser Online]*. Accessed February 20, 2023, from https://www.who.int/standards/classifications/international-classification-of-functioning-disability-and-health#:~:text=ICF%20is%20the%20WHO%20framework,and%20measure%20health%20and%20disability

Part II
Education for Sustainability and SDG

EU GREEN: How Peripheral Regional Universities Created a European Alliance Based in SDGs

Gemma Delicado, Rocío Yuste, Raquel Pérez-Aloe, Giorgio Pelosi, and Cesaltina Pires

Abstract This chapter describes how nine peripheral universities located far away from the center of their countries launched a consortium that focused on a particular set of sustainability-related practices. The consortium has been one of the four alliances funded in the European Union Call for European Universities in 2022. This alliance was built out of the conviction that these universities can serve their communities, students, territories, and planet by working together to scale up and promote our sustainability and SDG-related activity. The name of the alliance is EU GREEN: University Alliance for Sustainability: Responsible GRowth, Inclusive Education and Environment. "GREEN" is a term that covers sustainability and reflects their mission to contribute to a more rightful society, equalized economy, and sustainable environment. This chapter will describe the method and tools used to relate the different stages and spheres of the creation process of the alliance that will be based on sustainability and address the major societal challenges identified in the 17 SDGs of the UN 2030 Agenda. The results will suggest recommendations

G. Delicado (✉)
Departamento de Filología inglesa, University of Extremadura, Cáceres, Spain
e-mail: gdelpue@unex.es

R. Yuste
Departamento de Ciencias de la Educación, University of Extremadura, Cáceres, Spain
e-mail: rocioyuste@unex.es

R. Pérez-Aloe
Catedrática de Tecnología electrónica, Departamento de Ingeniería Eléctrica, Electrónica y Automática, University of Extremadura, Badajoz, Spain
e-mail: raquel@unex.es

G. Pelosi
Department of Chemical, Life and Environmental, Sustainability Sciences, University of Parma, Parma, Italy
e-mail: giorgio.pelosi@unipr.it

C. Pires
Departamento de Gestão, University of Evora, Évora, Portugal
e-mail: cpires@uevora.pt

C. Rego et al. (eds.), *Transitioning to a Circular Economy*, Studies on Entrepreneurship, Structural Change and Industrial Dynamics,
https://doi.org/10.1007/978-3-031-77661-8_7

for potential new networks that will navigate the challenging journey to becoming a European University Alliance.

Keywords Alliance · European Universities · Peripheral · Higher education · SDGs

1 Introduction

1.1 What Are "European Universities"?

In September 2017, Emmanuel Macron gave a historic and profoundly pro-European speech entitled "New Initiative for Europe" at the Sorbonne, where he called for a transformation of the European Union to protect it from external threats and Europhobic populisms within the continent (Macron, 2017; Grove, 2019). The French President alleged that "the Europe we know is too weak, slow and ineffective" and added that "only Europe can give us a capacity for action in the face of the great contemporary challenges." The President claimed that "fragmentation is only superficial" and diversity should be Europe's "greatest opportunity." Macron used the example of the different languages of Europe, which should be seen as an asset instead of a barrier, pledging that students should be able to speak at least two European languages by 2024. Furthermore, instead of lamenting the divergences among states, connections among them should be reinforced, and in his vision, by 2024 half of the European students "should have spent at least six months in another European country by the time they are 25, whether they are university students or learning a trade". To accomplish this plan, the French President projected the foundation of new constructions in the European higher education landscape:

> I believe we should create European Universities—a network of universities across Europe with programs that have all their students study abroad and take classes in at least two languages. These European Universities will also be drivers of educational innovation and the quest for excellence. We should set for ourselves the goal of creating at least 20 of them by 2024. However, we must begin setting up the first of these universities as early as the next academic year, with real European semesters and real European diplomas. (Macron, 2017)

It was in this historic speech that the term "European Universities" was coined and a new era for European Higher Education started.

According to Andrew Gunn (2020), the idea of creating a supranational university is not new and it originated in the late 1940s. However, despite various attempts, the aspiration remained unfulfilled for decades. After President Macron's speech, a new impulse to pursue the supranational university philosophy resulted in the launch of the European Universities Initiative pilot phase. Concretely, only 2 months later at the Gothenburg summit in November 2017, European leaders, through Donald Tusk, brought Macron's plan to the meeting and called for changes to strengthen their universities by making them pillars of European values. The aspiration sought a globally innovative, competitive, and attractive European Research Area, which, after the Bologna process, could promote unhindered transnational cooperation, in addition to equality and inclusion, with the ultimate goal of transforming the alliance's institutions (European Commission/EACEA/Eurydice, 2018; Birk, 2019; Cesaer, 2019; Cippitani & Gatt, 2009). At the summit, it was agreed to launch the first call for the "European Universities" strategy, which aimed to create 20 alliances. This number has increased up to 40 comprising more than 400 higher education institutions. The European Council published the conclusions of the Gothenburg Summit that included a call for Member States, the Council, and the Commission:

> Strategic partnerships across the EU between higher education institutions and encouraging the emergence by 2024 of some twenty 'European Universities', consisting in bottom-up networks of universities across the EU which will enable students to obtain a degree by combining studies in several EU countries and contribute to the international competitiveness of European universities. (EU Council, 2017)

This strategy was also supported by the Education Committee of the Council in May 2018. The Committee reinforced the emergence of European Universities as a flagship for a European Education sphere (EU Council, 2018). The initiative tried to bring higher education institutions beyond the existing models of cooperation in higher education in order to gradually achieve the long-term consolidation of "European Universities." It is clear that the complex plan for the European Universities Initiative will redefine internationalization as we now understand it, as well as national legal frameworks, if the original intentions materialize (de Wit & Deca, 2020; Claeys-Kulik, 2019).

Alliances among Europe are framed in the European Universities Initiative, which was launched in 2019 as one of the flagship programs of the European Union with the aim of building a solid European education area co-developed by higher education institutions, student organizations, associated members, Member States and the European Commission (EU Commission, 2019). This action is probably the most important strategic action of the European Union for education in the last 20 years. The idea is that European Universities are transnational alliances that will become the universities of the future, mainly focused on promoting European values in European Higher Education. In summary, the EUI is a Flagship initiative of the European Commission to achieve:

- European Interuniversity campuses
- Transnational and multidisciplinary teams working together to face social challenges
- Creating worldwide known Universities
- Building bridges for a common future (EU Commission, 2019)

1.2 First Call (Pilot)

This first Call launched in 2019 was a pilot call to test this initiative. The Commission received 54 applications but only 17 European Universities were funded. These 17 consortia involved 114 higher education institutions from 24 Member States. According to the Commission, the evaluation was carried out by 26 independent "external experts, including rectors, professors and researchers, appointed by the Commission (EU Commission, 2019). The consortia selected by the Comission were the following described in Table 1:

1.3 Second Call (Pilot)

Most of the consortia decided to try again in 2020. For this second pilot call, 62 consortia applied and only 24 were selected to become European Universities. This new Call involved 165 higher education institutions from 26 Member States and other countries participating in the Erasmus+ program. I need to find a chart for the second call (EU Commission, 2020).

1.4 Third Call

The third call was the first of the three that was not on a pilot mode. The results of the 2022 call bring to 44 the total number of European Universities involved in the initiative, which now comprises 340 higher education institutions in 31 countries, including all member states as well as Iceland, Norway, Serbia and Turkey. **44 European Universities** are presently ongoing, and their **1300 associate partners** will increase the quality and impact on European Higher Education (EU Commission, 2022).

1.5 How Did EU GREEN Emerge?

In this context, the EU GREEN—European Universities Alliance for Sustainability: Responsible for GRowth, Inclusive Education and Environment (https://www.eugreenalliance.eu/)—was selected with a score of excellence by the European Commission in the third call. The alliance emerged from academic debates between three southern European universities that have a long trajectory of collaboration: Extremadura, Evora and Parma. Extremadura and Evora are neighbor partners from 2 close but at the same time different countries, Spain and Portugal. Both institutions are very similar and located in regions with similar social and academic challenges. Their closeness and similar problematics made collaboration extremely fluent for decades. Collaboration among researchers is clear in the numerous papers published together. In research, also a great number of projects were jointly conducted.

Table 1 Universities funded in the 2019 pilot call

Reference number	Country	Applicant organization	Project title	Maximum EU grant €	% co-funding
612247-EPP-1-2019-1-ES-EPPKA2-EUR-UNIV	ES	UNIVERSIDAD DE GRANADA	ARQUS European University Alliance	5,000,000,00	35.38%
612361-EPP-1-2019-1-SI-EPPKA2-EUR-UNIV	SI	UNIVERZA VLJUBLJANI	European Universities Transforming to an Open, Inclusive Academy for 2050	4,986,791,01	80.00%
612442-EPP-1-2019-1-DE-EPPKA2-EUR-UNIV	DE	UNIVERSITAET POTSDAM	European Digital UniverCity	4,997,910,74	80.00%
612461-EPP-1-2019-1-DE-EPPKA2-EUR-UINV	DE	TECHNISCHE UNIVERSITAT DARMSTADT	University Network for Innovation, Technology and Engineering	5000,000,00	71.84%
612468-EPP-1-2019-1-ES-EPPKA2-EUR-UNIV	ES	UNIVERSIDAD DE CADIZ	The European University of the Seas	5000,000,00	78.67%
612489-EPP-1-2019-1-DE-EPPKA2-EUR-UNIV	DE	JOHANNES GUTENBERG-UNIVERSITAT MAINZ	Fostering Outreach within European Regions. Transnational Higher Education and Mobility	4,995,587,94	80.00%
612511-EPP-1-2019-1-FR-EPPKA2-EUR-UNIV	FR	UNIVERSITE DE STRASBOURG	European Partnership for an Innovative Campus Unifying Regions	5,000,000,00	80.00%
612518-EPP-1-2019-1-HU-EPPKA2-EUR-UNIV	HU	MAGYAR KEPZOMUVESZETI EGYETEM	ALLIANCE FOR COMMON FINE ARTS CURRICULUM	5,000,000,00	80.00%
612521-EPP-1-2019-1-NL-EPPKA2-EUR-UNIV	NL	UNIVERSITEIT TWENTE	ECIU University	5,000,000,00	80.00%

(continued)

Table 1 (continued)

Reference number	Country	Applicant organization	Project title	Maximum EU grant €	% co-funding
612531-EPP-1-2019-1-FR-EPPKA2-EUR-UNIV	FR	COMMUNAUTED UNIVERSITES ET ETABLISSEMENTS UNIVERSITE PARIS-SACLAY	European University Alliance for Global Health	5,000,000,00	80.00%
612546-EPP-1-2019-1-ES-EPPKA2-EUR-UNIV	ES	UNIVERSITAT DE BARCELONA	CHARM European University (Challenge-driven, Accessible, Research-based, Mobile) (CHARM-EU)	5,000,000,00	76.24%
612574-EPP-1-2019-1-NL-EPPKA2-EUR-UNIV	NL	UNIVERSITEIT MAASTRICHT	Young Universities for the Future of Europe	5,000,000,00	76.19%
612599-EPP-1-2019-1-FR-EPPKA2-EUR-UNIV	FR	UNIVERSITE DE LA ROCHELLE	European University for Smart Urban Coastal Sustainability	4,516,063,00	80.00%
612621-EPP-1-2019-1-FR-EPPKA2-EUR-UNIV	FR	SORBONNE UNIVERSITE	The 4EU+ Alliance	5,000,000,00	32.32%
612648-EPP-1-2019-1-FR-EPPKA2-EUR-UNIV	FR	UNIVERSITE D'AIX MARSEILLE	CMS—a European civic university alliance	5,000,000,00	67.68%
612661-EPP-1-2019-1-BE-EPPKA2-EUR-UNIV	BE	KATHOLIEKE UNIVERSITEIT LEUVEN	UNA Europa	5,000,000,00	80.00%
612663-EPP-1-2019-1-FR-EPPKA2-EUR-UNIV	FR	INSTITUT D'ETUDES POLITIQUES DE PARIS	CMCA—The European University in social sciences	4,944,805,07	80.00%

Source: EU Commission: Directorate-General for Research and Innovation

Hidronatura-Gestión eficiente de los recursos hidronaturales, is one of the most noteworthy ones. It was a transnational campus for the proper management of inland water resources that covered the spheres of teaching, research, and mobility. Also in the academic field, there are already existing double degrees in nursing, modern languages, history and historical patrimony, hispanic studies, and business. The interchange of students, faculty and administration staff is also very smooth.

The collaboration with Parma, the third southern partner, was mainly grounded in international projects and internationalization affairs. The cooperation between the 2 offices for international services framed a flowing mobility not only for students but also for faculty and staff, resulting in further cooperation in research and teaching especially in agriculture, humanities, and health. One of the newest projects is SMALLDERS, where UEx and Parma are partners for a common objective: basic research activities and industrial development in agriculture and reduction of water consumption for small landowners.

Along with Parma and Évora, the other universities that make up this consortium have joined: Université D'Angers (France), Atlantic Technological University (Ireland), University of Gävle (Sweden), Universitatea din Oradea (Romania), Otto von Guericke Universität Magdeburg (Germany), and Wrocław University of Environmental and Life Science (Poland). All nine partners share existing collaborations (mobility of students, staff and faculty, research papers and projects) and, most important of all, similar strategies and concerns.

In numbers, EUGREEN are nine universities that all together have 158,000 students and 18,000 staff on 24 campuses. The total population of the nine European regions in which they are located is around 5,250,000 people. It is remarkable the important number of associated partners that exceeds 100.

1.6 Universities in the Periphery

The 2022 European Cohesion Report underlines the role that universities have in less developed regions: boosting economic growth, diminishing regional disparities, providing our graduates with the necessary skills for the job market, and investing in R&D.

As underlined previously, our population is less well-educated than in other regions, and our graduates have a tendency to seek better job opportunities elsewhere. This resulting brain-drain effect weakens the region further, diminishing its active and entrepreneurial population and impacting its age and gender structure. In addition, due to the ongoing consolidation of university hubs in capitals and large urban centers, universities located in peripheral regions face major challenges in securing adequate funding and have little impact on the definition of relevant national and European strategies. This, in turn, makes it harder to attract and retain talent in both research and innovation.

However, all policymakers agree that universities such as ours have a pivotal role to play in the social and economic development of their regions. They are critical "assets," even more so in less-favored regions where the private sector is often weak or relatively small, with low research and development intensity. As underlined by the European Commission's Practical Guide to Connecting Universities to Regional Growth, the successful mobilization of a university's resources can have a disproportionately positive effect on their regional economies and the achievement of comprehensive regional strategies.

They are a key pillar when defining regional smart specialization strategies, via a rigorous assessment of a region's knowledge base, assets and capabilities, including those embedded internally within the university or externally in the local business tissue. They support companies through applied research and the creation of research and learning environments for the continuous development of new technologies and business models for Industry 4.0 (Benešova & Tupa, 2017). Most importantly, they train the local students and act as beacons to attract external talent to the region (Fig. 1). In the following figure the universities that form the consortium are located in a map of Europe that show the geograhical balance of the alliance.

In summary, this chapter will explain the steps for the creation of the alliance, the political context where the alliance has been born, the missions and concerns of the alliance, the impact of the alliance in their territories, the weaknesses and recommendations for higher institutions concerned about SDG.

Fig. 1 Map of the EUGREEN partners. Source: elaborated by EU GREEN Alliance

2 Objectives

The general objective of this research is to find a common *leitmotiv* that serves as an umbrella term and concept for the 9 universities of the network.

The specific objectives are:

- To analyze previous interactions and collaborations/ collaboration networks within the Alliance
- To identify the main research strengths and weaknesses of the consortium
- To map the alliance research output to the SDGs

3 Methodology

In this section the research analysis will be described. The method used for the research: a bibliometric analysis of the research output from all nine member universities has been performed. According to scientific literature, a bibliometric analysis is a rigorous and popular method that is used to analyze a large scientific amount of data (Broadus, 1987). Bibliometric analysis has great acceptance, especially in the business field. This reputation can be credited to the development and accessibility of bibliometric software tools: Gephi, Leximancer, VOSviewer, and scientific databases such as Scopus and Web of Science, and the cross-disciplinary pollination of the bibliometric methodology from information science to business research (Donthu et al., 2020, 2021). More importantly, the popularity of bibliometric analysis in business research is not a fad but rather a reflection of its utility for handling large volumes of scientific data and producing high research impact (Table 2).

The data analysis was collected from Bibliographical Database (Scopus®) and H2020 Projects were collected from Community Research and Development Information Service—CORDIS https://cordis.europa.eu/, through the UNiCS platform owned by SIRIS Academic. SIRIS Academic was a consultancy firm with expertise in supporting universities in this kind of analysis, hired to do it from an external and non-biased perspective.

Table 2 Sources of information gathering for specific objectives

Specific objectives	Information gathering
To analyze previous common interactions and collaborations/collaboration networks within the Alliance.	– Scopus analyses – H2020 projects
To identify the main research strengths and weaknesses of the consortium.	– Bibliometric analysis
To map the alliance research output to the SDGs	– Bibliometric analysis

Source: Scopus, EU Commission: H2020 Projects, Directorate-General for Research and Innovations

4 Results

The results of common interactions are (Fig. 2):

The results of the analysis of the total number of Scopus publications by year (2014 to 2020) and project analysis for each institution are in Table 3:

The results of the analysis by Scopus clusters of research are the following (Fig. 3):

The result of the analysis by fields of research is (Fig. 4):

The results in Chart 3 show the absolute number of publications. However, to identify the relative specialization fields it is important to compare the relative importance of each scientific area with the European average (comparing the absolute number of publications does not consider the fact that publication rates differ a lot among areas). Through this comparative analysis, we conclude that EU GREEN is highly specialized in life sciences (all areas) and health sciences (almost all areas) and somewhat specialized in environmental science, chemical engineering, and decision sciences.

A similar comparative specialization analysis was done regarding the mapping of research on SDGs. The results show that the partners' research is more specialized than the European average in the following SGDs (by decreasing order of the degree of specialization):

SDG 2-Zero hunger
SDG 15-Life on land
SDG 8-Decent work and economic growth
SDG 12-Responsible consumption and production
SDG 4-Quality education
SDG 11-Sustainable cities and communities

On the other hand, the consortium research is less specialized than the European average in the following SDGs:

SDG 7-Affordable and clean energy
SDG 13-Climate action

Besides providing information on the SDG where the consortium has a specialization degree above the European average, the analysis performed allows us to estimate how far (above or below) the Alliance is from the European average. For instance, for the Alliance to become specialized in SDG 7, it would require 256 more publications. Another example is that although the Alliance is specialized in SDG 11, it is only slightly above the European average, as it has only 31 publications more than the required for the average specialization.

Based on the **publications data** (Scopus) and **projects** (CORDIS) here are the results of EU GREEN specialization on the SDGs: 2 (zero hunger), 8 (decent work and economic growth), 9 (industry, innovation and infrastructure), 11 (sustainable cities and communities), 12 (responsible consumption and production), 15 (life in land).

Collaborations between members
(as per publications listed in SCOPUS)

# of collaborations in SCOPUS publications (i.e. publications with affiliation to more than one member) between Alliance members, 2014-2020

University	Gâvle	IT Carlow	Magdeburg	Extremadura	Oradea	Angers	Évora	Parma	Waterford	Wrocław
Gâvle	-	0	0	0	0	0	1	2	0	0
IT Carlow	0	-	1	0	0	1	1	0	15	0
Magdeburg	0	1	-	2	0	0	1	24	1	3
Extremadura	0	0	2	-	1	1	116	18	0	0
Oradea	0	0	0	1	-	0	0	0	0	0
Angers	0	1	0	1	0	-	0	7	0	0
Évora	1	1	1	116	0	0	-	5	1	0
Parma	2	0	24	18	0	7	5	-	0	5
Waterford	0	15	1	0	0	0	1	0	-	0
Wrocław	0	0	3	0	0	0	0	5	0	-

Collaboration areas:

- **Évora-Extremadura**: Computer Science, Earth and Planetary Science, Medicine and Environmental Science, and to a minor extent in Agriculture and Business & Management
- **Magdeburg-Parma**: Medicine
- **Extremadura-Parma**: Agriculture, Computer Science, Environmental Science..

Fig. 2 Collaboration between members. Source: SIRIS Academic and Scopus

Table 3 Scopus publications and H2020 projects of each of the universities in the alliance

Institution	Number of Scopus publications	Number of H2020 projects
Otto von Guericke University of Magdeburg (Germany)	12,029	28
Hogskolan i Gavle (Sweden)	1684	2
Università di Parma (Italy)	13,105	56
Universite d'Angers (France	5198	9
Universidad de Extremadura (Spain)	8065	10
University of Évora (Portugal)	3946	39
Universitatea din Oradea (Romania)	1851	1
Institute of Technology Carlow (Ireland)	242	2
Wrocław University (Poland)	3980	10
Waterford Institute of Technology (Ireland)	1078	32

Source: SIRIS Academic and UEx

It can be compared the EU GREEN specialization in SDGs and other alliances SDGs presence (Table 4):

5 Discussion

Based on the preliminary results shown in the previous section and after focal group discussions taken over several months, it was decided that the leitmotiv of the network should be "Peripheral universities for the SDGs" with a special focus on education as the truly transformative impact of the consortium should be put on students and their training for SDGs and sustainability. However, to guarantee such a transformative experience for the students, all the pillars of the university have to also be aligned with the SDG and sustainability as all of them contribute to enriching the student pathway. For instance, research is organized in six research clusters in different sustainability areas with the aim of achieving research excellence by working together. The areas of these research clusters are also the same where the alliance is going to develop innovation and entrepreneurship expertise and where collaboration with society and local and regional communities is going to be fostered; and, more importantly, these are the educational areas where the Alliance wants to create joint study programs. The research bibliometric analysis was an important input in the Alliance decision process.

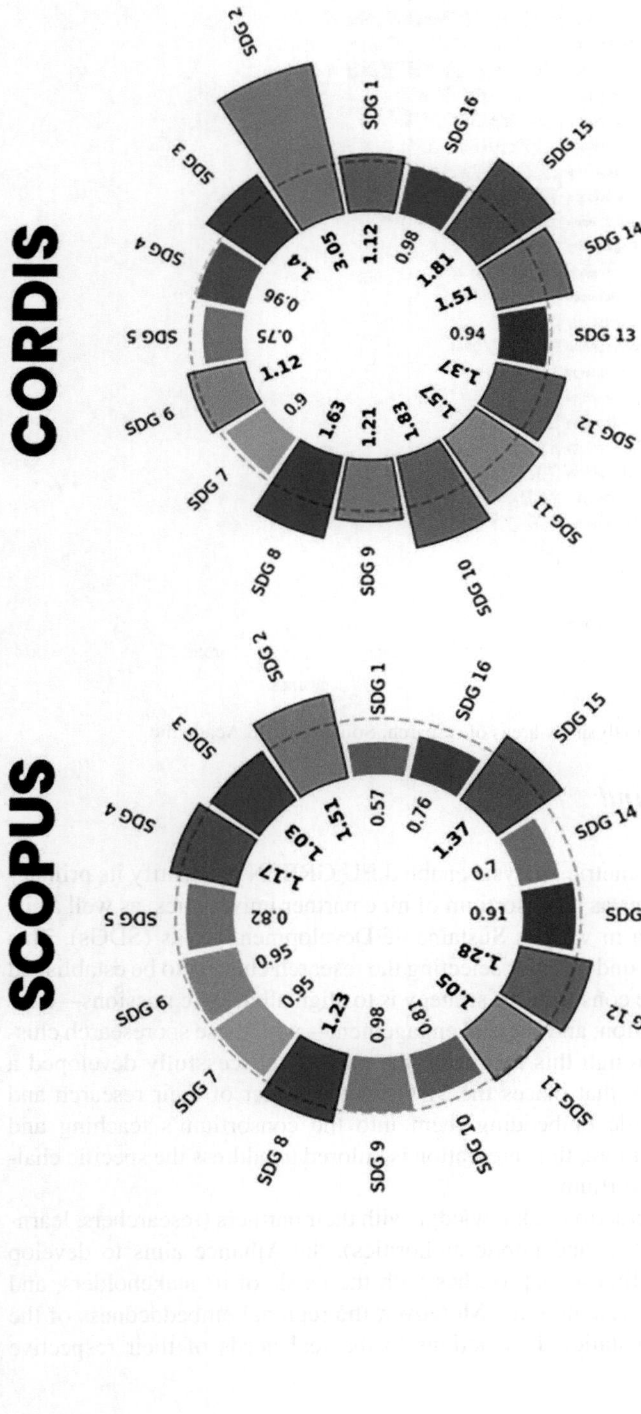

Fig. 3 Scopus and Cordis analysis. Source: SIRIS Academic

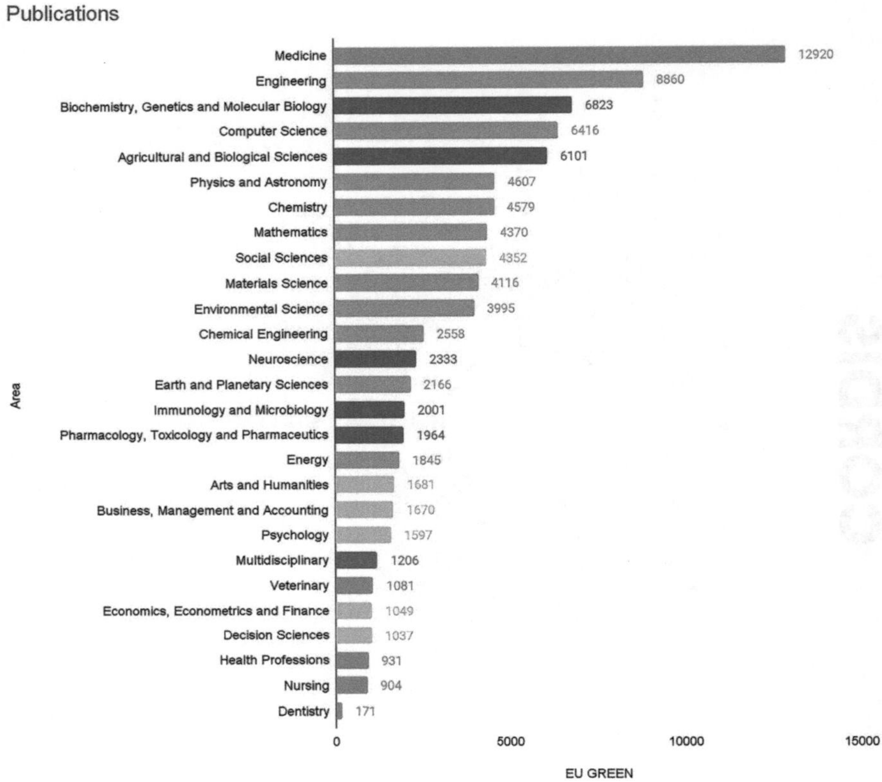

Fig. 4 Scopus and Cordis analysis by areas of research. Source: SIRIS Academic

5.1 Leitmotiv Found

The results of the bibliometric analysis enabled EU GREEN to identify its primary strengths and weaknesses as a consortium of nine partner universities, as well as its degree of specialization in various Sustainable Development Goals (SDGs). This analysis served as the foundation for selecting the research clusters to be established within the Alliance. The consortium's strategy is to align all its core missions—education, research, innovation, and societal engagement—with these six research clusters and the SDGs. Through this approach, the partners successfully developed a collaborative framework that places the SDGs at the center of their research and innovation agendas while embedding them into the consortium's teaching and learning model. Furthermore, this integration is tailored to address the specific challenges faced by the consortium.

By emphasizing co-creation of knowledge with their partners (researchers, learners, industry, civil society, and public authorities), the Alliance aims to develop novel methodologies, align its approaches with the needs of its stakeholders, and reinforce its long-term sustainability. Moreover, the regional embeddedness of the partner enables them to tailor their actions to the real needs of their respective

Table 4 Other alliances specialization in SDGs

Existing EUAs	How they define their positioning in terms of sustainability and SDGs	Presence of research-intensive universities and urban regions
AURORA	Sustainability pioneers/education/campus plan	Yes
CHARM-EU	The planet and the Sustainable Development Goals	Yes
Circle-U	Pooling resources for the sustainability agenda	Yes
EC2U	SDG3/SDG4/SDG11	Mid-sized cities
ECIU	SDG 11	Yes
ENGAGE EU	Focus on SDG with digital technologies	Yes
ENLIGHT	SDG11	Yes
EUNIWELL	SDGs in Well-Being & Health. Environment, Urbanity & Teacher Education	Yes
EURECA-PRO	SDG 12	Peripheral regions, technological focus
INVEST	SDG 2/SDG 3/SDG 4/SDG 6/SDG 11/SDG 13/SDG 14	Rural focus
ULYSSEUS	SDGs in general	Yes
UNITA	Commitment to the SDGs	Mountain regions
UNIVERSEH	SDGs as a priority	Yes
EU GREEN	All SDGs: we address the major societal challenges identified in the 17 SDGs We aim to scale up and promote *sustainability encompassing the economic, societal, cultural and environmental dimensions* of the term and their impact on one another across all four of our core missions: education, research, innovation and service to society.	Peripheral regions

Source: SIRIS Academic

territories and thus contribute to the development of other regional ecosystems across Europe, ensuring a true transformative impact on society, using as a compass the EU's commitment to implement the European Green Deal and reach its climate objectives "in a fair, cost effective and competitive way."

EU GREEN members aim to be at the vanguard of this venture, by embracing a task of actively doing good and by placing societal responsibility at the heart of their mission as a guiding belief that delineates the way we teach, the focus of our research, innovation and engagement activities and our contribution to the sustainable development of our regions. Only by educating citizens on sustainability and giving them the tools to implement it will societies become truly sustainable. These principles are also inextricably linked with the main objectives of the European Strategy for Universities, which places global leadership and green transformation as hallmarks of the European identity in education and leadership. EU GREEN thus means to operate and be influential not only in Europe but also globally.

6 Conclusions

EU GREEN found that implementing this bibliometric analysis was a powerful tool to find a common leitmotiv that was not obvious for or taken by already existing alliances. Although the partner had many characteristics in common (most universities were generalist, medium-sized, connected to their regions and far from national power places), before doing the bibliometric analysis, it was not obvious what be the alliance leitmotiv should be. In this sense, the general objective of this research, which is to find a common *leitmotiv* that served as an umbrella term and concept for the 9 universities of the network, was successfully achieved.

Several axioms/mottos were brought up before the analysis was done based on mutual trust, friendship after decades of collaboration, perceived similarities, and perceived weaknesses, but not based on scientific data. Topics like food (cheese and wine), agriculture (in a broad scope), and some others were discussed, but we were unable to reach a consensus. The objective analysis based on the alliance publications and projects allowed the identification of the consortium strengths, which enlightened our choice of leitmotiv and research clusters.

Regarding the achievements of the specific objectives, results show that:

Specific Objective 1: To analyze previous interactions and collaborations/collaboration networks within the Alliance.
Based on the analysis, previous proven collaborations were mainly done between Evora and Extremadura, Magdeburg and Parma and Extremadura and Parma. It should be noted that, although data is not shown here, the consortium partners had previous mobility agreements and mobility flows among them and collaborations in internationalization existed as well.

Specific Objective 2: To identify the main research strengths and weaknesses of the consortium.
Based on the analysis done, the strengths of the alliances are the fields of physical sciences, life sciences, health and social sciences. The results showcase that EU GREEN is highly specialized in Life sciences (all Scopus subject areas) and Health sciences (almost all Scopus subject areas), as well as other relevant areas such as environmental science and chemical engineering and decision sciences. For this reason, the network was connected to sustainability and SDGs.

Specific Objective 3: To map the alliance research output to the SDGs
Considering the SDG analysis, there are 3 SDG where the Alliance is highly specialized: SDG 2 (Zero hunger), SDG 15 (Life on Land) and SDG 8 (Decent work and economic growth). Moreover, the Alliance has a specialization above the European average in SDG 12 (Responsible consumption and production), SDG 4 (Quality of Education) and SDG 11 (Sustainable cities and communities). Each of these SDG includes many different issues organized around a particular sustainable development goal. For instance, SDG 2 (Zero Hunger) includes topics such as sustainable agriculture, agricultural research, and undernourishment to improve soil quality and promote local seed banks.

The recommendation for future networks that want to create a super campus together is to take the time, space, the right method, tools, and a deep debate to find the common grounds that will guide them for the next 10 years. For EU GREEN the analysis proved that their previous trajectory, common strengths, their accomplishment of SDGs enabled them to tailor their actions to the real needs of their respective territories and thus contribute to better training of students for sustainability, the development of other regional ecosystems across Europe, ensuring a true transformative impact on society, using as a compass the EU's commitment to implement the European Green Deal and reach its climate objectives "in a fair, cost-effective, and competitive way."

It is difficult to imagine in 2023 a supercampus composed of several European universities under the umbrella of the same leitmotiv and sharing a common long-term strategic plan. However, this is Europe's most important educational venture in the last decades. Among those already awarded in the 2 previous calls there were 40, and in 2022 only 4 were awarded, one of them being the one led by the University of Extremadura. The Alliance and its partners, led by the University of Extremadura, focus on promoting sustainability and addressing regional disparities. Together they require a deep shift in thinking away from a focus on growth and towards a holistic perspective that considers the economic, societal, cultural and environmental dimensions of sustainability and their impact on one another within each of our regions. Their approach is closely aligned with the SDGs across all four of our core missions: education, research, innovation, and service to society. Together, they have built an inclusive and collaborative partnership that places these SDGs at the heart of our research and innovation approaches and integrates them into their unique teaching and learning model, while adapting them to the specific challenges they face.

7 Limitations

It should be clarified that ATU was not yet a member when this analysis was done. In the analysis, IT Carlow was a partner. Once IT Carlow merged with Waterford and both turned into SETU, they joined another alliance where Waterford had collaborated as an associated partner. Then, EU GREEN searched for a partner with similar characteristics to Carlow's. The limitation of this study is that it includes the original consortia.

References

Benešova, A., & Tupa, J. (2017). Requirements for education and qualification of people in industry 4.0. *Procedia Manufacturing, 11*(2017), 2195–2202.

Birk, K. (2019). *European universities: balancing excellence, exclusiveness and inclusion*. EAIE blog. Accessed from https://www.eaie.org/blog/european-universities-balancing-excellence-exclusivenessinclusion.html

Broadus, R. N. (1987). Toward a definition of "bibliometrics". *Scientometrics, 12*, 373–379.

Cesaer. (2019, October 28). *Position on European Universities*. Accessed from https://www.cesaer.org/content/5-operations/2019/20191028-position-on-european-universities.pdf

Cippitani, R., & Gatt, S. (2009). Legal developments and problems of the Bologna Process within the European higher education area and European integration. *Higher Education in Europe., 34*(3–4), 385–397.

Claeys-Kulik, A. (2019). *European Universities initiative – Chances and challenges*. University World News, 6th November 2019. Accessed from https://www.universityworldnews.com/post.php?story=20191106144804387

De Wit, H., & Deca, L. (2020). Internationalization of higher education, challenges and opportunities for the next decade. In *European higher education area: Challenges for a new decade*. Springer.

Donthu, N., Kumar, S., & Pattnaik, D. (2020). Forty-five years of Journal of Business Research: A bibliometric analysis. *Journal of Business Research, 109*(1), 1–14.

Donthu, S. K., Pandey, N., & Gupta, P. (2021). Forty years of the *International Journal of Information Management*: A bibliometric analysis. *International Journal of Information Management, 57*, 102307.

EU Commission. (2019). *First 17 "European Universities" selected: A major step towards building a European Education Area, Brussels, Press release 26 June 2019*. Accessed from https://europa.eu/newsroom/content/first-17-%E2%80%9Ceuropean-universities%E2%80%9D-selected-majorstep-towards-building-european-education_en

EU Commission. (2020). *24 new European Universities reinforce the European Education Area, Brussels, Press release 9 July 2020*. Accessed from https://ec.europa.eu/commission/presscorner/detail/en/ip_20_1264

EU Commission. (2022). *Erasmus+: 44 European Universities to cooperate across borders and disciplines, Press release 27 July 2022, Brussels*. Accessed from https://ec.europa.eu/commission/presscorner/detail/en/ip_22_4702

EU Council. (2017). *European Council meeting 14 December*. Accessed from https://www.consilium.europa.eu/media/32204/14-final-conclusions-rev1-en.pdf

EU Council. (2018). *Draft Council conclusions on moving towards a vision of a European Education Area Brussels, 8 May*. Accessed from http://data.consilium.europa.eu/doc/document/ST-8701-2018-INIT/en/pdf

European Commission/EACEA/Eurydice. (2018). *The European higher education area in 2018: Bologna Process implementation report*. Publications Office of the European Union. Accessed from https://doi.org/10.2797/63509

Grove, J. (2019). *Defer Macron's dream for European universities, mobility experts say*. Times Higher Education, 26th September. Accessed from https://www.timeshigh-ereducation.com/news/defermacrons-dream-european-universities-mobility-experts-say

Gunn, A. (2020). The European universities initiative: A study of alliance formation in higher education. In A. Curaj, L. Deca, & R. Pricopie (Eds.), *European higher education area: Challenges for a new decade*. Springer. https://doi.org/10.1007/978-3-030-56316-5_2

Macron, E. (2017). *Speech on new initiative for Europe, 26 September*. Accessed from https://www.elysee.fr/emmanuel-macron/2017/09/26/president-macron-gives-speech-on-new-initiative-for-europe

Designing New Forms of Business Education at Circular Economy and Sustainable Business: An Experience Report on Stakeholders' Engagement Through Co-Creation

Denise Aparecida Pereira Siloto, Thiago Gomes de Lima, Lays Alves Pereira, and Thais da Silva Barcelos

Abstract The circular economy and sustainable business have been gaining ground as significant pathways to facing the challenges of the green transition. Despite an increased interest in those concepts, many concerns about disseminating them as a new business paradigm still need to be addressed. Higher education institutions are fundamental in this endeavor and can play an important role as enablers of multi-stakeholder engagement. In this way, this chapter addresses the problem of how to design a business education course that focuses on the circular economy through co-creation. This experience report takes place in Macaé/RJ, Brazil's oil capital and deeply dependent on productive activities linked to fossil fuels. The prototype course occurred in November 2022 as an extension course, and stakeholders' engagement was essential from the project's conception and implementation to its closure and evaluation. The Lawshe content validity method was applied to evaluate students' perceptions of the teaching topics. The results showed that the introduction module for circularity had higher scores regarding essentiality, highlighting the relevance of the fundamentals of circular economy and sustainability. Finally, adopting the co-creation approach for designing an executive education course proved to be a powerful way to strengthen the engagement of participants and insert content aligned with the region's needs. Some final recommendations to practitioners were mentioned.

Keywords Business education · Circular economy · Sustainable business · Co-creation · Stakeholder

---

D. A. P. Siloto (✉) · T. G. de Lima · L. A. Pereira · T. da Silva Barcelos
Macaé, Brazil
e-mail: denise.siloto@ufrj.br; thiago.gomes@ufrj.br

C. Rego et al. (eds.), *Transitioning to a Circular Economy*, Studies on Entrepreneurship, Structural Change and Industrial Dynamics,
https://doi.org/10.1007/978-3-031-77661-8_8

1 Introduction

In recent decades, few topics in higher education have attracted as much attention as the concept of sustainability. Since the publication of the Brundtland Report in 1987, educational systems worldwide have been pressured to incorporate programs that contribute to a more sustainable society (Figueiró & Raufflet, 2015).

The Sustainable Development Goals (SDGs), adopted by the United Nations (2015), reinforce, on more than one front, the importance of entrepreneurship, technological changes, and education to find lasting solutions to social, environmental, and economic current challenges. SDG 8, for example, is about decent work and economic growth; one of its targets is to improve resource efficiency in consumption and production and to decouple economic growth from environmental degradation. SDG 9 promotes investment in innovation and scientific research as crucial drivers of sustainable economic growth. SDG 11 highlights the impact of higher education on integrating local and regional contexts into their research (e.g., to solve local problems) and curricula. It is worth mentioning that the 17 SDGs are integrated, and actions in one area will affect the outcomes in others.

One of the instrumental strategies discussed for achieving sustainable development is the transition to a circular economy (CE). The "take-make-consume-dispose" linear economy system is replaced by a circular economy model, which emphasizes redesigning processes and cycling of materials (Murray et al., 2017).

Although the social dimension of sustainability has yet to be strongly developed in the CE (Murray et al., 2017), its practices can contribute directly to achieving a significant number of SDGs. Strong relationships exist between CE and the targets of SDG 6 (Clean Water and Sanitation), SDG 7 (Affordable and Clean Energy), SDG 8 (Decent Work and Economic Growth), SDG 12 (Responsible Consumption and Production), and SDG 15 (Life on Land) (Schroeder et al., 2019).

Higher education institutions (HEIs) are uniquely positioned to be at the forefront of sustainability issues. They can be strategic agents in supporting the circular economy and sustainable business through knowledge creation (research), dissemination (teaching), and community development (outreach activities) (Johnston, 2007). As outlined in SDG 17, multi-stakeholder partnerships and cooperation are required to advance with the SDGs, without mentioning the efforts on skills training, capacity-building programs, and technology development (Schroeder et al., 2019).

HEIs contributions to a green transition include but are not limited to collaborating with industry, assisting policymakers, building human and intellectual capital, influencing campus culture, and promoting an inclusive sustainability discourse (Salas et al., 2021). According to Bugallo-Rodríguez and Vega-Marcote (2020), HEIs play a pivotal role in integrating sustainability and circular economy concepts into teacher training programs, enabling future educators to champion these principles in their practices.

As for business schools globally, educators have launched numerous courses, programs, and projects for sustainable management. The main idea behind those initiatives is to transform the way of thinking of future business leaders and entrepreneurs, contributing to increasing companies' awareness of their imperative role and responsibility in building more sustainable societies (Figueiró & Raufflet, 2015).

Acting collaboratively with multiple stakeholders, especially in executive education, provides an intense and constant exchange of knowledge between universities, private companies, public services, and the community, which favors finding solutions to complex challenges with more assertiveness and agility.

According to Holmlid et al. (2015), the collaborative design suggests that people can act and achieve results together. In this sense, offering services in an HEI (courses, workshops, or lectures) is inherently a co-creative practice, as subjects collaborate to co-create and deliver value. Thus, co-creation can be defined as a creative and interactive process that values the different views of stakeholders and seeks to combine professional and local experiences to generate solutions to meet various demands (Cottam & Leadbeater, 2004).

The co-creation approach in HEIs can provide better experiences to teachers, researchers, and students, which contributes to developing new skills, more significant commitment, and innovation in teaching and learning practices. Research bases have revealed different studies considering co-creation in the field of education. Researchers are observed exploring co-creation approaches applied to teaching and learning processes, Bovill (2020), Könings et al. (2021), Bovill et al. (2016). There are researchers adopting co-creation with an emphasis on different demands of higher education, such as demands related to the processes of learning and teaching, Sterenberg et al. (2018), related to curriculum reform, Bovill et al. (2013), related to positive outcomes for higher education, Tarı Kasnakoğlu et al. (2022), associated with the co-creation of games and projects with different stakeholders, Baxendale et al. (2021), Gkogkidis (2020), and among others, such as those aimed at the innovation of services offered by universities, Mann (2020).

Given this, this work intends to present an experience report in which executive education was used to engage stakeholders in a short course (continuing education) focused on circular economy and sustainable business. The course's overall objective was to sensitize business leaders and regional entrepreneurs to the benefits, challenges, and implications of the circular economy and to support the development of strategies so that they could model their organizations in this direction. The first version of the course was co-created by different social actors and launched as a prototype so that the syllabus for future editions could be validated and enhanced.

2 Circular Economy and Sustainable Business

The Circular Economy (CE) theory is the result of an attempt to reconcile, on the one hand, the pressure of increasing production and consumption and, on the other, the evident limitation of natural resources (Ohde & Mattar, 2018). Since the Industrial Revolution, the evolution of the global economy has been dominated by a linear "take-make-consume-dispose" model. Although the industry has experienced significant advances in efficiency gains, a system whose foundation is consumption assumes significant losses along the entire value chain (MacArthur, 2015).

Circular Economy was created to contrast with the Linear Economy model, in which raw materials are used and discarded without concern for the impacts generated. The idea of the circular economy is to promote a healthy and regenerative economy, which aims to keep each and every resource at its highest level of utility and value, that is, which provides for the reuse of material that has already been extracted from nature (Ohde & Mattar, 2018).

According to Morseletto (2020), the economic model in which the circular economy is inserted considers essential aspects, among them, the efficient use of resources, the depreciation of waste, the perspective of long-term value, the devaluation of primary resources, closed cycles of products, in addition to ensuring environmental and socioeconomic concerns. In summary, the Circular Economy aims to reduce the ecological and economic damage caused by solid waste and contributes to reducing waste generation (Mancini et al., 2021).

Several countries worldwide are developing reduction strategies to promote remanufacturing and reuse, including South Korea, China, and the United States. In Europe, the first call for proposals aimed at the Circular Economy originated in the Horizon 2020 program in 2014. In this universe, it is worth considering the Ellen MacArthur Foundation, which, since 2010, has been expanding the dissemination of the Circular Economy among executives and public policy managers. Although, since 1990, the concept has been applied timidly in some eco-industrial parks, such as the Kalundborg Symbiosis in Denmark (Stahel, 2016).

Global data warns about the importance of reviewing waste disposal and minimizing the emission of greenhouse gases. Andrew (2018) warns that 8 million tons of plastic are dumped annually into the oceans, and greenhouse gas emissions are out of control. Moreover, to make matters worse, it is known that the global demand for resources is expected to double by 2050 (United Nations Environment Program, 2017). With this, scientists advise that resources must be managed more responsibly and under a long-term vision (Haas et al. 2015).

According to Geng et al. (2019), the circular economy will be able to operate on four levels: products (designed to be recyclable and reusable), companies (creating new business models), networks (articulation of companies and customers networks) and policies (aiming to support markets). It is known that companies are becoming aware that remaining in the linear system could bring negative impacts such as increased exposure to risks, especially resource prices; interruptions in the supply chain; higher competition; and stagnant demand for some sectors (MacArthur, 2013).

In the case of Brazil, de Oliveira et al. (2018) present some challenges to be overcome in the transition from linear to circular business models since they are still incipient, unidentified, and need more articulation. It is known that Brazil has a vast territory and regional diversity, which contribute to the promotion of the circular economy still being a complex challenge (de Oliveira et al., 2018). Thus, the need to disseminate the concepts of circularity and its importance at all national, state, regional, and municipal levels increases to make it known in the country's various locations.

Adapting business models to circular business models had become a priority agenda worldwide. For Stahel (2016), circular business models can be divided into two groups: those that promote reuse and extend useful life (repairs, remanufacturing, upgrades, and retrofits) and those that transform old goods into new resources (recycling materials). Both options can be considered by entrepreneurs and contribute to strategic direction. As a consequence, new markets and business models may emerge; for example, vehicle manufacturers can withdraw cars and update parts so that they work longer (Geng et al., 2019).

Given the importance of this topic, several researchers are developing studies within this universe and reinforcing the importance of involving stakeholders and academia in transitioning business models. Barreiro-Gen and Lozano (2020) stressed that the circular economy should also be implemented beyond the boundaries of organizations in a more systematic way and through collaborative strategies in partnership with the various stakeholders. Mancini (2021) reinforced that society is increasingly concerned about the environmental impacts of people's behaviors and requires industries to provide products that promote waste prevention and a transition to a circular economy. Research is crucial to contribute to the technological changes in many segments.

Additionally, knowledge of the circular economy is currently concentrated in large industries and far from small and medium-sized enterprises (SMEs); therefore, introducing it into academic and professional training could be a valid strategy (Stahel, 2016). For this, researchers must collaborate with society, industry, and non-profit organizations, as well as in disciplines such as sociology, economics, engineering, and science, to expand the dissemination of the Circular Economy (Geng et al., 2019).

At the interface between a sustainable society (which includes the Circular Economy) and higher education are the different spaces, structures, subjects, norms, activities, and behaviors that make up the ecosystem of a community. The theme of sustainability has been gaining relevance as the perception spreads that, as a species, human beings are threatened by the conflict between the planet's natural systems and the economy. The view that there will not be enough natural resources to provide new entrants to the market with a way of life similar to that of the world's middle class causes concern and fear (Siloto et al. 2021).

Faced with this crisis scenario, Nascimento (2012) describes some possible responses that society can develop: the first would be the technological solution. This path bets on the inventiveness of human beings to solve the challenge of scarce resources. Another answer would be a radical change in production and consumption patterns, pointing to what economists have agreed to call "economic degrowth." Sustainable development is another possible response to the crisis, shaping the idea of a balance between the social, economic, and environmental spheres.

Furthermore, it is essential to mention that the term sustainability must always be seen as a discourse, and precisely for this reason, it will never be naive or disinterested (Acselrad, 1999). Talking about sustainable business is, above all, debating the inconsistencies inherent in uniting such divergent concepts.

One of the significant interpretive matrices that make up the debate on sustainability comprises the unfolding of the original proposal of the Brundtland Report. It is a discourse that understands that the market economy can lead the transition to sustainability through changes in the production process, containment of population growth, and ecologically oriented consumption. According to this view, the dynamics of the capitalist system can adapt to environmental demands and transform them into new incentives for competitiveness (Lima, 2015).

In order to meet the growing demand for businesses that are considered sustainable, companies must develop plans and structures to deal with ethical considerations, i.e., codes, values, or policies to guide business behavior. These documents will determine what is acceptable inside and outside the company; therefore, they should be periodically reviewed to identify and eliminate weaknesses (Ferrell et al., 2017).

Faced with the scenario presented, the university has an essential role in developing competencies in the various layers of education, whether at the undergraduate level, continuing education, specialization courses/MBA, master's degree, or Ph.D. programs. This chapter shares the experience in continuing education (business education) that adopted the co-creation approach in the conception processes, involving different interested parties to generate ideas. Section three in this chapter contextualizes the co-creation approach and brings a broader reading about the applicability of co-creation practices in executive education.

3 Co-creation

Initially, it is worth mentioning that the term co-creation has been used in different contexts. In this work, co-creation focuses on design practices in which different social actors collaborate to create new desired futures (Holmlid et al., 2015). Mattelmäki et al. (2011) suggested four co-design modes. In the first mode, users are given a voice, and their suggestions are considered in the design process. In the second mode, user participation is facilitated with co-creative tools provided by designers/researchers. In the third mode, the designer is not just a facilitator; he participates in collective creation. In the fourth modality, designers and researchers support and facilitate a collaborative process with multiple stakeholders, not just users.

Regarding the design processes, the literature shows different models that explain and describe the step-by-step. This work has considered the model of lenses by Holmlid et al. (2015), which presents perspectives that direct the process toward innovations. This model contemplates four lenses, namely (1) Insight Generation, (2) Concept exploration and development, (3) Converging Towards a Specification, and (4) Transformation and Implementation, as shown in Fig. 1.

Synthetically, the activities of each lens can be:

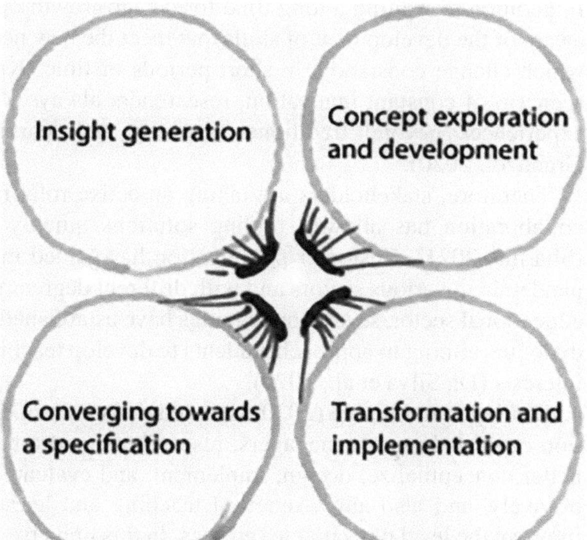

Fig. 1 The four lenses for viewing co-creative practices (Holmlid et al., 2015)

- Insight Generation: Raises needs, wants, and possibilities and therefore deals with exploration and curiosity about what users experience and might experience in the desired future.
- Concept exploration and development: In this lens, interest is focused on exploring possible concepts that may suit the needs and desires.
- Converging Toward a Specification: In this lens, it is expected to develop the specification of a service so that interested parties have a shared understanding of the service and its specificities.
- Transformation and Implementation: This lens is related to project implementation, although prototyping can take on new challenges and formats when used as a means of transformation and implementation.

This is not a linear progression, as different lenses share common elements and meaningful connections. Therefore, co-creative practices can contribute through different lenses. Sometimes, the co-creative tools become everyday processes to facilitate interactions between the organization and the client. In the case of business education, adopting co-creative processes can be a proactive strategy for innovation.

3.1 *Co-creation Applied to Business Education*

The engagement of professionals has always been considered a challenging issue in the universe of executive education (Naim, 2021). It is also known that constant changes in the business sphere have a direct impact on the career plan of executives.

In addition to waiting a long time for career growth opportunities, they need to be aware of the development of skills that meet the new needs identified by the market, which change constantly, in short periods of time (Krishnamurthy, 2020). In this scenario of constant innovation, researchers always warn about designing quality experiences beyond traditional teaching and learning methods (Grandzol & Grandzol, 2020).

Therefore, stakeholders are taking an active role in co-creation activities. This collaboration has allowed finding solutions quickly to the installed challenges (Sharma, 2021). Customer participation has gained much strength throughout the pandemic in various sectors and with different degrees of service co-creation. In the educational sector, several universities have established a favorable environment for dialogue, aiming to approach students to develop teaching plans that meet collective interests (Da Silva et al., 2020).

Dollinger and Lodge (2020) signaled that co-creation processes in higher education could help guide managers, researchers, and other university stakeholders to better conceptualize, design, implement, and evaluate activities developed collaboratively, and also allow renewed teaching and learning experiences, as well as improve the level of available services. In this universe, students can contribute with feedback, opinions, and skills that stimulate the mutual generation of value and interact and work collaboratively with teachers (Dollinger et al., 2018).

Wilden et al. (2019) reveal that for HEIs, co-creation processes strengthen relationships with companies, students, and employees by developing new skills and improving services based on opportunities for collaboration between the parties involved (Maxwell-Stuart et al., 2018). It was also observed that paying students felt more satisfied when participating in co-creation activities (Maxwell-Stuart et al., 2018). In the case of executive education, some researchers have recommended that activities should be designed to favor the intense exchange of experiences and combined practices. There is a tendency to promote intercultural and inter-institutional exchanges, including content related to developing soft skills and creating classrooms in a hybrid format and face-to-face in-field activities (da Silva et al., 2022).

In this work, the design process of an executive education course will be explored, which involves different stakeholders who contributed to the final product through the generation of insights. It included doctoral researchers, undergraduate students, HEIs professors, HEIs advisors in the extension area, and market professionals. According to co-design mode one, those involved were given a voice, and their thoughts were considered in the design process.

4 Methodology

One of the specific objectives established for this project involved creating a syllabus for a course that addressed the circular economy, where it was possible to converge knowledge in the area with the challenges and opportunities in the region. The co-creation method, which guided the journey from project planning to close, is a method that develops in a non-linear way but which can be observed through four perspectives: generation of insight, development and exploration of concepts, convergence towards a specification, and finally, implementation of the innovative process (Holmlid et al., 2015). Specifically, regarding the generating insights perspective, several actions were taken to get ideas and involve different stakeholders. This journey will be reported in the results section.

In the design process, there was an active effort to collect insights from multi-stakeholders: from those enrolled in the course, from professors, from students who were members of the project team, and from Federal University of Rio de Janeiro (UFRJ) advisors who are responsible for continuing education courses.

While conceiving and designing the course, the project coordinators conducted an extensive survey of circular economy courses with other higher education institutions. The survey results were shared with potential lecturers to discuss new insights and align the syllabus for each module. The meetings with the teachers took place through individual sessions.

UFRJ advisors, who are responsible for approving new continuing education courses, were essential in analyzing the structure of the project, its objectives, and connections with teaching, research, and extension activities. Communication with them took place through the university's electronic channels. Insights from students, who were project team members, were debated through alignment meetings involving project coordinators and professors.

Insights from participants enrolled in the course were collected before classes through the following strategic questions in the registration form:

1. Does the company where you work already include sustainability and/or circular economy actions in its operations? If so, describe these actions.
2. What would you like to learn about circular economy and sustainable business?
3. Cover letter for registration.

These questions were of great value for reflecting on the contents, aligning expectations, and identifying good practices in the region. The selected participants' profiles and expectations were shared with the teachers to enrich the content. The enrollment letters contributed to the knowledge of the subjective aspects that motivated the participants to seek the course.

At the end of the last module of the course, an online form was made available to students. The link directed the student to a reaction semi-structured questionnaire, where she or he could evaluate, one by one, all the topics worked on by the lecturers. This feedback helped to identify the essential items covered along the journey, and the results will be helpful for the design of future offerings of this course.

The semi-structured questionnaire, built on the Google Forms platform, aimed to analyze the essentiality of the subjects covered throughout the Circular Economy and Sustainable Business course. The feedback included 40 assertions corresponding to all topics worked on in classes. For each statement, the specialist (the student who attended the course) should judge whether that topic was "not important," "important, but not essential," "essential," or "I do not know." In addition, they had to answer two open questions:

1. What were the strengths of the course? What should we keep?
2. What are our opportunities for improvement? What can we do differently?

Lawshe (1975) proposed a highly appropriate method for validating research instruments, and his method was chosen to analyze the quantitative data on the form. The experts invited for the questionnaire validation are people related to the context in which the items are to be validated. Therefore, they may be clients, students, researchers, and specialists in the subject, among others. In this sense, the specialists invited to apply their opinions should analyze and classify each item. After the answers from these experts, the tabulation of data is developed, aiming to identify the level of essentiality of each item under evaluation. Subsequently, it is necessary to quantify the responses to verify whether each item deviates from the unit and approaches 50%.

Given this, for Lawshe (1975), an item that is evaluated as "essential" by more than half of the interviewees has its content considered valid, and the greater the tendency of the interviewees, the essentiality of the item, that is, more than 50%, the greater the degree of validity of the content. Therefore, to classify the consensus among specialists, Lawshe (1975) suggests a Content Validity Ratio (CVR) method, as shown in the equation:

$$\text{CVR} = \frac{n_e - (N/2)}{(N/2)} \tag{1}$$

Equation 1—Content Validity Ratio (CVR). Source: Lawshe (1975).

n_e is the number of evaluators who classified the item as "essential," and N is the total number of respondents.

The results are interpreted as follows: (1) Negative CVR: less than half of the experts judged the item as "essential"; (2) Zero CVR: half of the experts judged the item as "essential" and the other half as "non-essential"; and (3) Positive CVR: more than half of the experts judged the item as "essential."

In this framework, Lawshe's Method has been considered by several specialists in order to validate questionnaire items and criteria validation. Freitas (2013) adopted Lawshe's Method to identify success evaluation criteria in implementing ERP systems. Lima (2015) and Silvério (2016) adopted Lawshe's Method to analyze the criteria for evaluating the quality of services in the hotel sector according to guests' perceptions. De Souza et al. (2020) applied Lawshe's method to analyze customers' perceptions of the products and services offered by a supermarket in Rio de Janeiro.

The analysis of qualitative information was completed through a critical reading of the answers to the two open questions regarding the course's strengths and weaknesses. Next, excerpts from the text were highlighted and categorized for interpretation and search for convergences in the students' answers.

5 Results

The purpose of creating a circular economy and sustainable business short course in the city of Macaé was to establish a dialogue between an HEI (Federal University of Rio de Janeiro—UFRJ) and a group of entrepreneurs, managers, analysts, professors, and researchers from the region to disseminate knowledge and experiences on those themes and to foment and drive the green transition in a city traditionally focused on the oil & gas industry.

The design of a sustainability-focused business course is inherently transdisciplinary, as it must offer solutions to complex and multidimensional problems. Defining the course modules was challenging since they addressed environmental sciences, business management, economy, engineering, design, innovation, and creativity topics. NEGEP, a laboratory for studies in management, ergonomics & projects, and NUPEM, Institute of Biodiversity and Sustainability, both from UFRJ, led the initiative.

The way to face the proposed challenge included a benchmark survey with other higher education institutions offering short- and long-term courses on the same subject. The institutions consulted were Cambridge Judge Business School (Executive education course: "Circular Economy and Sustainability Strategies"), MIT Professional Education (Online course: "Circular Economy—Transition for Future Sustainability"), TU DELFT (Online course: "Circular Economy—An Introduction"), UFRGS (Specialization course: "Circular Economy and Innovation") and FGV (Short-course: "Circular Economy and New Business Models"). After critically reading the courses' syllabuses, a group of professors with different backgrounds selected the most repeated modules and those that best spoke to the region's needs.

The next step was to establish contact with potential lecturers and discuss the amendment of each module being designed. The course had professors from Polytechnic Institute—UFRJ Macaé, Institute of Biodiversity and Sustainability—UFRJ Macaé, UFF, FEMASS, and Mackenzie. Automobile and cosmetics industry managers were also invited to participate in the case studies discussion.

The ten modules that made up the course syllabus were organized in topics by the lecturers as follows (Table 1):

Once the ten modules and corresponding teachers were defined, the first version of the short course, a prototype, was launched in November 2022. The course was planned to be free of charge, conducted face-to-face, at night (classes between 6:30 pm and 10:30 pm), and the ten planned modules would be presented in 5 days (2 modules per day). The course also required readings recommended by the lecturers and completing a final assignment.

The participant's profile to whom the course was aimed was adults interested in professional updating courses, with an interest and/or activity in sustainability and circular economy. Initially, 50 participants were selected, but only 31 met the attendance requirements at the end of classes.

Table 1 "Circular economy and sustainable business" course syllabus

Modules	Topics
1. Introduction to circularity	What is circular economy and why is it important?
	Differences between linear and circular economy
	Circular economy fundamentals & key concepts
	Circular economy: Challenges and trends
	Climate change and greenhouse gas emissions
2. Sustainability and business management	Sustainability definition and its historical context
	Waste is nutrients Waste-recovery management—cradle to cradle (C2C)
	Case studies C2C
3. Modelling circular business	How to circularize a linear business model
	Innovation and business benefits of transitioning to a circular business model
4. Sustainability culture	State of the art—what is culture and why is it important?
	How to build a sustainable culture in organizations?
	Planning for organizational culture changes
	Case study in sustainability culture
5. Circular business financial analysis	Risk sources
	Green finance
	Environment, social and governance (ESG)
	How business is measured and evaluated?
6. Marketing and sustainability	Assessment of approaches to marketing and sustainability
	Ethical and purposeful marketing
	Marketing and corporate social responsibility (CSR)
	Stakeholders and strategic partnerships
7. Sustainable supply chain	What is supply chain?
	Traditional supply chain
	Circular supply chain
	Sustainable supply chain and the importance of collaboration
	Supply chain planning in circular economy
	Case studies in supply chain
	Reverse logistics and other sustainable initiatives in the oil & gas industry
	Integrating the ESG concept into supply chain
	Supply chain key performance indicators (KPI)
8. EcoDesign and product design	Creativity and product design
	Materials circularity
	Circularity measures
9. Innovation strategies	Sustainability and innovation
	Innovation initiatives that improve circularity

(continued)

Table 1 (continued)

Modules	Topics
	Innovation to sustainable strategies
	Intellectual and industrial property in circular economy
	Case study—sustainable innovations
10. Case studies	EC in real life: Automotive and beauty industry executives

Source: Author's elaboration

Among the 31 students of the course who participated in the research, it is worth highlighting some relevant information. Concerning academic education, 42% have completed a degree or are specialists, while 25% have a master's or Ph.D., corroborating the interest in continuing education. Regarding gender, 55% are female, which is in line with research that suggests that the presence of women in companies may help them to perform better than their peers in corporate sustainability (McElhaney & Mobasseri, 2012). Biologists, engineers, administrators, and economists, who work in both the public and private sectors, attended the classes, which enriched the experience of a course that, since its inception, was intended to be transdisciplinary.

Upon completing the course, students received a link to a feedback form, whose main objective was to assess the importance of each of the subjects covered in the ten modules of the course. The questionnaire also provided two open questions to capture qualitative information.

Following a quantitative approach, the structured feedback of the 31 participants was analyzed using the Lawshe method. For each topic covered in the ten modules, it was calculated the Content Validity Radio (CVR). Items whose CVRcritical value is higher than the CVRcalculated must be excluded or reformulated in new course editions. The data will be presented by module; in this sense, each table will describe the corresponding topics in descending order of essentiality.

Only two topics were classified as "to be excluded": the topic "State of the art—what is culture and why is it important?" from Module 4 (Sustainable Culture) and "Traditional Supply Chain" from Module 7 (Sustainable Supply Chain). Both were planned as opening themes to introduce students to the subject and highlight the differences with circular models. For future editions of the course, they could be suppressed or absorbed by other topics, being presented more superficially.

The topics of module 1, together, were considered the most essential by the students who participated in the research. This module, as seen in Table 2, was designed to cover key concepts and fundamentals of the circular economy, distinguish between linear and circular business models, and discuss challenges and trends on this theme. Climate change and the emission of greenhouse gases were presented as aggravating factors that make a transition in the forms of production and consumption even more urgent.

The intention of module 2 was to promote a debate regarding the definition of the concept of sustainability and discuss the incompatibilities between sustainability and economic growth. Students were introduced to the ecological perspective of

Table 2 Module 1—introduction to circularity

1. Introduction to circularity	n_e	N	CVRcalculated	CVRcritical	Decision
What is circular economy and why is it important?	31	31	1.000	0.352	Accept
Circular economy fundamentals & key concepts	31	31	1.000	0.352	Accept
Circular economy: Challenges and trends	31	31	1.000	0.352	Accept
Differences between linear and circular economy	29	31	0.871	0.352	Accept
Climate change and greenhouse emissions	29	31	0.871	0.352	Accept

Source: Author's elaboration

Table 3 Module 2—sustainability and business management

2. Sustainability and business management	n_e	N	CVRcalculated	CVRcritical	Decision
Sustainability definition and its historical context	30	31	0.935	0.352	Accept
Waste-recovery management—cradle to cradle (C2C)	30	31	0.935	0.352	Accept
Case studies C2C	30	31	0.935	0.352	Accept
Waste is nutrients	27	31	0.742	0.352	Accept

Source: Author's elaboration

sustainability and the term's origins in international multilateral organizations. The Cradle-to-Cradle vision was introduced with the idea that there is no such thing as waste and that all waste should be seen as nutrients. According to Table 3, all module items were considered essential, with the lowest value attributed to the topic "Waste are nutrients."

Module 3 presented a step-by-step on how to circularize a linear business model and aimed to describe all the business benefits of transitioning to the circular economy. Both topics were considered essential by the students (Table 4).

The importance of building a sustainable culture in organizations was discussed in Module 4. The topics described in Table 5 dealt with the importance of organizational culture to carry out lasting transformations oriented towards sustainability and how to incorporate it into the business strategy. Actual cases of cultural changes were discussed with students. The only topic considered non-essential was the "State of the art—what is culture and why is it important?" It should be reformulated or deleted in future course editions.

Module 5 proposed to describe the ESG as a metric to measure organizations' sustainability level. It also introduced the concept of green finance and described the characteristics of sustainable finance. Students learned to develop circular scenarios and recognize risks and make plans to avoid potential pitfalls in this context. According to Table 6, all topics were considered essentials, especially the ESG discussion.

Module 6 presented the contradiction between a part of the marketing objectives and the claims of sustainability by the companies. Ethical Marketing and Marketing

Table 4 Module 3—modeling circular business

3. Modelling circular business	n_e	N	CVRcalculated	CVRcritical	Decision
How to circularize a linear business model	29	31	0.871	0.352	Accept
Innovation and business benefits of transitioning to a circular business model	28	31	0.806	0.352	Accept

Source: Author's elaboration

Table 5 Module 4—sustainability culture

4. Sustainability culture	n_e	N	CVRcalculated	CVRcritical	Decision
How to build a sustainable culture in organizations	30	31	0.935	0.352	Accept
Planning for organizational culture changes	28	31	0.806	0.352	Accept
Case study in sustainability culture	27	31	0.742	0.352	Accept
State of art—what is culture and why is it important?	16	31	0.032	0.352	Exclude

Source: Author's elaboration

Table 6 Module 5—circular business financial analysis

5. Circular business financial analysis	n_e	N	CVRcalculated	CVRcritical	Decision
Environment, social and governance (ESG)	31	31	1.000	0.352	Accept
How business is measured and evaluated?	29	31	0.871	0.352	Accept
Risk sources	28	31	0.806	0.352	Accept
Green finance	27	31	0.742	0.352	Accept

Source: Author's elaboration

Table 7 Module 6—marketing and sustainability

6. Marketing and sustainability	n_e	N	CVRcalculated	CVRcritical	Decision
Assessment of approaches to marketing and sustainability	31	31	1.000	0.352	Accept
Ethical and purposeful marketing	30	31	0.935	0.352	Accept
Marketing and corporate social responsibility (CSR)	29	31	0.871	0.352	Accept
Stakeholders and strategic partnerships	28	31	0.806	0.352	Accept

Source: Author's elaboration

with Purpose were introduced to the students, as well as the importance of networking and strategic partnerships to achieve sustainable goals. The idea of Corporate Social Responsibility (CSR) was presented, and the class discussed how CSR and Marketing could share joint initiatives. As shown in Table 7, all items were considered essentials by the students.

Table 8 Module 7—sustainable supply chain

7. Sustainable supply chain	n_e	N	CVRcalculated	CVRcritical	Decision
Supply chain planning in circular economy	30	31	0.935	0.352	Accept
Sustainable supply chain and the importance of collaboration	30	31	0.935	0.352	Accept
Integrating the ESG concept into supply chain	30	31	0.935	0.352	Accept
Reverse logistics and other sustainable initiatives in the oil & gas industry	29	31	0.871	0.352	Accept
Circular supply chain	29	31	0.871	0.352	Accept
Supply chain key performance indicators (KPI)	29	31	0.871	0.352	Accept
Case studies in supply chain	28	31	0.806	0.352	Accept
What is supply chain?	24	31	0.548	0.352	Accept
Traditional supply chain	14	31	−0.097	0.352	Exclude

Source: Author's elaboration

Table 9 Module 8—EcoDesign and product design

8. EcoDesign and product design	n_e	N	CVRcalculated	CVRcritical	Decision
Circularity measures	30	31	0.935	0.352	Accept
Materials circularity	30	31	0.935	0.352	Accept
Creativity and product design	25	31	0.613	0.352	Accept

Source: Author's elaboration

The supply chain is crucial in achieving sustainability goals in the production process. Module 7 introduced the term, differentiated between linear and circular models, guided planning for a circularized supply chain, presented indicators to be implemented and monitored, and brought examples of reverse logistics in the Oil & Gas industry. Except for the topic "Traditional Supply Chain," all others were considered essential by the research participants (see Table 8).

Module 8 connected creativity, technological changes, and the circular economy. The design of products and services oriented towards sustainability and the intelligent use of materials was debated among the students. Circularity measures and actual industry examples illustrated the potential for action in this area of knowledge. Table 9 shows that all items were considered essential, emphasizing "Circularity Measures" and "Materials Circularity."

Module 9, on innovation strategies for sustainability, discussed initiatives that contribute to achieving sustainability in organizations, presented possible strategies to be implemented, and discussed examples of innovation in companies. Students were also introduced to the importance of intellectual and industrial property in the circular economy. Table 10 shows that all the topics were classified as essentials.

The last module of the course allowed students to debate with executives from the automotive industry and the beauty industry about initiatives aimed at sustainability and circular economy that have been implemented in the respective sectors. The issues of electric cars, the transition from combustion engines to batteries, and

Table 10 Module 9—innovation strategies

9. Innovation strategies	n_e	N	CVRcalculated	CVRcritical	Decision
Innovation to sustainable strategies	30	31	0.935	0.352	Accept
Case study—sustainable innovations	30	31	0.935	0.352	Accept
Sustainability and innovation	28	31	0.806	0.352	Accept
Innovation initiatives that improve circularity	28	31	0.806	0.352	Accept
Intellectual and industrial property in circular economy	24	31	0.548	0.352	Accept

Source: Author's elaboration

Table 11 Module 10—case studies

10. Case studies	n_e	N	CVRcalculated	CVRcritical	Decision
EC in real life: Automotive and beauty industry executives	30	31	0.935	0.352	Accept

Source: Author's elaboration

Table 12 Lowest CVRcalculated values

Module	Topic	n_e	N	CVRcalculated	CVRcritical
7. Sustainable supply chain	Traditional supply chain	14	31	−0.097	0.352
4. Sustainability culture	State of art—what is culture and why is it important?	16	31	0.032	0.352
7. Sustainable supply chain	What is supply chain?	24	31	0.548	0.352
9. Innovation strategies	Intellectual and industrial property in circular economy	24	31	0.548	0.352
8. EcoDesign and product design	Creativity and product design	25	31	0.613	0.352
2. Sustainability and business management	Waste is nutrients	27	31	0.742	0.352

Source: Author's elaboration

the usefulness and disposal time of the new materials were discussed. Returning packaging and the recycling challenge were also debated since companies depend on their consumers' engagement and active participation. As shown in Table 11, 30 out of 31 students considered this topic essential.

Another revealing perspective for organizing and presenting the results was the selection of the six items with the highest and lowest CRVcalculated values, as shown in Tables 12 and 13. One of the insights these tables reveal concerns the importance of dedicating time and efforts to introductory topics and topics focused on observing the application of circular practices in products, finance, and businesses. On the other hand, more general topics, which worked to introduce the theme or to contrast the circular and linear economy, were considered less essential. Particular topics, such as intellectual property, also received lower scores.

Table 13 Highest CVRcalculated values

Module	Topic	n_e	N	CVRcalculated	CVRcritical
1. Introduction to circularity	What is circular economy and why is it important?	31	31	1.000	0.352
1. Introduction to circularity	Circular economy fundamentals & key concepts	31	31	1.000	0.352
1. Introduction to circularity	Circular economy: Challenges and trends	31	31	1.000	0.352
5. Circular business financial analysis	Environment, social and governance (ESG)	31	31	1.000	0.352
6. Marketing and sustainability	Assessment of approaches to marketing and sustainability	31	31	1.000	0.352
8. EcoDesign and product design	Circularity measures	31	31	1.000	0.352

Source: Author's elaboration

The qualitative information, coming from the two open questions of the questionnaire, also offers interesting directions for interpreting the course performance. The connection between theory and practice, combined with the presentation of case studies, was mentioned as one of the strengths of the course by 39% of the students. Another item that stood out among the positive feedback (mentioned by 16% of students) was the diversity of lecturers and students with different academic backgrounds, offering plural and sometimes antagonistic perspectives on sustainability and the reach of the circular economy.

Feedback on opportunities for improvement brought some insights into what can be done differently in new courses on the same topic. The duration of the course was mentioned by 25% of the students, who stated that they would like the course to last longer so that it would be possible to delve deeper into each of the topics presented; this return signals an opportunity to develop long-term courses in the region focused on the circular economy.

Another revealing issue the interviewees mentioned was using disposable plastic cups during the course. This feedback is in line with the importance of changing the culture of the university campus to express sustainability. As mentioned earlier, HEIs must contribute not only to the creation and dissemination of sustainability-related content but must attempt to make their culture and operations sustainable as well.

The inclusion of more in-depth material on greenhouse gas emissions and carbon pricing was mentioned by 2 of the 31 students. Following the quantitative results, topics from module 1, introductory to the concept of circular economy, were flagged as topics to be expanded and deepened.

6 Conclusions

The present chapter focuses on a project to develop a short-term, no-cost course centered on Circular Economy and Sustainable Business. Feedback obtained from participants attested to the course's alignment with the requirements of professionals within the region, thus establishing a conducive environment for future iterations.

The project was developed through the co-creation approach and, for its evaluation, incorporated qualitative strategies and quantitative questionnaires based on the content validation methodology suggested by Lawshe.

The assessment of the course syllabus from the students' standpoint holds potential value for other HEIs interested in dedicating themselves to the circular economy subject. In the current scenario, it is imperative to foster executive education programs that exhibit a profound connection with the Sustainable Development Goals (SDGs), given that such endeavor represents a strategic path for cultivating skills and building capacities that possess the potential to contribute towards the realization of a more sustainable society.

The results of the quantitative evaluation showed that the introductory module on circularity garnered superior ratings, thereby underscoring the significance of instilling a firm foundation in the fundamentals of circular economy and sustainability, particularly within the Brazilian context, wherein these concepts remain incipient and confined to a limited cohort of individuals and organizations.

The high index of contents validated as essential by the students confirms the effectiveness of the co-creation processes employed during the course. Thus, the results obtained through Lawshe's method emphasized the importance of engaging multi-stakeholders through different strategies.

Some lessons learned:

(i) The co-creation approach in designing executive education courses was instrumental in engaging participants throughout the project lifecycle.
(ii) Ensuring alignment between the course content and its various facilitators proved imperative to minimize redundancies and favor the logical sequencing of topics.
(iii) Considering the requirement for hybrid facilitation strategies, alternatives for teacher-participant interaction were worth considering, whether through questioning, practical activities, or virtual collaborative methodologies.
(iv) Technological advances have eliminated geographic restrictions, including lecturers from different regions of the world. In this context, meticulous attention had to be paid to factors such as time zones, content delivery mode (recorded or remote), and quality of internet connectivity.

(v) Encouraging the execution of practical projects in which students applied the acquired concepts was a fundamental approach, even more so because the course was dedicated to emerging themes that still need to be disseminated.

Finally, the case expanded the possibilities for dialogue and cooperation with multi-stakeholders, a requirement to advance with the sustainability agenda outlined in SDG 17, focusing on facing the complex challenges of a green transition.

6.1 Study Limitations

The present study could contribute to co-creation practices beyond the design of continuing education courses. For example, it could be interesting to apply an evaluation of the co-creation design process aimed at understanding how the participants felt in contributing to the generation of ideas, and even that would allow for evaluating the quality of the collaboration.

6.2 Future Studies

For future studies, exploring the different lenses of the co-creation process and adopting different co-creative practices in each lens is suggested. Other co-creation models and co-design modes can be considered. Executive education requires a close look at the needs of the local and global markets. Therefore, involving executives from different levels of organizations and sectors can be favorable in gathering insights.

References

Acselrad, H. (1999). *Discursos da sustentabilidade urbana*.

Andrew, M. (2018). *Co-creating learning and teaching: Towards a shared understanding*. [Master's thesis, University of Edinburgh].

Barreiro-Gen, M., & Lozano, R. (2020). How circular is the circular economy? Analyzing the implementation of circular economy in organizations. *Business Strategy and the Environment, 29*(8), 3484–3494.

Baxendale, C., MacKinnon, K., & Goldfinch, J. (2021). Co-creating the curriculum: Opportunities and challenges in higher education. *Teaching in Higher Education, 26*(7–8), 1021–1035.

Bugallo-Rodríguez, A., & Vega-Marcote, P. (2020). Circular economy, sustainability and teacher training in a higher education institution. *International Journal of Sustainability in Higher Education, 21*(7), 1351–1366.

Bovill, C. (2020). *Co-creating learning and teaching: Towards relational pedagogy in higher education*. Critical Publishing.

Bovill, C., Cook-Sather, A., & Felten, P. (2013). *Engaging students as partners in learning and teaching: A guide for faculty*. Jossey-Bass.

Bovill, C., Cook-Sather, A., Felten, P., Millard, L., & Moore-Cherry, N. (2016). Addressing potential challenges in co-creating learning and teaching: Overcoming resistance, navigating institutional norms, and ensuring inclusivity in student–staff partnerships. *Higher Education, 71*(2), 195–208.

Cottam, H., & Leadbeater, C. (2004). *Open welfare: Designs on the public good.* Design Council.

Gkogkidis, V. (2020). *Co-creation in higher education: Students and academics co-creating learning and teaching*. [Doctoral dissertation, University of Sussex].

da Silva, M., da Silva, L. C. A., & Brambilla, F. R. (2020). Value co-creation: A study of life quality at an university in the South Brazil. *Independent Journal of Management & Production, 11*(3), 833–852.

da Silva, S. S., Spers, R. G., & de Azevedo Amendomar, A. (2022). COMO SERÁ A EDUCAÇÃO EXECUTIVA DO FUTURO? *Encontro Internacional de Gestão, Desenvolvimento e Inovação (EIGEDIN), 6*(1).

De Oliveira, F. R., França, S. L. B., & Rangel, L. A. D. (2018). Challenges and opportunities in a circular economy for a local productive arrangement of furniture in Brazil. *Resources, Conservation and Recycling, 135*, 202–209.

De Souza, J. C., Alves, M. C. C., Machado, F. D. A. L., Shimoya, A., & Luna, P. S. P. (2020). Aplicação do Método de Lawshe para Validação de Itens de um Questionário sobre a Satisfação de Clientes de um Supermercado em Campos Dos Goytacazes, RJ. *Exatas & Engenharias, 10*(27), 42–53.

Dollinger, M., & Lodge, J. (2020). Student-staff co-creation in higher education: An evidence-informed model to support future design and implementation. *Journal of Higher Education Policy and Management, 42*(5), 532–546.

Dollinger, M., Lodge, J., & Coates, H. (2018). Co-creation in higher education: Towards a conceptual model. *Journal of Marketing for Higher Education, 28*(2), 210–231.

Ferrell, O. C., Bacellar, C., & Hartline, M. D. (2017). *Estratégia de marketing: teoria e casos.* Cengage Learning Edições Ltda.

Figueiró, P. S., & Raufflet, E. (2015). Sustainability in higher education: A systematic review with focus on management education. *Journal of Cleaner Production, 106*, 22–33.

Freitas, R. J. D. (2013). *Um estudo sobre critérios de avaliação de sucesso na implementação de sistemas ERP*.

Geng, Y., Sarkis, J., & Bleischwitz, R. (2019). How to globalize the circular economy. *Nature, 565*(7738), 153–155.

Grandzol, J. R., & Grandzol, C. J. (2020). Best practices for online business education. *The International Review of Research in Open and Distance Learning, 7*(1), 1–18.

Haas, W., Krausmann, F., Wiedenhofer, D., & Heinz, M. J. (2015). How circular is the global economy?: An assessment of material flows, waste production, and recycling in the European Union and the World in 2005. *Journal of Industrial Ecology, 19*, 765–777.

Holmlid, S., Mattelmäki, T., Visser, F. S., & Vaajakallio, K. (2015). Co-creative practices in service innovation. In *The handbook of service innovation* (pp. 545–574). Springer.

Johnston, A. (2007). Higher education for sustainable development: Final report of international action research project. In *Forum for the future action for a sustainable world.*

Könings, K. D., Mordang, S., Smeenk, F., Stassen, L., & Ramani, S. (2021). Learner involvement in the co-creation of teaching and learning: AMEE Guide No. 138. *Medical Teacher, 43*(8), 924–936.

Krishnamurthy, S. (2020). The future of business education: A commentary in the shadow of the Covid-19 pandemic. *Journal of Business Research, 117*, 1–5.

Lawshe, C. H. (1975). A quantitative approach to content validity. *Personnel Psychology, 28*(4), 563–575.

Lima, T. G. (2015). *Analysis of guests' perceptions: A study on criteria for evaluating the quality of services in hospitality* (Master's thesis, Instituto Candido Mendes, Campos dos Goytacazes, Brazil).

MacArthur, E. (2013). Towards the circular economy. *Journal of Industrial Ecology, 2*(1), 23–44.

MacArthur, E. (2015). *Rumo à Economia Circular: O racional de negócio para acelerar a transição*. Ellen MacArthur Foundation.

Mancini, S. D., de Medeiros, G. A., Paes, M. X., de Oliveira, B. O. S., Antunes, M. L. P., de Souza, R. G., et al. (2021). Circular economy and solid waste management: Challenges and opportunities in Brazil. *Circular Economy and Sustainability, 1*(1), 261–282.

Mann, S. (2020). *The research interview: Reflective practice and reflexivity in research processes*. Palgrave Macmillan.

Mattelmäki, T., Brandt, E., & Vaajakallio, K. (2011). On designing open-ended interpretations for collaborative design exploration. *CoDesign, 7*(2), 79–93.

Maxwell-Stuart, R., Taheri, B., Paterson, A. S., O'Gorman, K., & Jackson, W. (2018). Working together to increase student satisfaction: Exploring the effects of mode of study and fee status. *Studies in Higher Education, 43*(8), 1392–1404.

McElhaney, K. A., & Mobasseri, S. (2012). *Women create a sustainable future—Center for Responsible Business*. Haas School of Business, University of California, Berkeley.

Morseletto, P. (2020). Targets for a circular economy. *Resources, Conservation and Recycling, 153*, 104553.

Murray, A., Skene, K., & Haynes, K. (2017). The circular economy: An interdisciplinary exploration of the concept and application in a global context. *Journal of Business Ethics, 140*, 369–380.

Naim, A. (2021). Application of quality matters in digital learning in higher education. *Texas Journal of Multidisciplinary Studies, 1*(1), 3–12.

Nascimento, E. P. D. (2012). Trajetória da sustentabilidade: do ambiental ao social, do social ao econômico. *Estudos Avançados, 26*, 51–64.

Ohde, C., & Mattar, H. (2018). *Economia circular: um modelo que dá impulso à economia, gera empregos e protege o meio ambiente*.

Salas, D. A., Criollo, P., & Ramirez, A. D. (2021). The role of higher education institutions in the implementation of circular economy in Latin America. *Sustainability, 13*(17), 9805.

Schroeder, P., Anggraeni, K., & Weber, U. (2019). The relevance of circular economy practices to the sustainable development goals. *Journal of Industrial Ecology, 23*, 77–95.

Sharma, P. (2021). Customer co-creation, COVID-19, and sustainable service outcomes. *Benchmarking: An International Journal, 28*(7), 2232–2258.

Siloto, D. A. P., Leal, G. F., & Schottz, V. (2021). Escolhas alimentares e sustentabilidade: contribuições de um estudo de caso em uma feira livre. *Tematicas, 29*(58), 74–101.

Silvério, A. S. (2016). *Levantamento bibliométrico nas bases scielo e scopus dos itens para avaliação da qualidade em serviços: um estudo de caso na hotelaria de campos dos goytacazes/rj*. Dissertação do Programa de Mestrado em Engenharia de Produção. Universidade Cândido Mendes.

Stahel, W. R. (2016). The circular economy. *Nature, 531*(7595), 435–438.

Sterenberg, G., Barrett, M. J., & McMillan, B. (2018). Co-creating curriculum with students: A narrative inquiry into elementary school teachers' experiences. *Journal of Curriculum Studies, 50*(5), 670–685.

Tarı Kasnakoğlu, B., Aydın, G., & Demir, M. (2022). Co-creation of learning and teaching in higher education: A systematic literature review. *Educational Research Review, 35*, 100419.

United Nations. (2015). *Transforming our world: The 2030 agenda for sustainable development*. United Nations.

United Nations Environment Program. (2017). *Towards a pollution-free planet: Background report*. United Nations Environment Program.

Wilden, R., Gudergan, S., Akaka, M. A., Averdung, A., & Teichert, T. (2019). The role of co-creation and dynamic capabilities in service provision and performance: A configurational study. *Industrial Marketing Management, 78*, 43–57.

Teaching Sustainable Development Goals from Theory to Praxis in Universities Based on the Theory of Graphs

Gloria María Isaza Zapata, Byron Enrique Portilla Rosero, and Juan Santiago Calle Piedrahita

Abstract The teaching of the Sustainable Development Goals (SDGs) in universities serves as a foundation for theoretical knowledge, yet often lacks practical application. Our study employed a methodology centered around graph theory and in-depth interviews with instructors from four universities (two private and two public) during the first semester of 2022. The findings revealed that instructors predominantly approach the teaching of SDGs from a theoretical standpoint, emphasizing the need to raise awareness about the pressing real-world challenges our planet faces in the absence of intervention. While the SDGs are integrated into various subjects' curricula, little effort is made to actively contribute to sustainability. In conclusion, instructors primarily teach the SDGs within their respective courses but fail to adopt a cross-disciplinary approach to foster widespread awareness. While teaching the SDGs is crucial in understanding the limitations of our resources, it is equally essential to do so in a practical manner, employing real-life examples from society to facilitate effective learning and application.

Keywords Sustainable Development Goals · Education · Teachers · Graphs

1 Introduction

Universities are developing strategies to promote the teaching of the Sustainable Development Goals (SDGs) in various courses. In Colombia, educational programs are integrating the SDGs into their syllabi and evaluating student comprehension through final projects. Instructors are guided by the objectives outlined in the work plan, aiming to bridge the gap between SDGs theory and practice. Societal

---

G. M. Isaza Zapata (✉) · B. E. Portilla Rosero · J. S. Calle Piedrahita
Medellín, Colombia
e-mail: gloria.isaza@upb.edu.co; byronpor@teamconectados.com; juan.callepi@amigo.edu.co

C. Rego et al. (eds.), *Transitioning to a Circular Economy*, Studies on Entrepreneurship, Structural Change and Industrial Dynamics,
https://doi.org/10.1007/978-3-031-77661-8_9

transformation also fosters the formulation of new SDG-related policies. However, education serves as the fundamental pillar for fostering awareness and driving the implementation of these new goals. It plays a crucial role in catalyzing change. Despite the international community's commitment to significant agreements forged in various United Nations summits and conferences to define the key axes of sustainable development, most signatory countries have consistently fallen short of meeting these objectives.

While the principles of economics have been utilized to study the SDGs (Lun et al., 2022; Marinova et al., 2017; Peel, 2006), education is perceived as a humanistic component (Koon, 2021) but is integrated into the development of organizations with a focus on psychological studies (Sanchez et al., 2020; Sánchez-García et al., 2018). However, there remains a paucity of studies dedicated to effectively integrating the SDGs into teaching and their practical application. The SDGs are rooted in the needs of the past, which have disrupted social, environmental, and ecological progress in detrimental ways (Gast et al., 2017). Consequently, strategies supported by indicators are necessary to mitigate future damages from a forward-looking perspective. In other words, we must devise approaches that contribute to timely intervention to ensure an environmentally sustainable future (Dotres Zuniga & Sanchez Paz, 2020; Lysaght et al., 2022) and foster sustainable economic development. Teaching the SDGs in educational institutions plays a vital role in raising awareness, establishing environmentally conscious institutions, promoting community social development, and fostering equitable prosperity (Marinova et al., 2017).

Furthermore, education plays a crucial role in fostering connectivity and linking the objectives of 2030. When basic education is complemented with the experiential knowledge of organizations driven by environmental and social ideologies, it contributes to the assimilation of new sustainable knowledge. Consequently, the educational process itself must undergo a transformation to adopt innovative perspectives on the reality of finite resources. The question arises: How can we effectively teach the Sustainable Development Goals (SDGs) to cultivate future professionals with genuine environmental, social, and economic ideas that can reshape their perception of resources?

The educational journey begins by comprehending fundamental theories and challenging prevailing resource management paradigms. Educators hold a pivotal role in imparting novel approaches to the utilization of scarce resources. It becomes imperative to adopt a fresh outlook on resource management and cultivate a comprehensive, inclusive, and diverse pedagogical approach to teaching SDGs, enhancing both knowledge and attitudes toward sustainability. However, the teaching of SDGs must also shift focus towards practical thinking. This necessitates educational reform aimed at promoting social, environmental, and economic development.

Existing literature highlights three primary categories under which the teaching of Sustainable Development Goals can be classified: sustainability in teaching, sustainable development, and human-social aspects. Sustainability in teaching encompasses themes related to students, UNESCO, assessment methods, economic approaches, and training. Sustainable development pertains to planning, design, engineering education, and biodiversity. The third category, human-social, covers topics associated with psychology, capacity building, adult education, and human experiences.

It is also important to teach the new generation about the decrease in forest areas in Colombia. In 1990, the country's forest area covered 56.3% of its surface, but after 20 years, it decreased to 54.5% (Faber-Langendoen, 1992). Although the country has made significant progress in human rights, increasing its score from 0.56 in 1980 to 0.72 in 2012 (Ladrach & Wright, 1995; Villarraga-Flórez et al., 2016), it still lags behind other countries in this aspect.

The objective of the research was to identify the teaching of the SDGs in the classes from a theoretical reality towards praxis. In other words, the current interest in generating knowledge of this topic in the classroom permeates the understanding of the reality of the SDGs in their execution. In addition, it ensures that communities can participate in the development of strategies to implement the SDGs on a daily basis. In short, the classroom provides the opportunity to learn what the SDGs are and apply them in context.

2 Sustainability Theory

"Sustainability is not only about caring for the environment but also a need to redefine key concepts" (Zamora-Polo & Sánchez-Martín, 2019, p. 4224). However, there are several visions that make it difficult to understand and address sustainability challenges (Albareda-Tiana et al., 2018a). Project-Oriented Learning (POL) and Cross-disciplinary Workshops provide evidence for developing competencies in sustainability (Albareda-Tiana et al., 2018b) by using different teaching methods. Additionally, "the use of Massive Open Online Courses (MOOCs) to teach SDGs and promote sustainable development in teacher training provides a new vision for professional and personal life" (Zamora-Polo et al., 2019, p. 4225). However, there may be significant differences between what students think and what they actually do.

Recent research endeavors have been dedicated to promoting initiatives on how to effectively teach sustainability within higher education institutions (Leal Filho et al., 2019). García-González et al. (2020) conducted a study demonstrating that although students were familiar with the term "sustainability and the Sustainable Development Goals" (SDGs), there was no significant improvement in their comprehension before and after training. However, Sonetti et al. (2021) revealed a discrepancy in the perceived relevance of SDGs between the students' own perspectives and the university's perspective. In addition, Tamrat (2022) highlighted a significant gap in current engagement patterns, with teaching, course provision, and a few operational and managerial aspects receiving attention, while research and community service, the core functional areas, are often overlooked. This neglect is a matter of concern. To address these challenges, colleges and universities must adopt a transdisciplinary approach to teaching the SDGs, enabling students to acquire essential knowledge and skills for future endeavors (Chang & Lien, 2020) (see Fig. 1).

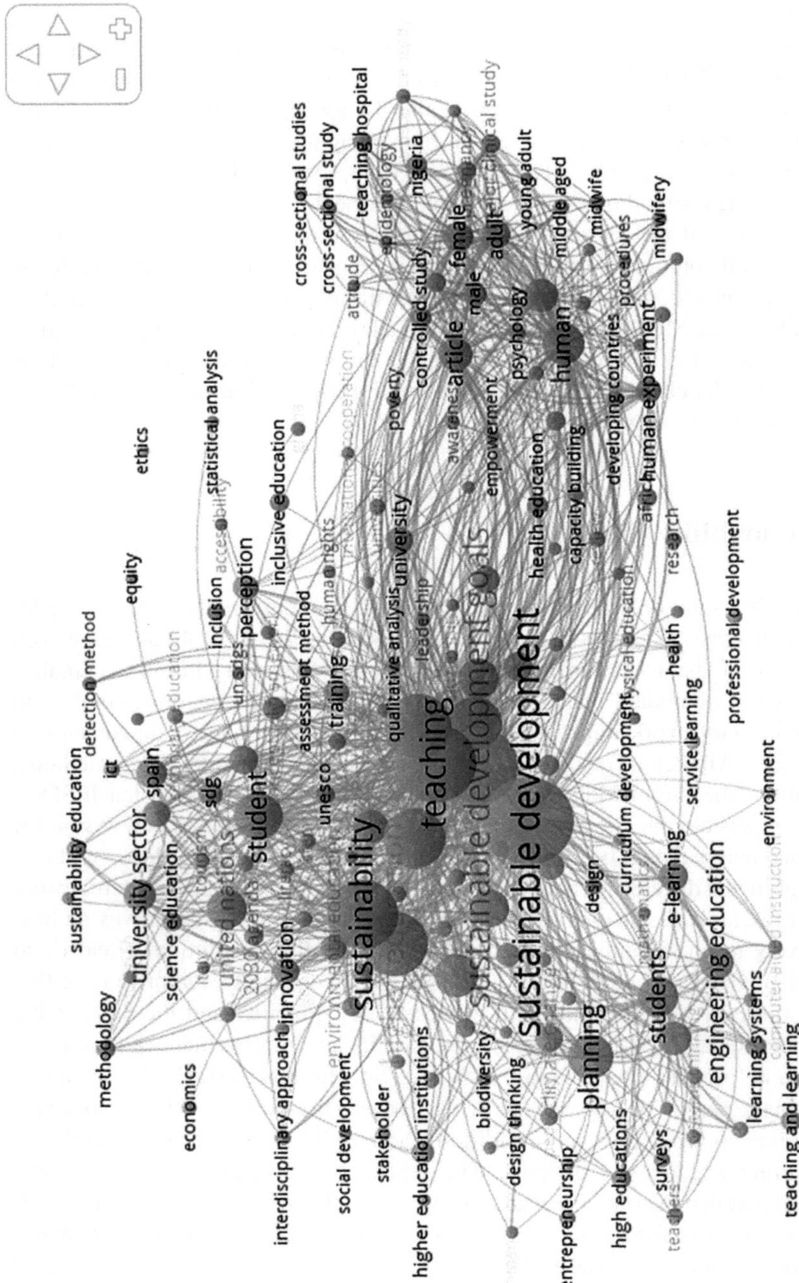

Fig. 1 Terms used in Scopus. Source: Author's elaboration VosViewer Scopus

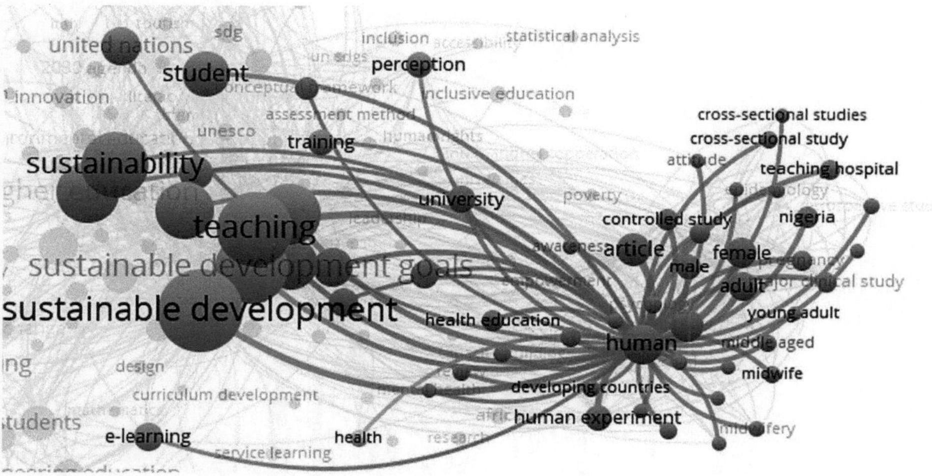

Fig. 2 Terms related to human and SDGs. Source: Author's elaboration VosViewer Scopus

The relationship between human beings and the advancement of sustainability hinges upon the effective teaching of sustainable development objectives. It is through education and training that individuals develop an improved perception of both positive and negative environmental events, thereby shaping their understanding of sustainability. The sustainability of organizations, in turn, relies on how the reality of sustainability is imparted within their specific context. Furthermore, educational reengineering is necessary to comprehend the environmental landscape and devise new policies that can effectively influence the implementation of pollution reduction plans. The teaching of SDGs adopts a human-centered approach, drawing from fields such as psychology and human health. Additionally, it is crucial to effectively communicate the negative consequences to transform them into positive outcomes through curriculum-based strategies (See Fig. 2).

Similarly, the integration of teaching and the sustainable development of the Sustainable Development Goals (SDGs) entails educational planning, particularly through e-learning, with a focus on the human element and fostering international cooperation. Teaching students to think critically about the application of the SDGs is a valuable contribution to fostering sustainability by shaping their perception of innovative approaches. This equips students with the necessary skills to apply theoretical knowledge in practical contexts. Moreover, curriculum design plays a crucial role in emphasizing the significance of understanding the SDGs and their corresponding indicators. In essence, it is essential to incorporate the SDGs into the curriculum to cultivate long-term awareness and facilitate their effective implementation (see Fig. 3).

Since 2019, there has been a growing emphasis on the human element in sustainability, leading to the incorporation of sustainable development into university-level education through scientific studies. In 2020, research on sustainability and the Sustainable Development Goals (SDGs) emerged in the context of education in

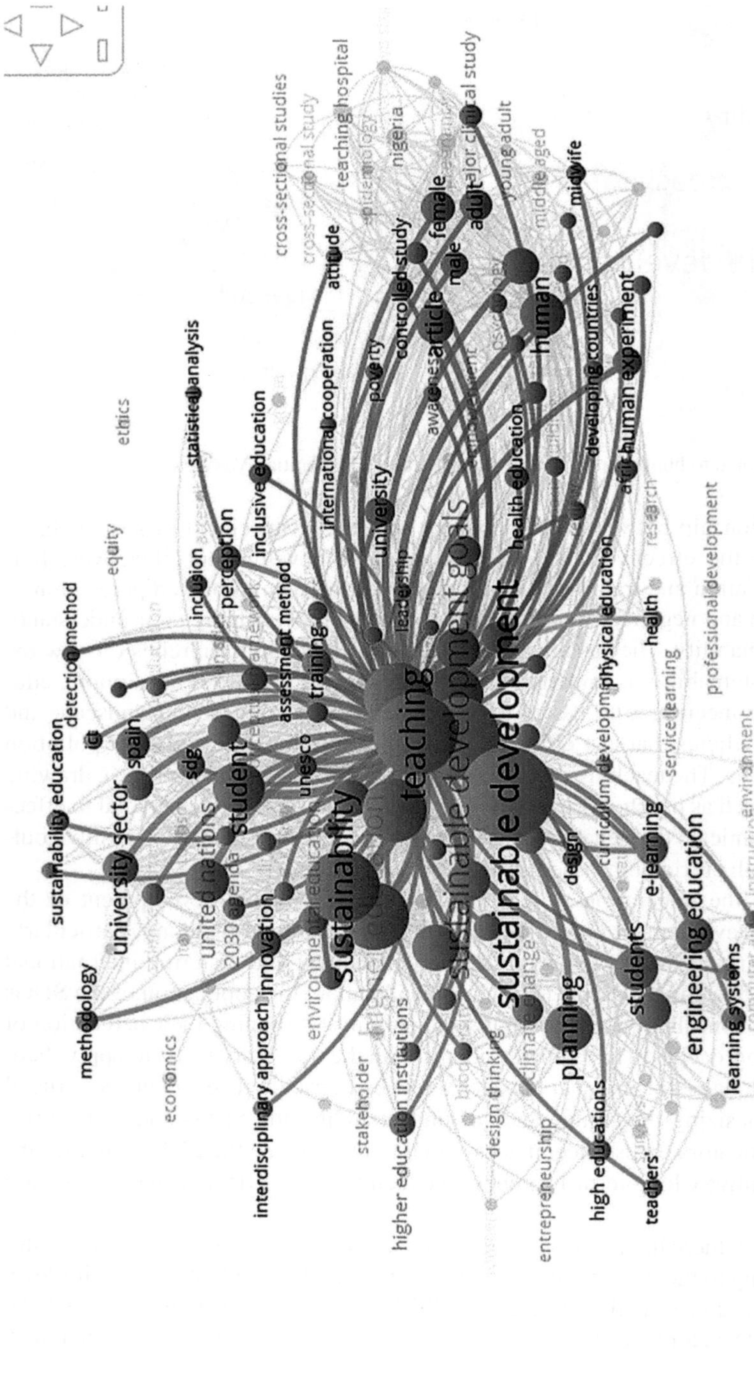

Fig. 3 Terms relate to sustainability development and teaching. Source: Author's elaboration VosViewer Scopus

Spain, focusing on learning systems, higher education, methodologies, and surveys. Teaching sustainability entails careful resource planning and the transmission of knowledge from teachers to students, utilizing e-learning, article readings, case studies, poverty understanding, and promoting educational inclusivity. The teaching of SDGs adopts a case-based methodology, analyzing negative impacts and encouraging critical thinking to develop solutions to problems. Additionally, an interdisciplinary approach is employed to view the complexity of sustainability challenges in a holistic and systematic manner (Fig. 4).

The term "SDGs" focuses on defining the human condition based on its capacity to build economic development while exploiting social and environmental resources. Furthermore, studies focus on assimilating poverty as not only a lack of economic development but also a lack of social development. This approach configures the reality of teaching to observe problems as opportunities for innovation by detecting resource shortages.

3 Theory of Graphs

Graph theory is a mathematical structure that facilitates the representation of what is being studied. It allows for the representation of a single type of relationship between the variables defined in the initial search terms. In this case, the search focuses on the SDGs in education. Graph theory contributes to the abstract visual representation of an image linked by nodes that relate to the search keywords. Graph theory is used to graphically interpret and solve problems in various fields, that is, the mathematical union of simplifying facts in the form of colors or connections of nodes or vertices. Graphs are used to predict search factors (Carmesin, 2022; Li, 2008; Liu et al., 2022). Graphs contribute to the optimization of data for information interpretation (Zolkin et al., 2022). The analysis of the diagrams focuses on the items consolidated in the search in greater detail (Nath, 2022). Society is focused on reducing data in a way that allows for agile and systemic interpretation for decision-making (Eddoujaji et al., 2023).

Graph theory is a mathematical framework that enables the representation of relationships between variables in a given inquiry. This method utilizes nodes and edges to connect different terms and establish the relationship between them. Graph theory is a powerful tool that can be applied to a wide range of problems, including those related to sustainable development (Carmesin, 2022; Li, 2008; Liu et al., 2022). By simplifying complex data sets through color coding and node clustering, graphs provide an effective means of data visualization and interpretation. This allows for the prediction of search factors, optimization of data, and identification of patterns and trends (Zolkin et al., 2022; Nath, 2022).

Additionally, graphs aid in representing various combinations of data to schematize connections and flow of information (Gao et al., 2023). Individual data points alone are insufficient for informed decision-making; it is the combination of data that leads to the generation of useful information. Graphs organize and simplify

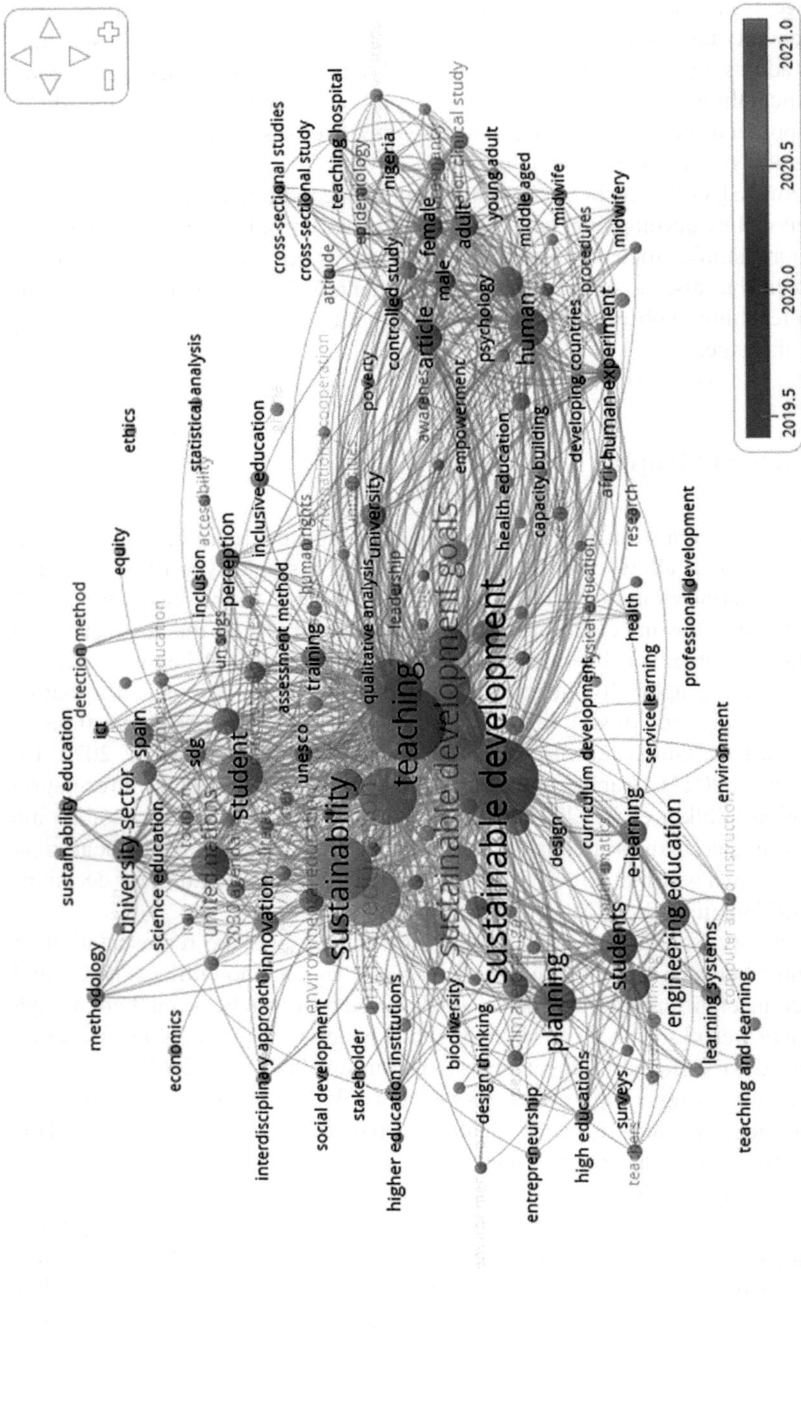

Fig. 4 Chronological development of the terms SDGs and teaching. Source: Author's elaboration VosViewer Scopus

data in a coherent way with respective nodes, generating a diagram that visually represents intersections between nodes (Rivera et al., 2020). Graph theory considers data connectivity in the context of certain constraints based on the interpretation model. These restrictions arise from the type of data being analyzed and patterns of data repetition (Cheng, 2019).

Education contributes to teaching the SDGs by interacting with data in the form of information and making decisions in favor of the environment. The teacher configures a diffusion study to explain the importance of the SDGs based on norms (Velasquez & Alvarez-Alvarado, 2021) that contribute to achieving innovation (Meijer et al., 2018) and permeate the graphic image of the data (Worden et al., 2020) to grant structural decision-making benefits (Christidis & Losada, 2019) and implement dynamic interpretation schemes (Yang et al., 2008). This permeates information optimization (Mo et al., 2022) and analysis of prediction schemes (Kyriakou et al., 2019). The analysis of graphs in education emphasizes the importance of successfully assimilating the indicators of the SDGs. The professional context defines a diffuse dynamic of the characteristics of the SDGs based on the legitimacy of the indicator. Graphs contribute to improving the interpretation of different events as one event. Graphs pay tribute to the graphic interpretation of the subject studied based on an innovative instrument (Katambi et al., 2002). In summary, graph theory involves the interaction of nodes from the interpretation of data sets in the form of an image to optimize information.

4 Methodology

The methodology used for this research involved a qualitative approach, using in-depth interviews with four professors from different professional areas (education, engineering, and administration) at four universities (two private and two public) during the first semester of 2022. The first professor T1PL, T1 is a professor and PL Public University, interviewed on February 25, 2022, at 2:00 pm, interviewed for 2 h at the end of the business sustainability management class. Professor T1PR, T1 is a teacher and PR is a private institution, in the teacher's office at 9:00 am on March 9, 2022, for 1 h and 45 min, with an audio recording of the teacher with her respective authorization. The third teacher (T2PR) and the second teacher (T2) belong to a private institution (PR). They were interviewed on March 22, 2022 at 10:00 am, in the teachers' room for 3 h. The teacher of the public university T2PL was interviewed on April 7 at 3:00 pm for 1 h. The teachers were in charge of a course focused on SDG issues, with note-taking, because they did not authorize recording.

The interview was based on three guiding questions per category: (1) Teaching SDGs: How are the SDGs taught theoretically? Why is teaching the SDGs important? Why are the SDGs taught? (2) Teaching praxis: What kind of examples do you use to teach the SDGs? What methodology do you use in the classroom to teach the SDGs? How and why is the curriculum integrated into the teaching of the SDGs during the class? (3) Motivation to understand the importance of the SDGs: What kind of motivational strategies are used in class and how are their effects? How do you incorporate ICT to motivate learning about the SDGs? Which of the SDGs generates the greatest learning motivation and why in students?

In addition, the research was based on articles retrieved from Scopus and Web of Science, analyzed using Vos Viewer and Tree of Science (TOS) with the keywords "Sustainable Development Goals" and "teaching."

To begin, contact was made with university administrators to identify professors who would teach courses related to the topics of the SDGs and business sustainability. Once the teachers were identified, a meeting was held to explain the purpose of the research and informed consent was obtained. The interviews with each teacher were conducted on the university campus and lasted between 1 and 3 h on average. Notes and recordings of the responses of each teacher were taken according to their authorization and the data was categorized according to three themes: teaching the SDGs, teaching praxis and motivation to understand the importance of the SDGs. Finally, the data was analyzed through a cross-examination of the categories and interpreted based on the objectives of the study.

The selection of the Universities is due to the fact that these are the most recognized in the academic environment of the city for the implementation of the SDGs in the teaching processes in the classroom according to state exam tests. The two public universities have more than 100 years of accreditation in the academic environment and the two private ones are the ones that currently permeate the knowledge of promoting sustainable entrepreneurship.

5 Results

Traditionally, higher education institutions have primarily focused on designing curricula that are specific to particular professions. However, there has been a noticeable shift towards integrating topics related to the Sustainable Development Goals (SDGs) into courses, with the aim of fostering long-term awareness. Education plays a vital role in exploring research methods that enable teachers to impart theory in a practical manner, facilitating better understanding and application of SDG concepts. While universities are increasingly promoting the incorporation of SDGs into their curricula, public universities typically introduce SDG-related topics in their administration programs during the final year of study. Private universities with a focus on education may teach specific SDG topics in select courses, but without a comprehensive integration throughout the curriculum, resulting in limited student awareness regarding SDGs. Additionally, education teachers often lack formal training on SDGs, which can lead to deficiencies in pedagogical approaches to transforming the paradigm of environmental care, contributing to social initiatives, and enhancing the overall well-being of individuals.

Teacher T1PL has stated that "… although students learn about SDGs in their courses, they often fail to meet the objectives after leaving the university. …" On the other hand, T1PR believes that "... teaching from an SDG perspective helps students learn to conserve scarce resources, but the lack of collaboration from parents who have not been trained in SDGs slows down conscious knowledge."

Teachers make significant efforts to instill in their students the importance of the Sustainable Development Goals (SDGs). It is equally crucial for parents to indirectly receive training on these matters to enhance awareness in both the short and long term. During meetings, teachers engage in discussions regarding pedagogical approaches to persuade students about the significance of learning about the SDGs. They also contribute to fostering comprehension of business sustainability, which encompasses environmental, social, and economic dimensions. By familiarizing themselves with social issues, students can identify ways to enhance education, alleviate poverty and hunger, and improve maternal health. From an environmental perspective, teachers can cover topics such as water conservation, recycling, and tree planting. Simultaneously, from an economic standpoint, teachers can emphasize the generation of shared values.

During meetings, teachers engage in discussions about didactic and pedagogical approaches to convince students of the significance of learning about the SDGs. They also contribute to understanding the management of business sustainability, encompassing the environmental, social, and economic dimensions. By teaching about social issues, they contribute to improving education, reducing poverty, hunger, and maternal mortality rates. From an environmental perspective, teachers cover topics such as water conservation, recycling, and afforestation. From an economic standpoint, they focus on generating shared value.

T2PR stated, "...the economic has always been prioritized over the social and environmental concerns... this has led to the destruction of the planet... it is time to prioritize the environmental aspect followed by the social and then the economic."

T1PL suggested "... creating a book club about the SDGs to teach their importance..." This would be an opportunity to teach the SDGs in a different way, not just by reading indicators but also by generating debates about their significance. Such a discussion would help analyze the problems of the SDGs and their approach to teaching them. In other words, talking about the SDGs from a positive perspective would help students understand them better than teaching them about the negative impacts. Teachers employ various pedagogies to disseminate the importance of the SDGs. For instance, T1PL uses "games to teach how to separate waste from a recycling perspective. ..."

The T1PR teacher stated, "... readings on the subject and case studies of SMEs so that students can understand that it is not only the large companies that should make efforts but also small ones can contribute to the environment."

The T2PR teacher said, "... the SDGs are taught from a business perspective, linking corporate social responsibility, organizational management, and sustainability."

The management of business sustainability course offered at the public university provides an opportunity to integrate the Sustainable Development Goals (SDGs) into various aspects, including business sustainability, value chain analysis, process mapping, and financial evaluation of projects. This course takes a management-oriented approach to SDGs. On the other hand, the private university's focus is primarily on teaching the SDGs as a topic, without placing significant emphasis on their relevance within organizational contexts. Furthermore, while the SDGs are primarily taught from a United Nations perspective, it is crucial to incorporate their teachings in the context of both urban and rural environments. Professors at public universities often possess a better understanding of the syllabus as they are more acquainted with economic environments. In contrast, private universities specializing in education tend to concentrate solely on the SDGs as a subject rather than offering a comprehensive course. Their emphasis lies in connecting the SDGs with the public from a perspective of sustainable economic development.

The T1PR teacher stated, "The SDGs are taught in the last semester so that students can link all the effects in the organization. Case studies are presented, and there are debates in round tables and talks with businessmen."

These talks with businessmen allow students to understand the theory in organizational reality and the importance of understanding entrepreneurs' rationing to identify new sources of renewable resources. Business case studies contribute to generating debate on the paradigm shift of the SDGs from a continuous improvement opportunity approach. Virtual interaction games enable students to learn from their devices, such as interactive banking based on learning the economic theme of the SDGs. Likewise, the debates in round tables leverage the identification of the reality of a specific topic of the 17 SDGs.

The teacher T1PR expressed, "The motivation to teach the SDGs is based on finding process innovation, environmentally friendly products, and quality of social innovation."

Teachers in the field of education argue that the focus of education indicators should not solely be on increasing the population of students but should also include strategies for achieving this goal.

The T2PL teacher expressed, "Educational indicators contribute to the inclusion of people. It is difficult to teach respect for others and prevent bullying that leads to dropouts. It promotes learning that we are all different and must be respected, generates ethical and moral values, promotes being capable of accepting that there are differences in thought and attitudes, and renews the ideology of being an economic man."

However, the teacher T2PL stated that "… education is the means to understand the organizational complexity of submitting statistics so that it can be dynamic… it is time to learn to unlearn."

The utilization of educational games plays a vital role in fostering critical thinking about environmental and social issues. It serves as a tool to encourage individuals to consider solutions that encompass both social and environmental perspectives. Consequently, the economic aspect emerges as a consequence of addressing the social and environmental aspects.

The Sustainable Development Goals (SDGs) have become an integral component of the educational process, cutting across various disciplines. As a result, educational institutions require well-prepared teachers who can effectively impart strategies that enable students to tackle the challenges presented by an ever-changing world of work. This entails fostering collaboration between the industry and providing quality education that equips students with the necessary skills to navigate the complexities of the contemporary labor and social landscape. The integration of theoretical knowledge from different fields within the educational environment should be accompanied by a practical application that promotes problem-solving abilities.

Education about SDGs is an opportunity to learn how to use technology to generate diverse knowledge. The teacher T2P2 stated, "Teaching the SDGs is an opportunity to change attitudes and skills to prevent social and environmental disasters. It is an opportunity to learn to identify solutions, understand the environment, and submit strategies to take care of it. It is time to act and change habits."

The teacher has the opportunity to transmit appropriate messages to simultaneously apply the 17 SDGs in the short term for long-term effects.

The T2PL teacher said, "... if we don't teach the SDGs today, they won't be applied in 2030 as a solution but as a new problem."

The implementation of the SDGs is not just an industrial problem but an educational one as well. It is important to understand how industry and education can collaborate to provide solutions and integrate them into applied reality in companies. Learning to identify opportunities is key to finding long-term solutions. Higher education institutions must incorporate systemic thinking that considers the reality of the SDGs.

The T2PL teacher also stated, "... the university is the place to teach how to live with social and environmental well-being, not environmental destruction."

Universities have a responsibility to generate debates that raise awareness of the damage caused by humanity and to exemplify the practice of implementing SDG indicators.

Incorporating the Sustainable Development Goals (SDGs) into educational processes requires a comprehensive examination of the curriculum, its implementation, and the policies that support flexible actions and community involvement. Communities must actively seek new avenues for integrating SDGs into their daily lives and creating a lasting legacy for future generations, fostering better sustainable development. It is crucial for individuals to take conscious and active steps toward contributing to society.

This process involves preventive measures that encourage reflection, systematic thinking, and the adoption of new perspectives on the environment and its reality. When it comes to intervention, it is essential to consider the specific context of the environment, its inhabitants, their values and representations, and the factors that have influenced their presence in that particular environment rather than others. Moreover, individuals must recognize their responsibility towards the environment and actively engage in its conservation.

The convergence of promotion, prevention, and intervention within the framework of the SDGs allows for the establishment of new perspectives on social reality within the educational community. This strengthens community bonds, fosters greater support, acknowledges individuals and their life experiences, appreciates cultural heritage, promotes sustainable practices, and establishes a sense of shared responsibility in society towards self-care, environmental stewardship, and the well-being of others. These approaches become ingrained ways of experiencing life within society. Consequently, teachers play a critical role in nurturing critical thinking and fostering creativity skills among students, intensifying their efforts towards these objectives.

Teachers play a pivotal role in promoting social development and enhancing the quality of life. They have the opportunity to cultivate students' creative thinking skills and empower them to find innovative solutions to emerging social and environmental challenges. By encouraging students to think differently and view problems as opportunities, teachers can foster the development of innovative solutions. Effective teaching involves conveying clear and coherent messages based on factual information that offer practical solutions. For instance, teaching students about how artificial intelligence can contribute to curing future diseases. Leveraging AI as a teaching tool allows students to analyze and address future challenges. However, it is crucial to program AI effectively to avoid unintended negative consequences. Therefore, teachers need to be prepared to embrace and utilize technology to address social and environmental issues sustainably.

The integration of the Sustainable Development Goals (SDGs) is essential in the education and training of future professionals. Embracing this approach enables the continual improvement and adaptation of responses to the evolving needs of society. Strategies such as project-based learning can be employed to apply the SDGs, fostering collaboration, acknowledging the core principles of the respective fields, and providing opportunities to apply theoretical knowledge to real-world situations while developing practical skills.

Furthermore, incorporating technology in education facilitates access to information and educational resources, enhancing the learning experience and fostering motivation. Virtual platforms can serve as tools for promoting service-learning based on the SDGs, allowing students to engage with real community problems and apply their knowledge in practical contexts, thereby contributing to the development of their environment. This approach respects individual students' needs and different learning paces. Techniques like cooperative learning, case studies, and learning challenges can be leveraged to support this approach.

Therefore, addressing the SDGs, preparing students for the challenges of the workforce, and promoting lifelong learning helps them to discern how to receive information or take advantage of opportunities to actively participate in solving social problems and be an active part of practical actions that promote leadership, compassion, empathy, and a sense of responsibility for the common good.

6 Discussion

In private universities, the teaching of the Sustainable Development Goals (SDGs) typically begins from the sixth semester onward as part of the cross-curricular approach. In contrast, public universities incorporate the SDGs in the last semester of the business sustainability management course. In public education, the SDGs receive limited exposure in engineering and education degree programs in private

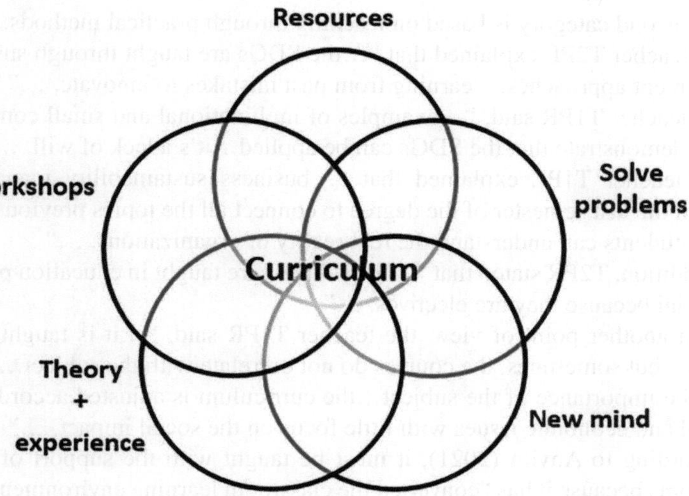

Fig. 5 Relationship between curriculum and SDGs. Source: Author's elaboration

universities, often as a distinguishing feature of the curriculum. However, students often associate sustainable development and sustainability solely with environmental aspects. It is important to emphasize that sustainable development encompasses not only environmental considerations but also a thriving economy and social equity. To address inequalities and poverty in the country, it is crucial to implement transversal strategies throughout the curriculum. Education serves as a vital means to teach sustainability and develop innovative strategies to promote holistic, sustainable development.

The first category was related to teaching SDGs. The SDGs are understood as "the process of improving non-renewable resources… is to generate real awareness and not just imagine… teaching people to see the environmental destruction caused by not thinking about the social and environmental impact" (T1PL). Likewise, T2PR exposed that "teaching for economic sustainability involves learning to manage resources, identifying innovation in the use of new materials, reducing poverty, and learning to care for the environment for a safer future."

The teacher T2PL stated, "…teach with examples of destruction from other countries… and teach to live in peace, respect gender, take care of nature (oceans, rivers, water sources, forests). …"

According to Lu et al. (2022), creative thinking must be taught as a strategy to offer a structured curriculum. In turn, activities must be integrated in accordance with government policy guidelines to maximize the performance of the SDGs (Adhikari & Shrestha, 2022). "Teaching students about the SDGs and incorporating sustainability into their business curriculum contributes to understanding the reality of organizational situations" (Fang & O'Toole, 2023, p. 1). Therefore, the topics of the SDGs must be included in the curriculum in a transversal way. "The ultimate goal is to help students become active citizens in their communities" (Cottafava et al., 2019, p. 223) (see Fig. 5).

The second category is based on teaching through practical methods.

The teacher T2PL explained that "...the SDGs are taught through sustainability management approaches...learning from past mistakes to innovate. ..."

The teacher T1PR said, "...examples of multinational and small companies are used to demonstrate that the SDGs can be applied...it's a lack of will. ..."

The teacher T1PL explained that "...business sustainability management is taught in the last semester of the degree to connect all the topics previously covered so that students can understand the real reality of organizations. ..."

In addition, T2PR stated that "...the subjects are taught in education programs as superficial because they are electives. ..."

From another point of view, the teacher T1PR said, "...it is taught in various courses...but sometimes, the courses do not correlate with the subject...this undermines the importance of the subject...the curriculum is adjusted according to new political and economic issues with little focus on the social impact. ..."

According to Anyim (2021), it must be taught with the support of intelligent classrooms because it has "converted the classroom learning environment into high-tech content, which creates an interactive classroom teaching efficiency for teachers and students in learning situations" (p. 5579). Furthermore, gender equality and good health and well-being are the SDGs that students and the public are most concerned about (Yuan et al., 2022, p. 1). However, "at the same time, academic and administrative staff must acquire environmental and egalitarian values that must be transmitted through their professional practice" (Álvarez-García et al., 2022, p. 53).

Institutions that prioritize the integration of Sustainable Development Goals (SDGs) within their environment not only make a positive impact on the environment but also gain recognition as entities that excel in adaptation and collaborative work. By promoting relevant skills, these institutions prepare their graduates for future challenges in their respective fields. Another distinctive characteristic is their emphasis on amplifying the voices of communities, facilitating adaptation and fostering active participation. These institutions become leaders in driving positive change. The ability to adapt and demonstrate resilience is a defining feature of their relationships with the environment, while inclusivity is consistently promoted in all its forms. By incorporating the SDGs into the educational sphere, students are equipped with innovative ideas that foster co-responsible participation and empower them to tackle global challenges.

The development of innovation skills in the training process enables individuals to see problems as opportunities for improvement (Rivera et al., 2020). That is, the scarcity of resources leads to thinking creatively to innovate the use of new inputs that are environmentally friendly. Teaching the language and terminology of the SDGs creates a response that allows the indicators to be assimilated in a coherent way with what humans do (Mambu, 2023). Rural teachers showed greater procedural knowledge of land use than urban teachers (Niens et al., 2021).

Therefore, the T1PR teacher expressed, "... in various regions, the SDGs are applied to the social, economic, and environmental areas to generate awareness."

The teacher T2PL said, "... it is taught to learn how to live in a community to solve community problems."

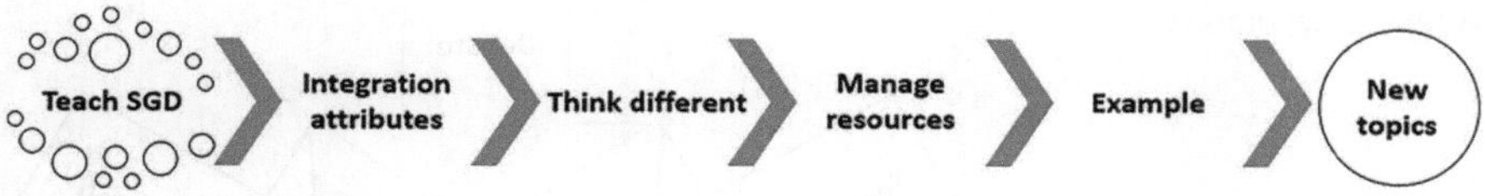

Fig. 6 SDGs based on curriculum. Source: Author's elaboration

In other words, the fundamental role of universities as engines of social transformation from different dimensions contributes to learning how to teach to learn to change attitudes before the events of the destruction of the earth (Chicharro & Campión, 2022) (see Fig. 6).

The third category was based on motivation to understand the importance of the SDGs.

The T1PL teacher expressed that "... teaching with games helps students assimilate the importance of the SDGs. ..."

T2PL said that "... teaching research methodology contributes to investigating examples of other research to make coherent decisions in relation to the environment and social care. ..."

The T2PR teacher expressed that "... teaching ethical values from childhood contributes to better assimilate the values of caring for the earth, the human being, and personal finances... teaching by example is important. ..."

The T1PR teacher said that "... the library contributes to learning about the topics related to the SDGs and with business examples given by company managers, students can better understand the importance of complying with the SDG indicators."

Teachers demonstrate a strong interest in emphasizing the importance of monitoring SDG indicators as a means to assess and improve conditions (Poza-Vilches et al., 2022). Teaching strategies play a crucial role in helping students understand the indicators as a control process, enabling timely interventions for improvement. Libraries serve as valuable resources for supporting student learning by providing relevant materials on the SDGs. Additionally, organizing training (Dei & Asante, 2022) can effectively communicate the significance of the SDGs. It is crucial to introduce the SDGs at an early age (Rodrigo et al., 2021) to instill values and behaviors that students can carry with them as they grow, becoming exemplary contributors to society.

The T1PR teacher said that "... the ability to teach with new virtual technologies in 3D contributes to understanding the policies that are implemented to care for the land and the environment... however, this should be a cross-curricular course and not just a course taught in the fifth year of the degree program. ..."

In other words, the SDGs can be explicitly incorporated into programs (Ramirez-Mendoza et al., 2020). The T2PR teacher expressed that "... the SDGs are incorporated as a learning process in various courses so that students can learn to integrate the themes of the courses to make decisions..." Cooperation and collaboration between schools (Chowdhury & Koya, 2017) can help to identify strategies that

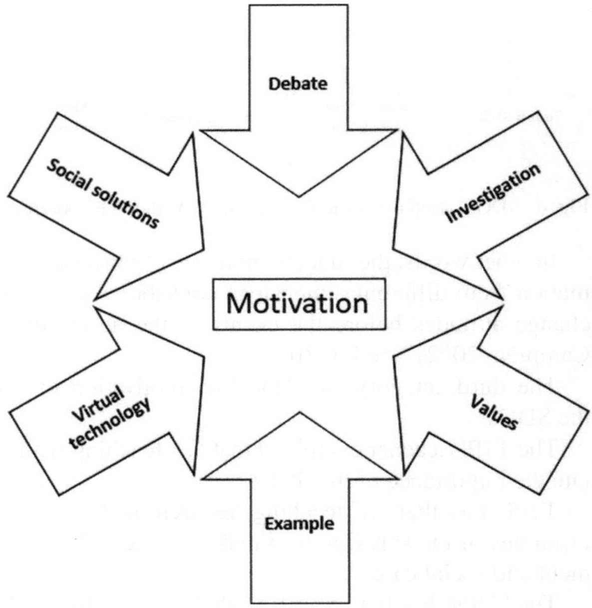

Fig. 7 SDGs based on motivation. Source: Author's elaboration

enable students, with the support of teachers, to better understand the importance of caring for the environment, contributing to society, and taking care of economic resources in the long term (see Fig. 7).

Teaching the SDGs through a virtual system (Schwanke, 2020) enables both students and teachers to access class recordings and increase awareness. Moreover, the experiences shared by academic teachers in class and their ability to respond contribute to learning about the SDGs by generating debates about resource scarcity and ways to promote care for them (Almuaqel, 2022).

Therefore, the T1PR teacher stated, "…the student learns by engaging in social and environmental activities…the teacher must accompany them to distinguish between what is favorable and not favorable to the implementation of the SDGs. …"

Furthermore, different models based on educational practices can leverage the interpretation of applicability (Baena-Morales et al., 2021) to configure a means of caring for the environment and promoting social activities in the community. The challenges posed by pedagogical tools supported by technology (Fernandez Rivas et al., 2020) make it possible to highlight the importance of the SDGs. However, committed teaching, research, and community participation (Perales Franco & McCowan, 2021) complement learning from the SDGs.

The T2PL teacher noted, "…community research is essential for students to learn how to separate waste and apply the 5Rs: Reject, Reduce, Reuse, Recycle, and Reincorporate…" It is an opportunity to teach that resources are not only economic but must also be transformed into positive support for society to take care of the environment.

The utilization of digital tools in social work education facilitates the collaborative construction of solutions for environmental care (Trubavina et al., 2021). Teaching about the SDGs and their indicators enables society to intervene and raise awareness regarding the challenges they present. Hence, the integration of teachers' global competencies (Yaccob et al., 2022) allows for the adoption of integrated teaching practices with a growth mindset (Narayana & Warrier, 2022) based on sustainable cooperation (Dlouhá & Pospíšilová, 2018). However, the decision to incorporate the study of SDGs often varies depending on the geographical location of the university (Westerman et al., 2020). Sustainability can be used to demonstrate the practical application of mathematics to students while instilling values (Lafuente-Lechuga et al., 2020) to enhance their knowledge.

Teaching about the SDGs in primary school helps universities understand their applications and definitions (Dieste et al., 2019). Teaching with understanding allows for a comprehensive assimilation of the SDGs (Leal Filho et al., 2021), and incorporating written narratives, films, and guest speakers (Addo et al., 2022) helps to assimilate technological developments that contribute to learning about the SDGs. Games and challenges also contribute to learning about the SDGs (Willicks et al., 2018), allowing students to understand the reality of the environment. In addition, the combination of basic and transversal skills, together with case studies, aims to learn from the application of the SDGs in any situation (Gillet & Voneche-Cardia, 2021). Various educational media contribute to teaching the reality of how to apply the indicators in any situation.

Teachers must break away from the paradigm of the difficulty of implementing the SDGs and have an open mind to facing problems and providing solutions. Students leverage the reality of the SDGs when faced with negative events. The curriculum is the teacher's guide to teaching the SDGs.

"Teaching the SDGs promotes seeing the problems accumulated over decades, providing an opportunity to rectify the damage done by humans, and offering a new way of understanding social and environmental problems," as expressed by T1PR teacher.

Teaching the indicators allows for the search for strategies for possible solutions, but at the beginning, it can be complex to incorporate them into the organization.

As the T1PL teacher said, "… they teach based on literature, but it can be too much at times, and they need to select what is believed to be taught. …"

The education of the SDGs is promoted from childhood so that by the time students reach university, they can better interpret them. Open systems facilitate the integration of the SDGs into the organization's structure, adjusting it to the curriculum to define their importance.

The T2P2 teacher expressed that "… the education of the SDGs is based on the understanding of their importance, identifying problems and their indicators with solutions, contributing in a different way, and organizing long-term strategies that are environmentally friendly. …"

From another perspective, teaching the SDGs allows for:

- Improving governance and transparency within the organization

- Enhancing the organization's contribution to the SDGs
- Engaging employees in a common project
- Becoming a pioneering company in sustainability
- Implementing Corporate Social Responsibility Actions

Teaching the SDGs aims to benefit the organization in the long run, by contributing to understanding the reality of their application and developing new strategies to leverage their implementation in the short term. It also involves a commitment from teachers to take responsibility for a joint project.

Teaching the SDGs motivates to solve future problems related to social, environmental, and economic aspects, fostering intrinsic motivation to learn from a diverse and complex environment. Additionally, extrinsic motivation rewards the prevention of problems in the implementation of the SDGs.

The T2PR teacher said, "… we learn from teachers and students, who contribute their points of view to generate constructive discussions… it allows us to see short-term solutions when innovating… it is important to learn from mistakes and identify solutions where they are not believed to exist."

Understanding the environment helps improve conditions for implementing didactic strategies that lead to solving problems, as expressed by the T2PL teacher: "… understanding the environment helps to improve the conditions to implement didactic strategies that lead to solving the problem." The teacher leverages the philosophy of incorporating the SDGs in class from theoretical reality to praxis. The T1PR teacher expressed, "… using gamification allows one to learn to see the consequences of not having policies to care for the environment and society… it is the opportunity to learn to differentiate a problem without a solution from a problem with a solution…" The teacher must be multifunctional to interpret the students' thoughts and incorporate them into the curriculum.

7 Conclusions

In conclusion, when teachers incorporate the teaching of SDGs in the classroom, they empower students to develop strategies that promote resource conservation and take responsibility for social, environmental, and economic policies that contribute to environmental care. This action has a significant impact on the future of global sustainability, as it depends on humanity's shift away from destructive attitudes toward the conservation of natural resources and the recognition that they are not inexhaustible. To support this, teachers should provide SDG bibliographies in the curriculum to encourage reading and stimulate reflection and the creation of new intellectual works. By engaging in research, teachers generate opportunities for society to recognize and explore the SDGs, fostering awareness of social, environmental, and economic well-being.

The perception of teaching SDGs varies across professions. Professionals in public and private university administration recognize the SDGs as part of management training, while professionals in engineering (public) and higher education (private) have limited knowledge of the SDGs as part of their academic development without integration into specific subjects. If we do not transform the way we educate our professionals, the future of humanity will face serious threats. Sustainable development serves a critical purpose, and teachers incorporate information and communication technology (ICT) to make SDG education appealing to the new generation of students. However, achieving this requires a serious commitment from society as a whole, including teachers and students as a community, rather than relying solely on individual actions.

Teaching supported by games, new tools, and virtual systems helps students grasp the importance of the SDGs. Teachers should incorporate real-life business experiences into their courses to illustrate the practical implementation of the SDGs. The SDGs should be integrated into the curriculum and taught in a cross-cutting manner, meaning their significance should be explored in each subject with relevant examples from everyday life.

Teachers transmit the reality of the SDGs based on what is outlined in the curriculum, but it is important to analyze the curriculum's scope. They teach SDG indicators theoretically and incorporate practical business examples to enhance understanding and application. Furthermore, affordable training courses for the community should be designed to improve efficiency in resource management.

The limitations of this study include restricted access to teachers who teach courses related to the SDGs, as some courses only cover specific SDG topics rather than comprehensively addressing all 17 goals. Additionally, teachers may be hesitant to share their teaching methods for fear of replication without proper credit. Time constraints also posed challenges during interviews and recording sessions.

Future research should focus on exploring the perspective of students regarding SDG education. It is also essential to extend the interview script to the international university setting and secondary education. Furthermore, incorporating questions that address the failures resulting from the delayed implementation of the SDGs would provide valuable insights.

References

Addo, R., Koers, G., & Timpson, W. M. (2022). Teaching sustainable development goals and social development: A case study teaching method. *Social Work Education, 41*(7), 1478–1488. https://doi.org/10.1080/02615479.2022.2112168

Adhikari, D. R., & Shrestha, P. (2022). Knowledge management initiatives for achieving sustainable development goal 4.7: Higher education institutions' stakeholder perspectives. *Journal of Knowledge Management*. https://doi.org/10.1108/JKM-03-2022-0172

Albareda-Tiana, S., Vidal-Raméntol, S., & Fernández-Morilla, M. (2018a). Implementing the sustainable development goals at university level. *International Journal of Sustainability in Higher Education, 19*(3), 473–497. https://doi.org/10.1108/IJSHE-05-2017-0069

Albareda-Tiana, S., Vidal-Raméntol, S., Pujol-Valls, M., & Fernández-Morilla, M. (2018b). Holistic approaches to develop sustainability and research competencies in pre-service teacher training. *Sustainability (Switzerland), 10*(10). https://doi.org/10.3390/su10103698

Almuaqel, I. A. (2022). Special needs, special efforts: Learners with intellectual disabilities and sustainable development goals (SDGs). *IEEE Transactions on Engineering Management, 2022*, 1–18. https://doi.org/10.1109/TEM.2022.3189318

Álvarez-García, O., Quintana-Murci, E., Forteza-Forteza, D., & Touza, C. (2022). Toward a faculty aligned with the sustainable development goals: Sustainability, equality and equity action plan (University of the Balearic Islands). In *World sustainability series* (pp. 45–64). Springer Science and Business Media Deutschland GmbH. https://doi.org/10.1007/978-3-031-04764-0_3

Anyim, W. O. (2021). Sustainable development goal on quality education: A review of E-Learning resources and pedagogy in the University System. *Library Philosophy and Practice*. Accessed from https://www.scopus.com/inward/record.uri?eid=2-s2.0-85108542764&partnerID=40&md5=c5e950e1c057d2e76e45a86aaa864ff0

Baena-Morales, S., Jerez-Mayorga, D., Delgado-Floody, P., & Martínez-Martínez, J. (2021). Sustainable development goals and physical education. A proposal for practice-based models. *International Journal of Environmental Research and Public Health, 18*(4), 1–18. https://doi.org/10.3390/ijerph18042129

Carmesin, J. (2022). Graph theory—A survey on the occasion of the Abel prize for László Lovász. *Jahresbericht der Deutschen Mathematiker-Vereinigung, 124*(2), 83–108. https://doi.org/10.1365/s13291-022-00247-7

Chang, Y.-C., & Lien, H.-L. (2020). Mapping course sustainability by embedding the SDGs inventory into the university curriculum: A case study from National University of Kaohsiung in Taiwan. *Sustainability, 12*(10). https://doi.org/10.3390/su12104274

Cheng, I.-H. (2019). Constructing an ecosystem for quality learning in the wake of un sustainable development goals: A social-entrepreneurship study of riverside school, India. *Contemporary Educational Research Quarterly, 27*(4), 1–35. https://doi.org/10.6151/CERQ.201912_27(4).0001

Chicharro, A., & Campión, M. J. (2022). What do law and mathematics have in common? The sustainable development goals as a transversal substrate of numbers and laws. In M. L. Sein-Echaluce, A. Fidalgo-Blanco, & F. J. Garcia (Eds.), *CEUR Workshop Proceedings* (Vol. 3129). CEUR-WS. Accessed from https://www.scopus.com/inward/record.uri?eid=2-s2.0-85129207970&partnerID=40&md5=3843496564110ae295f993fcaaa85c7f

Chowdhury, G., & Koya, K. (2017). Information practices for sustainability: Role of iSchools in achieving the UN sustainable development goals (SDGs). *Journal of the Association for Information Science and Technology, 68*(9), 2128–2138. https://doi.org/10.1002/asi.23825

Christidis, P., & Losada, A. G. (2019). Email based institutional network analysis: Applications and risks. *Social Sciences, 8*(11). https://doi.org/10.3390/socsci8110306

Cottafava, D., Cavaglià, G., & Corazza, L. (2019). Education of sustainable development goals through students' active engagement: A transformative learning experience. *Sustainability Accounting, Management and Policy Journal, 10*(3), 521–544. https://doi.org/10.1108/SAMPJ-05-2018-0152

Dei, D.-G. J., & Asante, F. Y. (2022). Role of academic libraries in the achievement of quality education as a sustainable development goal. *Library Management, 43*(6–7), 439–459. https://doi.org/10.1108/LM-02-2022-0013

Dieste, B., Coma, T., & Blasco-Serrano, A. C. (2019). Inclusion of the sustainable development goals in the curriculum of primary and secondary education in rural schools of Zaragoza. *Revista Internacional de Educacion para la Justicia Social, 8*(1), 97–115. https://doi.org/10.15366/RIEJS2019.8.1.006

Dlouhá, J., & Pospíšilová, M. (2018). Education for sustainable development goals in public debate: The importance of participatory research in reflecting and supporting the consultation process in developing a vision for Czech education. *Journal of Cleaner Production, 172*, 4314–4327. https://doi.org/10.1016/j.jclepro.2017.06.145

Dotres Zuniga, S., & Sanchez Paz, N. (2020). Integration of corporate social responsibility in the risk management in constructive investments. *Avances, 22*(2), 170–182.

Eddoujaji, M., Samadi, H., & Bouhorma, M. (2023). HDFS improvement using shortest path algorithms. *Lecture Notes on Data Engineering and Communications Technologies, 147*, 253–269. https://doi.org/10.1007/978-3-031-15191-0_25

Faber-Langendoen, D. (1992). Ecological constraints on rain forest management at Bajo Calima, western Colombia. *Forest Ecology and Management, 53*(1), 213–244. https://doi.org/10.1016/0378-1127(92)90044-A

Fang, J., & O'Toole, J. (2023). Embedding sustainable development goals (SDGs) in an undergraduate business capstone subject using an experiential learning approach: A qualitative analysis. *International Journal of Management Education, 21*(1). https://doi.org/10.1016/j.ijme.2022.100749

Fernandez Rivas, D., Boffito, D. C., Faria-Albanese, J., Glassey, J., Afraz, N., Akse, H., Boodhoo, K. V. K., Bos, R., Cantin, J., Chiang, Y. W., Commenge, J.-M., Dubois, J.-L., Galli, F., de Mussy, J. P. G., Harmsen, J., Kalra, S., Keil, F. J., Morales-Menendez, R., Navarro-Brull, F. J., et al. (2020). Process intensification education contributes to sustainable development goals. Part 1. *Education for Chemical Engineers, 32*, 1–14. https://doi.org/10.1016/j.ece.2020.04.003

Gao, J., Gao, J., Ying, X., Lu, M., & Wang, J. (2023). Higher-order interaction goes neural: A substructure assembling graph attention network for graph classification. *IEEE Transactions on Knowledge and Data Engineering, 35*(2), 1594–1608. https://doi.org/10.1109/TKDE.2021.3105544

García-González, E., Jiménez-Fontana, R., & Azcárate, P. (2020). Education for sustainability and the sustainable development goals: Pre-service teachers' perceptions and knowledge. *Sustainability (Switzerland), 12*(18). https://doi.org/10.3390/su12187741

Gast, J., Gundolf, K., & Cesinger, B. (2017). Doing business in a green way: A systematic review of the ecological sustainability entrepreneurship literature and future research directions. *Journal of Cleaner Production, 147*, 44–56. https://doi.org/10.1016/j.jclepro.2017.01.065

Gillet, D., & Voneche-Cardia, I. (2021). Promoting critical and design thinking activities to tackle sustainable development goals in higher education. In *Proceedings of 2021 World Engineering Education Forum/Global Engineering Deans Council, WEEF/GEDC 2021*, pp. 224–230. https://doi.org/10.1109/WEEF/GEDC53299.2021.9657466

Katambi, S., Jiming, G., & Xiangyuan, K. (2002). Applications of graph theory to gross error detection for GPS geodetic control networks. *Geo-spatial Information Science, 5*(4), 26–31. https://doi.org/10.1007/BF02826471

Koon, V.-Y. (2021). Bibliometric analyses on the emergence and present growth of humanistic management. *International Journal of Ethics and Systems, 37*(4), 581–598. https://doi.org/10.1108/IJOES-03-2021-0062

Ladrach, W. E., & Wright, J. A. (1995). Natural regeneration in a secondary Colombian rain forest. *Journal of Sustainable Forestry, 3*(1), 15–38. https://doi.org/10.1300/J091v03n01_02

Lafuente-Lechuga, M., Cifuentes-Faura, J., & Faura-Martínez, Ú. (2020). Mathematics applied to the economy and sustainable development goals: A necessary relationship of dependence. *Education Sciences, 10*(11), 1–12. https://doi.org/10.3390/educsci10110339

Leal Filho, W., Shiel, C., Paço, A., Mifsud, M., Ávila, L. V., Brandli, L. L., Molthan-Hill, P., Pace, P., Azeiteiro, U. M., Vargas, V. R., & Caeiro, S. (2019). Sustainable development goals and sustainability teaching at universities: Falling behind or getting ahead of the pack? *Journal of Cleaner Production, 232*, 285–294. https://doi.org/10.1016/j.jclepro.2019.05.309

Leal Filho, W., Frankenberger, F., Salvia, A. L., Azeiteiro, U., Alves, F., Castro, P., Will, M., Platje, J., Lovren, V. O., Brandli, L., Price, E., Doni, F., Mifsud, M., & Ávila, L. V. (2021). A framework for the implementation of the sustainable development goals in university programmes. *Journal of Cleaner Production, 299*. https://doi.org/10.1016/j.jclepro.2021.126915

Li, P. (2008). Moving towards creating the new pattern from the graph creativity to the education of design. In *2008 IEEE international symposium on knowledge acquisition and modeling workshop proceedings, KAM 2008*, pp. 428–431. https://doi.org/10.1109/KAMW.2008.4810516

Liu, H., Zhang, G., & Ma, J. (2022). Research progress of graph embedding algorithms. *Journal of Zhejiang University, Science Edition, 49*(4), 443–456. https://doi.org/10.3785/j.issn.1008-9497.2022.04.008

Lu, S.-Y., Wu, C.-L., & Huang, Y.-M. (2022). Evaluation of disabled STEAM-students' education learning outcomes and creativity under the UN sustainable development goal: Project-based learning oriented STEAM curriculum with Micro:bit. *Sustainability (Switzerland), 14*(2). https://doi.org/10.3390/su14020679

Lun, X., Wang, T., Zhang, M., & Wang, Y. (2022). Refinement economic management method based on management accounting. In *Discrete dynamics in nature and society*. Wiley. https://doi.org/10.1155/2022/5964276

Lysaght, R., Ghaderi, G., Milley, P., & Labelle, P. R. (2022). Best practices in evaluating work integration social enterprises for persons with intellectual disabilities: A scoping review. *Journal of Policy and Practice in Intellectual Disabilities*. https://doi.org/10.1111/jppi.12439

Mambu, J. E. (2023). Embedding sustainable development goals into critical English language teaching and learning. *Critical Inquiry in Language Studies, 20*(1), 46–76. https://doi.org/10.1080/15427587.2022.2099863

Marinova, D., Todorov, V., Blinov, A., & Safonov, E. (2017). Principles of ethical economics: A basis for transition to sustainability. In *Methods for sustainability research*. Edward Elgar. https://doi.org/10.4337/9781786432735.00023

Mo, X., Jin, X., Tian, J., Shao, Z., & Han, G. (2022). Research on the division method of signal control sub-region based on macroscopic fundamental diagram. *Sustainability (Switzerland), 14*(13). https://doi.org/10.3390/su14138173

Narayana, H., & Warrier, U. (2022). Building inclusive societies through emotional intelligence based education systems: Indian perspective to achieve sustainable development goals. *ECS Transactions, 107*(1), 8989–9001. https://doi.org/10.1149/10701.8989ecst

Nath, S. S. (2022). Evaluation of thermodynamic consistency of kinetic parameters in cyclic enzyme-catalyzed reaction networks. *Chemical Physics Letters, 804*. https://doi.org/10.1016/j.cplett.2022.139890

Niens, J., Richter-Beuschel, L., Stubbe, T. C., & Bögeholz, S. (2021). Procedural knowledge of primary school teachers in Madagascar for teaching and learning towards land-use-and health-related sustainable development goals. *Sustainability (Switzerland), 13*(16). https://doi.org/10.3390/su13169036

Peel, D. S. (2006). Economics of stocker production. *Veterinary Clinics of North America: Food Animal Practice, 22*(2), 271+. https://doi.org/10.1016/j.cvfa.2006.03.007

Perales Franco, C., & McCowan, T. (2021). Rewiring higher education for the sustainable development goals: The case of the intercultural University of Veracruz, Mexico. *Higher Education, 81*(1), 69–88. https://doi.org/10.1007/s10734-020-00525-2

Poza-Vilches, F., García-González, E., Solís-Espallargas, C., Velasco-Martínez, L. C., López-Alcarria, A., Estrada-Vidal, L. I., Jiménez-Fontana, R., Rodríguez-Marín, F., Puig-Gutiérrez, M., Tójar Hurtado, J. C., & Gutiérrez-Pérez, J. (2022). Greening of the syllabus in faculties of education sciences through sustainable development goals: The case of public Andalusian universities (Spain). *International Journal of Sustainability in Higher Education, 23*(5), 1019–1044. https://doi.org/10.1108/IJSHE-02-2021-0046

Ramirez-Mendoza, R. A., Morales-Menendez, R., Melchor-Martinez, E. M., Iqbal, H. M. N., Parra-Arroyo, L., Vargas-Martínez, A., & Parra-Saldivar, R. (2020). Incorporating the sustainable development goals in engineering education. *International Journal on Interactive Design and Manufacturing, 14*(3), 739–745. https://doi.org/10.1007/s12008-020-00661-0

Rivera, F. M.-L., Hermosilla, P., Delgadillo, J., & Echeverría, D. (2020). The sustainable development goals (SDGs) as a basis for innovation skills for engineers in the industry 4.0 context. *Sustainability (Switzerland), 12*(16). https://doi.org/10.3390/su12166622

Rodrigo, M. T., Diy, W. D. A., Favis, A. M. T., Amante, F. U., Castro, J. C. M., Herras, I. Y., Mallari, J. C. F., Mora, K. A., Torres, J. M. R., & Cuyegkeng, A. C. (2021). A RECIPE for teaching the sustainable development goals. In M. M. T. Rodrigo, S. Iyer, A. Mitrovic,

H. N. H. Cheng, D. Kohen-Vacs, C. Matuk, A. Palalas, R. Rajenran, K. Seta, & J. Wang (Eds.), *29th International Conference on Computers in Education Conference, ICCE 2021 - Proceedings* (Vol. 1, pp. 451–456). Asia-Pacific Society for Computers in Education. Accessed from https://www.scopus.com/inward/record.uri?eid=2-s2.0-85126634705&partnerID=40&md5=2b2c5949c636ea4dd993f36cb6b63ca6

Sanchez, F., Sandoval, A., Rodriguez-Pomeda, J., & Casani, F. (2020). Professional aspirations as indicators of responsible leadership style and corporate social responsibility. Are we training the responsible managers that business and society need? A cross-national study. *Journal of Work and Organizational Psychology-revista de psicologia del y de las organizaciones, 36*(1), 49–61. https://doi.org/10.5093/jwop2020a5

Sánchez-García, J. C., Vargas-Morúa, G., & Hernández-Sánchez, B. R. (2018). Entrepreneurs' well-being: A bibliometric review. *Frontiers in Psychology, 9*(SEP). https://doi.org/10.3389/fpsyg.2018.01696

Schwanke, K. (2020). Promoting the un sustainable development goals in teaching, research and democratic urban life: Approaches to lifelong education for sustainable development. *Training, Language and Culture, 4*(4), 51–61. https://doi.org/10.22363/2521-442X-2020-4-4-51-61

Sonetti, G., Sarrica, M., & Norton, L. S. (2021). Conceptualization of sustainability among students, administrative and teaching staff of a university community: An exploratory study in Italy. *Journal of Cleaner Production, 316*. https://doi.org/10.1016/j.jclepro.2021.128292

Tamrat, W. (2022). Addressing the higher education for sustainable development agenda at the College of Open and Distance Learning, St. Mary's University, Ethiopia: Opportunities and challenges. *International Journal of Sustainability in Higher Education, 23*(1), 177–195. https://doi.org/10.1108/IJSHE-08-2020-0316

Trubavina, I., Medvid, M., Cwer, A. M., Petryshyn, L., & Meshko, H. (2021). Substantiation of the advanced training program "Social work with military personnel and military-social work in the context of sustainable development goals". In S. Semerikov, S. Chukharev, S. Sakhno, A. Striuk, A. Iatsyshyn, S. Klimov, V. Osadchyi, T. Vakaliuk, P. Nechypurenko, O. Bondarenko, & H. Danylchuk (Eds.), *E3S Web of Conferences* (Vol. 280). EDP Sciences. https://doi.org/10.1051/e3sconf/202128004007

Velasquez, W., & Alvarez-Alvarado, M. S. (2021). Outdoors evacuation routes algorithm using cellular automata and graph theory for uphills and downhills. *Sustainability, 13*(9), 4731. https://doi.org/10.3390/su13094731

Villarraga-Flórez, L. F., Rodríguez-Piñeros, S., & Martínez-Cortés, O. G. (2016). Social science in forestry curricula: A case study of Colombia forestry programs. *Sustainability, 8*(1). https://doi.org/10.3390/su8010036

Westerman, J. W., Acikgoz, Y., Nafees, L., Depillis, E., & Westerman, J. (2020). The sustainable development goals and business students' preferences: An exploratory study. *Journal of Business Ethics Education, 17*, 99–114. Accessed from https://www.scopus.com/inward/record.uri?eid=2-s2.0-85104741930&partnerID=40&md5=93041cf8d0c483ae9bc72039018e262d

Willicks, F., Stehling, V., & Haberstroh, M. (2018). Sustainable development goals meet "Third mission": The engineers without borders challenge in Germany. In *ASEE Annual Conference and Exposition, Conference Proceedings, 2018-June*. Accessed from https://www.scopus.com/inward/record.uri?eid=2-s2.0-85051192723&partnerID=40&md5=05d7efca68c99b96ec6265636a4baad7

Worden, K., Bull, L. A., Gardner, P., Gosliga, J., Rogers, T. J., Cross, E. J., Papatheou, E., Lin, W., & Dervilis, N. (2020). A brief introduction to recent developments in population-based structural health monitoring. *Frontiers in Built Environment, 6*. https://doi.org/10.3389/fbuil.2020.00146

Yaccob, N. S., Yunus, M. M., & Hashim, H. (2022). The integration of global competence into Malaysian English as a second language lessons for quality education (fourth United Nations sustainable development goal). *Frontiers in Psychology, 13*. https://doi.org/10.3389/fpsyg.2022.848417

Yang, J., Yan, P., Xia, D., & Geng, Q. (2008). Analysis of spatial clustering optimization. *Geospatial Information Science, 11*(4), 302–307. https://doi.org/10.1007/s11806-008-0109-5
Yuan, X., Yu, L., Wu, H., She, H., Luo, J., & Li, X. (2022). Sustainable development goals (SDGs) priorities of senior high school students and global public: Recommendations for implementing education for sustainable development (ESD). *Education Research International, 2022*. https://doi.org/10.1155/2022/2555168
Zamora-Polo, F., & Sánchez-Martín, J. (2019). Teaching for a better world. Sustainability and sustainable development goals in the construction of a change-maker university. *Sustainability (Switzerland), 11*(15). https://doi.org/10.3390/su11154224
Zamora-Polo, F., Sánchez-Martín, J., Corrales-Serrano, M., & Espejo-Antúnez, L. (2019). What do university students know about sustainable development goals? A realistic approach to the reception of this UN program amongst the youth population. *Sustainability (Switzerland), 11*(13). https://doi.org/10.3390/su11133533
Zolkin, A. L., Munister, V. D., Domracheva, E. A., Faizullin, R. V., & Kovaleva, K. A. (2022). Application of machine learning for optimization of operational processes in industrial drum units. In I. Kovalev, A. Stupina, & A. Voroshilova (Eds.), *AIP conference proceedings* (Vol. 2467). American Institute of Physics. https://doi.org/10.1063/5.0092463

Environmentally Sustainable Conscious Students When Starting Up

Juan Santiago Calle Piedrahita, Julián David Colorado Agudelo, Victoria Blanquised Rivera, Laura Elena Zapata Jiménez, and Sandra Milena Malavera Pineda

Abstract Environmental sustainability presents a new challenge for organizations due to the scarcity of resources, and entrepreneurs recognize the need for action. This research aims to determine the characteristics of entrepreneurial students with regard to environmental sustainability. The study used a descriptive cross-sectional positivist paradigm and administered a questionnaire to 181 university students that focused on 6 variables related to environmental criteria. The hypothesis proposed that working students have a greater concern for environmental issues. The main finding was that students in training demonstrated a critical and positive awareness of environmental sustainability. Additionally, those who work exhibited a greater acceptance of engaging in activities that promote environmental sustainability. In conclusion, the research demonstrates that students who work recognize the importance of environmental sustainability in business processes and the conservation of both renewable and non-renewable resources necessary for ensuring the continuity of organizations. These students also acknowledge the impact of undergraduate training on decision-making across various areas of entrepreneurship.

Keywords Entrepreneurship · Sustainability · Environmental · Education · Environment

1 Introduction

The recognition of entrepreneurship as a catalyst for improving community well-being, ecosystem protection, and sustainability is growing. However, it is no longer sufficient to consider entrepreneurship as a mere technical option. It needs to be

J. S. Calle Piedrahita (✉) · J. D. Colorado Agudelo · V. Blanquised Rivera ·
L. E. Zapata Jiménez · S. M. Malavera Pineda
Medellín, Colombia
e-mail: juan.callepi@amigo.edu.co; julian.coloradogu@amigo.edu.co; victoria.blanquisedri@amigo.edu.co; laura.zapataji@amigo.edu.co; sandra.malaverapi@amigo.edu.co

C. Rego et al. (eds.), *Transitioning to a Circular Economy*, Studies on Entrepreneurship, Structural Change and Industrial Dynamics,
https://doi.org/10.1007/978-3-031-77661-8_10

established as a fundamental subject in the education of professionals across all disciplines. This will solidify entrepreneurship as a scientific phenomenon supported by theoretical knowledge, with clear boundaries and opportunities for development, evolution, and application. Such an approach ensures that entrepreneurship continues to gain credibility within academia. Throughout history, entrepreneurship has undergone various stages of evolution. Initially, the focus was on identifying the psychological and sociological traits of entrepreneurs. Later, the emphasis shifted towards creating financially driven companies to address job market gaps. Subsequently, entrepreneurship was studied as a step-by-step process.

Today, entrepreneurship is more oriented towards generating sustainable and community-beneficial alternatives while preserving and conserving the ecosystem. It also encompasses rational resource management and social development, aiming to promote growth that does not harm the environment and ensures equitable productivity for individuals. Socially conscious individuals, driven by a vision of continuous improvement and environmental preservation, seek to establish economic, environmental, and social capital through entrepreneurship. This involves fostering a rational and expanding economy based on renewable resources, as well as reducing the use of non-renewable resources. It requires managing scenarios and alternatives that promote responsible usage and social justice. The guidelines and frameworks for sustainable entrepreneurship arise from the United Nations Sustainable Development Goals (SDGs).

However, it is crucial for entrepreneurs to determine which SDG(s) to prioritize, as it is not feasible to fully address all 17 SDGs established by the UN. In the initial stages, entrepreneurship lacks the necessary administrative, technical, political, and social structures to fully align with these parameters. Furthermore, there is a lack of a definitive model or system of indicators to help entrepreneurs assess the correctness of their work. The global context presents clear challenges concerning waste generation, limited environmental regeneration capacity, and excessive use of non-renewable resources. Therefore, sustainable entrepreneurship is essential in mitigating these issues and promoting responsible and sustainable economic growth. According to a study by the Inter-American Development Bank (IDB), sustainable businesses can achieve long-term profitability while minimizing their environmental and social impact. On average, companies focused on financial sustainability outperform their competitors in terms of profitability and long-term growth (Banco Interamericano de Desarrollo (BID), 2018). These businesses can also enhance their reputation and attract customers and employees who value sustainability. A study by Cone Communications revealed that 63% of American consumers expect companies to take responsibility for improving the environment and society (Cone Communications, 2015). There have been several studies that have conducted environmental sensitivity diagnoses on university students, including that of Mediavilla et al. (2020), who analyzed dimensions such as environmental valuation, estimation of the seriousness of environmental problems, and environmental hyperopia.

Sustainable businesses also play a role in job creation and economic development. The European Commission's report on sustainable entrepreneurship states that it can foster job creation and economic growth at the local and regional levels

(European Commission, 2016). Additionally, sustainable businesses can reduce their carbon footprint and promote the circular economy. A study by the Ellen MacArthur Foundation concluded that adopting circular economy principles in production and consumption could reduce greenhouse gas emissions by 50% by 2030 (Ellen MacArthur Foundation, 2015).

Furthermore, undergraduate students exhibit favorable environmental attitudes, and it is expected that their environmental behaviors align with these attitudes. A study conducted in Spain by Fraj and Martínez (2005) discovered that verbal or intentional ecological commitments are more prevalent than actual ecological commitments. Several studies have indicated that intentions often surpass actions, aiming to transform academic populations into active agents of social change, fostering a society with more balanced relationships with nature and promoting socio-ecological transformation.

Another finding reveals that university students demonstrate a considerable level of concern and sensitivity toward environmental problems. This interest stems from the relevance of the issue and reflects their aspiration to contribute to effective solutions. It signifies constructive pessimism, wherein individuals possess a present-oriented mindset that prioritizes remediation in favor of the environment (Kaida & Kaida, 2015).

One effective academic strategy to address the lack of environmental awareness involves incorporating training processes that emphasize global issues rather than solely focusing on local ones. By investigating the causes, consequences, effects, and particular interests that arise within social or organizational contexts and analyzing the health implications, a more comprehensive and restorative understanding of environmental dynamics can be attained. This approach enables individuals to fulfill their commitments toward the Sustainable Development Goals (Chen, 2020).

Education serves as the most influential tool for fostering collective awareness. By incorporating theory, case studies, problem-based learning, and educational visits to businesses and local areas to analyze environmental issues, essential elements can be integrated into curriculum updates that align with the Sustainable Development Goals. This way, progress can be made toward establishing a socially sustainable, supportive, and environmentally conscious society that aligns with global environmental objectives (Schneiderhan & Bogner, 2020).

Entrepreneurship possesses the capacity for creativity, innovation, and the willingness to undertake risks, transforming these qualities into opportunities that drive organizational growth. It is important to note that not all entrepreneurs become business owners. Education can equip individuals with the necessary tools to become entrepreneurs without necessarily founding a business, as well as instill the ability to undertake endeavors without being business owners. The opportunity for socially and environmentally conscious entrepreneurship stems from the passion to engage in meaningful activities within an organization. While universities and governments often prioritize teaching entrepreneurship as a means of generating employment and financial resources, they occasionally overlook the social and environmental dimensions. The theoretical approach to entrepreneurship is often considered straightforward, yet practical experiences often surpass theoretical frameworks. Moreover, the literature has not extensively explored environmental concepts from an environmental perspective. However, interdisciplinary courses related to the Sustainable Development Goals are taught within education to identify the indicators organizations must meet. In summary, creativity contributes to learning from entrepreneurs' experiences, presenting a fresh perspective on environmental considerations (Filippin & Dreher, 2004; Gulla & Brasethvik, 2002).

Research-oriented courses within universities should not solely focus on presenting methodologies as structured steps but also emphasize their contribution to investigating the financial consequences of entrepreneurship from an environmental and social standpoint. Simultaneously, organizations must identify collaborators who can comprehend the micro and macro environment to promote eco-entrepreneurship or ecosystems (Brones et al., 2017; Camon Luis & Celma, 2020; Fernandez-Lopez et al., 2020; Hernandez & Munoz, 2022; Tatsuo, 2010). Entrepreneurial education provides a perspective that views problems as opportunities.

Environmental entrepreneurship centers around establishing leadership and governance policies, analyzing finances from a capital market and seed capital perspective, fostering a culture supported by social norms and successful case studies, focusing on services that leverage infrastructure and professional services with a fresh environmental and social vision, and understanding the priorities of NGOs. Additionally, environmentally-focused entrepreneurs rely on a talented workforce with an education rooted in social and environmental thinking. Lastly, they conduct market analyses from the customer and network perspective.

Education plays a crucial role in teaching the consequences of the first and second industrial revolutions, including the detrimental impact of natural resource exploitation. Scarce resources today can be attributed to waste and a disregard for the Earth and its derivatives, perpetuated by the false belief that natural resources are limitless. While students are taught to prioritize improving the quality of life from an environmental and social perspective, some individuals still hold anthropocentric beliefs that prioritize exploiting natural resources for personal satisfaction or needs (Liu & Chen, 2020). Embracing an entrepreneurial mindset can serve as a solution rather than a cause for preventing environmental degradation (York & Venkataraman, 2010).

Furthermore, sustainable entrepreneurship contributes to innovation and the development of new technologies and solutions. A report by the Global Entrepreneurship Monitor highlights how sustainable entrepreneurship can drive innovation in areas such as renewable energy, waste management, and sustainable agriculture (Global Entrepreneurship Monitor, 2016). Lastly, sustainable entrepreneurship has a positive impact on the health and well-being of local communities. A study conducted by the University of Exeter in 2017 concluded that sustainable businesses can enhance the health and well-being of workers and the community at large by implementing ethical and fair business practices (University of Exeter, 2017). Sustainable entrepreneurship offers numerous long-term economic, social, and environmental benefits. By prioritizing sustainability, businesses can discover new opportunities, enhance their reputation, and contribute to a fairer and more sustainable world. Therefore, it is proposed that this study aims to determine the characteristics of environmentally sustainable entrepreneurial students as an initial step toward their future professional endeavors.

The research aimed to determine whether students in the labor force possess a higher level of awareness regarding the challenges associated with long-term environmental sustainability management. This analysis greatly contributes to understanding the reality of students in relation to entrepreneurship, specifically emphasizing the crucial aspect of environmental sustainability. The aim of this research is to examine the importance of environmental sustainability, which involves recognizing the finite nature of our planet's resources and the need for their prudent conservation and management. Through the study of sustainability, we can delve into methods to minimize resource depletion and encourage sustainable practices. Furthermore, by exploring environmental sustainability, we can enhance our understanding of the causes and effects of climate change, and devise strategies to reduce greenhouse gas emissions and adapt to its impacts. Acquiring knowledge about environmental sustainability enables us to appreciate the significance of biodiversity, actively work towards its preservation, and prevent the loss of species and habitats. Moreover, studying environmental sustainability allows us to evaluate the repercussions of human activities on ecosystems, identify potential risks to human health, and develop sustainable solutions that promote both ecological integrity and human well-being. Lastly, it empowers us to explore balanced and sustainable approaches to achieve economic growth, social equity, and environmental protection.

2 Corporate Environmental Sustainability

Sustainability and entrepreneurship are closely intertwined concepts that are increasingly interconnected in today's world. Sustainable entrepreneurship is centered around creating profitable businesses that not only prioritize environmental preservation but also have a positive social impact. Sustainability, on the other hand, refers to the ability to strike a balance between economic development, environmental protection, and social well-being. Thus, sustainable entrepreneurship seeks innovative solutions to address environmental and social challenges while generating economic value.

In sustainable entrepreneurship, various practices can be employed, including the use of renewable energy, incorporating recycled materials, responsible waste management, and the implementation of fair and ethical labor practices, among others. Furthermore, sustainable entrepreneurship benefits not only the environment and society but also the entrepreneurs and their companies. Sustainable businesses tend to be more appealing to consumers and investors, leading to increased business opportunities and long-term profitability. In essence, sustainable entrepreneurship represents an innovative and responsible approach to creating companies that can generate economic benefits while making a positive impact on the environment and society.

Fornell and Larcker (1981) provide a framework for statistically measuring the variables that influence organizational performance, specifically when implementing ecological management. Cohen and Winn (2007) identify four types of market imperfections: inefficient firms, externalities, defective price mechanisms, and information asymmetries.

Entrepreneurship can offer a solution to environmental problems (Dean & McMullen, 2007) and contribute to sustainable development (Hall et al., 2010). While sustainability is typically examined in large companies in terms of their environmental impact and financial resources, small businesses often overlook the social complexities of sustainability (Hockerts & Wüstenhagen, 2010). Statistical indicators assist in interpreting the sustainability approach by considering economic integration, environmental contributions, and social factors (Henseler et al., 2015), thereby providing a more comprehensive understanding of the context. However, many organizations tend to prioritize economic management while neglecting the analysis of their social and environmental impacts. To achieve sustainable entrepreneurship, organizations must grasp the environmental reality and implement it operationally with managerial acceptance and commitment (Shepherd & Patzelt, 2011).

Competitiveness is attained by identifying the strategic conditions of the organization through internal and external analysis (Barney, 1991) and making well-informed decisions. However, the environmental impact remains an under-researched variable in strategic management. Thus, organizations should focus on establishing policies and strategies aligned with the green economy (Gast et al., 2017) to indirectly contribute to economic development (Biagi et al., 2021; Zhang & Kong, 2022; Portillo, 2017). In other words, companies must analyze and envision the future of their organization from an ecological and environmental management perspective (Bansal, 2000) while managing processes rationally (DiMaggio & Powell, 1983).

From another standpoint, organizations are confronted with resource scarcity because they fail to incorporate the finite nature of environmental resources into their management practices. Company managers must fundamentally change their approach and integrate the positive or negative impact of environmental effects into financial management. This can be achieved through disclosure standards that shed light on the reasons behind the lack of care for environmental resources (Parrish, 2010).

In turn, case studies can contribute to comprehending the reality of an organization and engender new discussions that offer solutions to other companies grappling with similar issues, aiming for tangible or conceptual resolutions (Eisenhardt, 1989). For instance, current research establishes the connections between entrepreneurship and environmental quality for organizational development, which ultimately leads to economic well-being (Philip et al., 2022).

Sustainable entrepreneurship can bring about a positive societal impact through job creation and the enhancement of working conditions. Sustainable companies often prioritize corporate social responsibility (CSR) by ensuring that their business practices adhere to ethical standards and fairness. This includes fair wage policies, fostering diversity and inclusion in the workplace, and implementing measures to enhance worker safety and health.

Another crucial aspect of sustainable entrepreneurship is innovation. Sustainable businesses often need to discover innovative approaches to reduce their environmental and social impact, leading to the development of more efficient technologies and processes (Lv et al., 2022). These innovations can have broader applications and be adopted by other businesses and organizations, thereby positively impacting society (Haveman & Nedzhvetskaya, 2022; Zhang & Kong, 2022; Pangarso et al., 2022) (See Fig. 1).

Sustainable entrepreneurship taught to university students has a greater tendency towards a sense of belonging that seeks to apply the benefits of sustainability. In addition, there has been a 58% increase in entrepreneurship from a sustainability perspective, as students have greater awareness of the social, economic, and environmental impact. However, this is complemented by the students' work experience, as the knowledge gained in their work contributes to the creation of sustainable entrepreneurship in both internal and external environments. The above indicates that social responsibility is a fundamental policy that must be incorporated into universities for the creation and dissemination of scientific knowledge to ensure the contribution of competencies, qualities, and capabilities in communities, thus bringing about sustainable development. The responsibility of universities evolves from generating programs of altruism or generosity to exercising a knowledge management process that commits them socially from the very exercise of their primary functions.

Therefore, it is important to highlight that universities are social organizations that generate processes of change to provide improvements in the community, which requires a shift from rhetoric to practice, seeking greater frequency in practical and real-time exercise. When the need for an academic program is established, it must respond to and impact the needs of the region where processes of change are expected to be developed. Therefore, a program is designed that explicitly incorporates the concept of fourth-sector enterprises, supported by social responsibility for

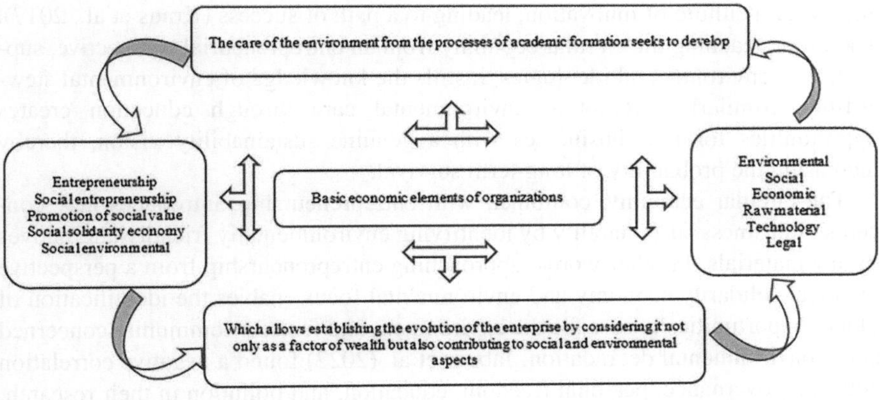

Fig. 1 Relationship of entrepreneurship in its context. Source: Author's elaboration

the creation of sustainable entrepreneurship that responds to local needs and is aligned with university principles.

Moreover, sustainable entrepreneurship can serve as an effective means of addressing major environmental and social challenges of our time, such as climate change, poverty, and inequality. Sustainable businesses can play a significant role in transitioning to a viable economic model (Errigo et al., 2022). In summary, sustainable entrepreneurship represents an innovative and responsible approach to establishing businesses that generate economic benefits while also making positive contributions to the environment and society. By prioritizing sustainability, businesses can discover new opportunities, enhance their reputation, and contribute to overall sustainability.

3 Circular Economy

Universities should incorporate education on the circular economy with a green focus to instill an understanding that resources are limited. This type of economy allows for non-renewable resources to coexist with human society by identifying new opportunities to undertake with environmental ideologies (Hockerts & Wüstenhagen, 2010; Soomro et al., 2020; Tsai, 2019) that contribute to environmental stewardship. Embracing an ecological orientation promotes long-term resource conservation and the generation of sustainability-oriented businesses, thereby achieving competitive and differentiated value. This entails obtaining products from a green economy (Andersén, 2022). Furthermore, entrepreneurs with education in social and environmental aspects have a greater potential to identify the development of green products that generate value (Gregori & Holzmann, 2022).

A new entrepreneur equipped with theoretical knowledge of sustainability has a higher likelihood of establishing companies with an environmental care focus (de Oliveira Lacerda et al., 2017). This means that courses on business sustainability, entrepreneurship, and motivation have an impact on environmental sustainability and foster a culture of innovation, leading to a path of success (Kraus et al., 2017). Therefore, teaching the circular economy from an entrepreneurial perspective, supported by environmental ideologies, instills the knowledge of environmental stewardship. Similarly, promoting environmental care through education creates opportunities for new businesses with a genuine sustainability vision, thereby increasing the probability of long-term survival.

The circular economy, combined with entrepreneurship, introduces new concepts of business sustainability by identifying environmentally friendly alternatives to raw materials. In other words, approaching entrepreneurship from a perspective of social solidarity economy and environmental focus enables the identification of global opportunities to care for the environment and foster a community concerned about environmental degradation. Jabeur et al. (2022) found a negative correlation between governance, personal freedom, education, and pollution in their research. Additionally, from a sustainability learning perspective, the circular economy supports the identification of factors where business orientation positively impacts environmental collaboration (Bouguerra et al., 2023).

Dynamic entrepreneurship contributes to the short-term development of organizations while maintaining a long-term vision aimed at attaining a competitive advantage and creating value that improves the quality of life (both socially and environmentally). Entrepreneurial students should analyze the environmental impact of their actions to establish circular economy policies. Understanding the effects of implementing the seven "R's" (reduce, reuse, recycle, repair, refurbish, redesign, and remanufacture) alters perceptions regarding the economic effects on social well-being. Universities must incorporate cross-cutting themes into courses that address environmental and social content, fostering motivation to care for the environment rather than perceiving it as an unlimited resource. Additionally, courses should focus on understanding the business history of the region to acquire knowledge from the challenges faced during their establishment, contributing to experiences that transform difficulties into eco-entrepreneurship (Nikolaou et al., 2018).

Similarly, the circular economy presents an opportunity to promote planetary care by recognizing that society contributes to its destruction. Therefore, universities that teach sustainability foster decision-making with strategic approaches to generate environmental and social value, leading to positive long-term economic value (Iqbal & Khizar, 2022). Rizos et al. (2015) examined the barriers and opportunities for small and medium-sized enterprises (SMEs) in transitioning to a circular economy, identifying the challenges they face and proposing solutions to help them adopt more sustainable practices. Kirchherr et al. (2018) also analyzed the issues in the European Union related to the implementation of the circular economy, highlighting factors such as the lack of appropriate incentives and policies, the lack of collaboration among different actors, and the lack of knowledge about the benefits it offers for sustainability.

The study conducted by Jayawardhena and Raluca (2016) investigated the influence of green entrepreneurship on environmental sustainability. Their findings suggest that green entrepreneurship has the potential to contribute significantly to the shift towards a more sustainable economy. By fostering innovation and facilitating the development of efficient and environmentally friendly technologies, green entrepreneurship can play a crucial role in promoting environmental sustainability.

Theoretical entrepreneurship serves as a catalyst for aspiring entrepreneurs and future business innovators, equipping them with essential elements to navigate the risks associated with creating environmentally and socially conscious companies. Teaching students about green economics enables entrepreneurs to identify crucial factors for achieving environmental sustainability (Omri & Afi, 2020). Additionally, incorporating examples of sustainable entrepreneurship with a focus on the environment, such as eco-friendly farms, green products, environmentally friendly packaging, clean production, and green initiatives, reinforces the significance of nature, resources (Yacob et al., 2019), and sustainable agriculture (Mupfasoni et al., 2018).

Elkington (1998) introduced the concept of the "triple bottom line," emphasizing the need to balance economic profitability with a company's social and environmental impact. According to Elkington, sustainable businesses can generate economic

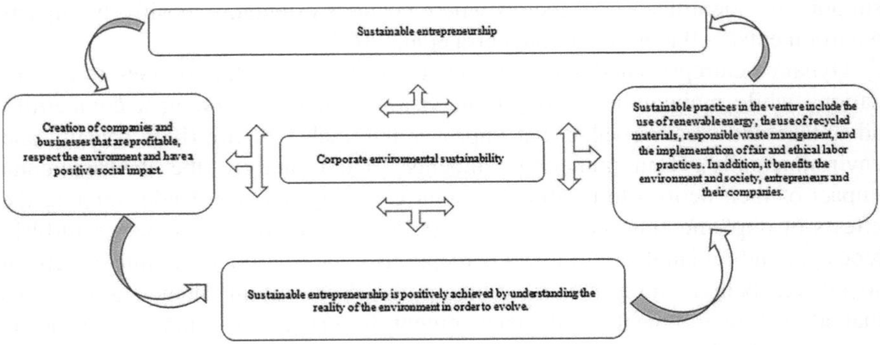

Fig. 2 Context of corporate environmental sustainability. Source: Author's elaboration

benefits while making positive contributions to society and the environment. Supporting this notion, Akkuş and Çalıyurt (2022) argue that sustainable businesses have the potential to contribute to the achievement of the United Nations' Sustainable Development Goals by addressing environmental and social challenges through innovation and corporate social responsibility.

Ho: Working students have a greater concern for environmental issues.

Environmental sustainability supported by entrepreneurs is generating involvement in the circular economy as an opportunity for new businesses that contribute to sustainability. Additionally, young people have a greater awareness of the negative effects of environmental destruction in favor of eco-entrepreneurial development. In other words, modern managers must conceptualize environmental and human issues alongside financial aspects. There is a transformation of behavior towards deeper new consumption habits, tastes, and needs (See Fig. 2). Entrepreneurs with a philosophy of environmental sustainability tend to make decisions that involve greater analysis of the impact of the organization's production on profitability generators, net present value, and weighted average cost of capital (WACC).

Therefore, the circular economy is an economic model aimed at minimizing waste and maximizing resource utilization through the design of regenerative systems. In this model, products, materials, and resources are kept in use for as long as possible, reducing waste and the need for new raw materials. Walker et al. (2021) proposed a framework for evaluating the sustainability of the circular economy, considering aspects such as resource efficiency, waste minimization, social equity, and economic viability. This evaluation framework can assist decision-makers in assessing the sustainability of circular economy practices and identifying areas for improvement. Thus, the circular economy, as a comprehensive approach, can address current environmental and social challenges. Adopting the circular economy can assist businesses and governments in reducing waste, increasing resource efficiency, and promoting innovation.

4 Methodology

In this study, we employed a quantitative, cross-sectional, and descriptive approach to examine behavior patterns among students undergoing entrepreneurship training in their final semesters of professional education. The research encompassed a statistical analysis conducted during the second quarter of 2022. The target population consisted of 341 enrolled students, and a non-probabilistic convenience sampling method was utilized, with a margin of error of 5.0%, a confidence level of 95%, and a presumed heterogeneity of 50%. This sampling strategy resulted in a sample size of 181 students who completed the analysis questionnaire, as indicated in Table 1.

Automation was employed to process and analyze information at every stage of the project. The statistical analysis utilized the SPSS program, where identified variables were subjected to descriptive analysis. For quantitative variables, statistical formulas such as mean, median, mode, minimum, maximum, and percentiles were employed. Qualitative variables were measured using percentages and represented through tables.

The primary hypothesis investigated in this study aimed to determine whether students who work exhibit a heightened awareness of environmental issues when engaging in entrepreneurial activities. To test this hypothesis, a bivariate statistical test, specifically the chi-square test, was conducted for each of the 33 variables outlined in the questionnaire. This examination directly explored the relationship between students with employment and their level of awareness.

In the final phase, a multivariate analysis was conducted using the logistic regression technique. This approach seeks to establish a relationship between students who hold jobs and their demonstration of environmental awareness while undertaking entrepreneurial endeavors.

The implementation of the research encountered challenges in gauging the students' willingness to complete the survey. Moreover, a feasibility analysis was conducted on the questionnaire by comparing the responses to odd and even questions within two groups on the same day, resulting in similar averages. Furthermore, reliability was assessed through a pre-posttest, which indicated that the instrument exhibited both structural integrity and positive significance. Regarding the composition of the participants, 53.6% were enrolled in the marketing course, 39.8% in the entrepreneurship course, and 6.6% in the research process courses (see Table 2).

The age range of the students in the study varied from 17 to 35 years, with an average age of 22.7 years. The participants' mean age was 22.7 years, accompanied by a standard deviation of 3.4. The age distribution showed a concentration of students between 21 and 24 years, with 23 years being the most prevalent age. Notably, the age range spanned from 17 to 35 years, encompassing both younger and older students. In terms of percentages, the majority of students (80%) fell within the 22–24 age range, placing them within the 24 and younger percentile (See Table 3).

Furthermore, the distribution of gender percentages shows an equal representation of men and women. Men account for 51.4% of the student population, which is equivalent to 93 individuals, while women make up 48.6% of the total.

Table 1 Questionnaire

#	Variables			
1	Country			
2	University you are studying at			
3	Faculty you are studying at			
4	Degree			
5	Year of study			
6	Age			
7	Gender			
	Select one of the following answer options	Negative	Neutral	Positive
8	Too much attention is paid to environmental problems			
9	For me, being a sustainable entrepreneur implies more advantages than disadvantages			
10	A career as a sustainable entrepreneur is attractive to me			
11	If I had the opportunity and the resources, I would like to set up a sustainable company			
12	Being a sustainable entrepreneur could be of great satisfaction for me			
13	Among the different employment options, I would like to be a sustainable entrepreneur			
14	I feel morally obligated to act in a way that respects the environment			
15	I would feel guilty if I did not act in a respectful way with the environment			
16	I would be a better person if I acted in a way that respects the environment			
17	Setting up a sustainable company could be easy for me			
18	If I wanted to, I could easily pursue a career as a sustainable entrepreneur			
19	I can control the processes of creating a new sustainable company			
20	I know the practical details necessary to set up a sustainable company			
21	Environmental pollution affects my health			
22	Environmental problems have consequences for my life			
23	I worry about environmental problems			
24	I can see with my own eyes that the environment is deteriorating			
25	Environmental problems will be a risk to the future of my children			
26	There is excessive talk about environmental problems			
27	The attention that is being given to the greenhouse effect is exaggerated			
28	Saving endangered species is an unnecessary luxury			

(continued)

Table 1 (continued)

#	Variables			
29	A better environment begins with oneself			
30	People who do not take the environment into account try to escape their responsibility			
31	I intend to start a company that solves sustainability problems in the next 5 years			
32	I am determined to create a sustainable company in the future			
33	My professional goal is to become a sustainable entrepreneur			

Source: Author's elaboration

Table 2 Number of students per course

University course	Students
Entrepreneurship	72
Research process	12
Marketing	97
Total	181

Source: Author's elaboration

Table 3 Age of the students

					Percentiles		
Mean	Median	Mode	Minimum	Maximum	40	60	80
22.7	22	23	17	35	22	23	24

Source: Author's elaboration

5 Results and Discussion

In terms of environmental variables, the surveyed students express a direct concern for addressing these issues, supported by *p*-values below 0.05. Interestingly, there are divergent opinions between employed and unemployed individuals regarding issues such as giving excessive attention and talking excessively about environmental problems, considering greenhouse effects, and saving endangered species as an unnecessary luxury. On average, 104 out of 181 respondents believed that work should continue to address these effects. Concerning moral and respectful sentiments toward working for the environment, 167 employed students responded positively, while others did not. As for environmental problems directly linked to health, 100% of the responses acknowledged the negative consequences for life and the future risk of sequelae. These findings indicate the urgent need to address these issues, with a *p*-value of 0.000. Based on Table 4, a significant statistical association is observed between variables indicating a conscious relationship with environmental issues among students who were employed in 2022. To identify causal factors, variables aligned with the Sustainable Development Goals (SDGs) concerning employment relationships are considered.

Regarding questions about establishing sustainable enterprises, no direct relationship is found for inquiries concerning interest in creating sustainable companies

Table 4 Factors associated with the conscious relationship with environmental problems when undertaking business of students enrolled in business administration and who have a labor relationship

		Employment relationship			
		No	Yes		
Variables	Categories	%	%	X^2	p
Too much attention is paid to environmental problems	Negative	47.8	63.2	4.09	0.04
	Neutral	0	0		
	Positive	52.2	36.8		
There is excessive talk about environmental problems	Negative	47.8	63.2	4.09	0.04
	Neutral	0	0		
	Positive	52.2	36.8		
The attention that is being given to the greenhouse effect is exaggerated	Negative	47.8	63.2	9.48	0.01
	Neutral	6	0		
	Positive	46.3	36.8		
Saving endangered species is an unnecessary luxury	Negative	53.7	63.2	1.56	0.21
	Neutral	0	0		
	Positive	46.3	36.8		
I feel morally obligated to act in an environmentally friendly manner	Negative	6	0	17.9	0
	Neutral	0	15.8		
	Positive	94	84.2		
I would feel guilty if I did not act in an environmentally friendly way	Negative	0	0	11.8	0
	Neutral	0	100		
	Positive	15.8	84.2		
I would be a better person if I acted in an environmentally friendly way	Negative	0	0	11.8	0
	Neutral	0	15.8		
	Positive	100	84.2		
People who do not take into account the environment try to escape their responsibility	Negative	0	0	11.8	0
	Neutral	0	15.8		
	Positive	100	84.2		
Environmental pollution affects my health	Negative	0	0	0	0
	Neutral	0	0		
	Positive	100	100		
Environmental problems have consequences for my life	Negative	0	0	0	0
	Neutral	0	0		
	Positive	100	100		
I care about environmental problems	Negative	0	0	0	0
	Neutral	0	0		
	Positive	100	100		

(continued)

Table 4 (continued)

Variables	Categories	Employment relationship No %	Employment relationship Yes %	X^2	p
I can see with my own eyes that the environment is deteriorating	Negative	0	0	0	0
	Neutral	0	0		
	Positive	100	100		
Environmental problems will be a risk to my children's future	Negative	0	0	0	0
	Neutral	0	0		
	Positive	100	100		
A better environment starts with yourself	Negative	0	0	0	0
	Neutral	0	0		
	Positive	100	100		
Being a sustainable entrepreneur could be of great satisfaction for me	Negative	0	0	11.8	0
	Neutral	0	41.1		
	Positive	100	58.9		
Among the different employment options, I would like to be a sustainable entrepreneur	Negative	0	0	11.8	0
	Neutral	0	41.1		
	Positive	100	58.9		
Starting a sustainable company could be easy for me	Negative	6	0	7.75	0.02
	Neutral	25.4	21.1		
	Positive	68.7	78.9		
I can control the processes of creating a new sustainable company	Negative	0	0	2.39	0.12
	Neutral	31.3	21.1		
	Positive	68.7	78.9		
If I wanted to, I could easily pursue a career as a sustainable entrepreneur.	Negative	6	0	10.1	0.01
	Neutral	25.4	15.8		
	Positive	68.7	84.2		
For me, being a sustainable entrepreneur implies more advantages than disadvantages.	Negative	0	0	11.8	0
	Neutral	0	15.8		
	Positive	100	84.2		
A career as a sustainable entrepreneur is attractive to me	Negative	0	0	11.8	0
	Neutral	0	15.8		
	Positive	100	84.2		
If I had the opportunity and the resources, I would like to start a sustainable company	Negative	0	0	11.8	0
	Neutral	0	15.8		
	Positive	100	84.2		
If I tried to start a sustainable company, I would have a high probability of success.	Negative	0	0	6.04	0.01
	Neutral	31.3	15.8		
	Positive	68.7	84.2		
I know the practical details necessary to set up a sustainable company	Negative	11.9	0	18.2	0
	Neutral	25.4	15.8		
	Positive	62.7	84.2		

(continued)

Table 4 (continued)

Variables	Categories	Employment relationship No %	Employment relationship Yes %	X^2	p
I intend to start a company that solves sustainability problems in the next 5 years	Negative	0	0	0.37	0.54
	Neutral	1.7	3.4		
	Positive	98.3	96.6		
I am determined to create a sustainable company in the future	Negative	0	0	0.37	0.54
	Neutral	1.7	3.4		
	Positive	98.3	96.6		
My professional goal is to become a sustainable entrepreneur	Negative	6.5	0	6.2	0.05
	Neutral	1.6	3.4		
	Positive	91.9	96.6		

Source: Author's elaboration

for the future, with p-values greater than 0.005. However, the majority of students surveyed demonstrate positive attitudes toward sustainable entrepreneurship, which also exhibits a statistically significant association with a p-value below 0.005. Ideas such as deriving satisfaction from creating sustainable businesses, constructing sustainable companies, and easily controlling their processes align well with the training received by business administrators. Moreover, these ideas present more advantages than disadvantages in such business ventures.

Based on the findings presented in Table 5, a multivariate analysis was conducted to examine the relationship between various factors. The independent variable chosen was the presence of a labor relationship with a company among business administration students by the year 2022. The dependent variables included several questions pertaining to environmental issues and the establishment of sustainable ventures. This analysis successfully established a significant statistical relationship between the variables, as evidenced by a p-value <0.005.

The results indicate that business administration students have displayed a keen interest in developing sustainable businesses. They possess knowledge about the requisite procedures, technical prerequisites, and administrative and managerial processes involved in creating and advancing sustainable ventures. Their responses reflect confidence in achieving business success while addressing environmental concerns. This demonstrates a noteworthy level of awareness among students regarding environmental issues and highlights the creation and sustenance of global sustainable companies as one of the potential solutions.

The conducted survey focused on specific inquiries related to environmental concerns, aligned with the objectives outlined in the Sustainable Development Goals. All participants (100%) responded affirmatively when asked about the impact of environmental issues on their lives, their worries about the environment, the observation of environmental degradation, and the potential risks it poses to future generations. Out of the total respondents, 114 individuals (63.0%) reported being employed in either public, private, or mixed organizations registered in

Table 5 Multivariate Logistic Regression analysis examining the relationship between business administration students' employment status and their consciousness of environmental problems in entrepreneurship

Variables		Category	Sig Anova	Sig coefficient
Presents employment relationship	Starting a sustainable company could be easy for me	Negative	0.001	0.032
		Neutral		
		Positive		
	If I wanted to, I could easily pursue a career as a sustainable entrepreneur.	Negative		0.003
		Neutral		
		Positive		
	I can control the processes of creating a new sustainable company	Negative		0.122
		Neutral		
		Positive		
	If I tried to start a sustainable company, I would have a high probability of success.	Negative		0.014
		Neutral		
		Positive		
	I know the practical details necessary to set up a sustainable company	Negative		0.000
		Neutral		
		Positive		

Source: Author's elaboration

Colombia, while the remaining 37.0% were not currently employed during the study period. The study reveals that among students without work experience, 37.0% believe that environmental deterioration will pose a threat to the future of the next generation if left unaddressed. On the other hand, among students with work experience, a larger percentage (63.0%) share the same perception regarding the potential risks to the environment in the future.

Regarding the question concerning excessive discussions about environmental problems in the survey, opinions were divided among the employed and unemployed population. One hundred and four respondents answered negatively, whereas 77 responded positively. Among the employed individuals, 36.8% believe that sufficient attention is being given to the environmental issue, whereas 52.2% of the unemployed share the same belief. Furthermore, 40.3% of the surveyed population expressed dissatisfaction with the efforts made to address the greenhouse effect.

Results indicating agreement or strong agreement (scoring higher than 4) suggest that students demonstrate sensitivity towards sustainability-related issues and have a self-perception of environmental consciousness by adopting pro-environmental beliefs and values. The majority of evaluated students display significant awareness of environmental sustainability. Conversely, students who are part of the workforce tend to prioritize quality of life from a social and environmental development perspective rather than solely focusing on economic aspects. They recognize the reality of limited resources and the negative consequences of their depletion, prompting them to promote long-term viability for their organizations through fostering social and environmental awareness. Therefore, organizations that engage students in their workforce development and provide them with university courses on environmental and social issues witness improvements in their organizational development.

It is worth noting that the non-employed individuals (37.0%), who are predominantly students, exhibit concerns about environmental problems. Conversely, the employed individuals (63.0%) express a higher level of awareness regarding environmental conservation due to their adherence to organizational policies, as they witness the deterioration of the planet. This implies that students are worried about the environmental situation, while those who are employed demonstrate greater awareness of the issue due to their involvement in organizational practices. However, 47.8% of the unemployed population consider environmental communication to be exaggerated, while 52.2% perceive exaggeration in advertising as a positive aspect since it keeps them informed about the detrimental impact humans have on the planet. Among the employed population, 36.8% share the view that advertising is exaggerated, yet acknowledge its positive contribution to encouraging actions within their organizations.

Currently, establishing the connection between entrepreneurship, the environment, and sustainability holds significant importance in generating alternative training options for undergraduate students and continuously improving quality of life, p-value <0.005. Students develop an awareness of environmental sustainability when they identify cross-cutting topics in their classes that contribute to the integration of sustainability principles throughout different curricular contents, including planning, organization, direction, and control, p-value <0.005. Although sometimes challenges may arise in integrating diverse knowledge systems, students learn to create synergy from these situations. Experienced teachers aim to bridge the gap between theoretical knowledge and practical applications, encouraging students to generate inquiries from an environmental and social perspective. In other words, they learn to identify constructive criticism to address environmental challenges given the limitations of resources, p-value <0.005.

Likewise, 63.2% of employed individuals acknowledge that advertising related to environmental matters is excessive, yet it does not impact their decision-making process significantly. Conversely, 36.8% of those who are employed perceive attention towards the greenhouse effect in a positive light, whereas 46.3% of the unemployed individuals consider it beneficial to highlight cases involving greenhouse effects. Among the surveyed workers, 63.2% view the exaggeration of greenhouse effect attention as unfavorable, while 47.8% of the non-employed population regard the presentation of greenhouse effects negatively.

The hypothesis stands confirmed: individuals who are employed exhibit a higher level of concern for environmental issues. The researchers' initial hypothesis, suggesting that students enrolled in entrepreneurship and marketing courses, who also have work affiliations, demonstrate a greater awareness of environmental matters. Employing the chi-square statistical formula p-value of <0.05 indicates a significant association between the variables under study. This indicates that awareness is cultivated during educational processes and further strengthened within the workplace through the execution of commercial activities and the recognition of environmental needs inherent to various positions held by the enrolled students. If the research has been conducted appropriately, the Null hypothesis can be rejected with a probability of committing a Type 1 error of only 0.5%, given that $p < 0.005$.

Students in the workforce exhibit greater awareness of environmental issues due to their exposure to corporate policies. Conversely, those who are not yet part of the workforce may have a lower perception of environmental issues, but they are aware of the need to intervene in future processes to mitigate environmental damage.

Students actively engage in conversations with teachers during class discussions on how to transform raw materials with an environmentally friendly philosophy. They also employ social media platforms to promote eco-friendly products and raise awareness for environmental care. Sustainable entrepreneurs in educational courses initiate debates that foster creative and innovative ideas contributing to a shift away from a purely economic perspective. Changing the economic mindset leads to sustainable growth, which ultimately promotes a favorable quality of life for both society and the environment. Entrepreneurs embracing environmental sustainability philosophies instill strategies in their collaborators that prioritize environmental care throughout the entire lifecycle of materials, from acquisition to use and reintegration into the organization. In this manner, a systemic conceptualization of the environment is fostered, indirectly promoting economic development.

6 Conclusions

In conclusion, higher education plays a vital role in promoting sustainable practices and cultivating responsible and environmentally conscious business leaders. It is crucial for higher education institutions to establish mechanisms that facilitate environmental education, which should be dynamic, creative, effective, and efficient, creating collaborative spaces for engagement with the government, society, and businesses.

Students who receive entrepreneurship classes demonstrate a heightened awareness of environmental issues. They exhibit concerns for environmental protection and actively seek renewable raw materials for their ventures. While conducting financial feasibility analysis, they incorporate environmental considerations into various scenarios, emphasizing the importance of factoring in environmental and social variables in investment decision-making. Entrepreneurial students advocate for environmental care among various stakeholders and initiate discussions on sourcing short-term renewable and biodegradable resources.

Greater knowledge of environmental and social issues leads to their practical application within organizations. Students who are conscious of environmental conservation promote financial activities that reward environmentally friendly practices as a form of appreciation. However, if solely viewed from a financial perspective, this may potentially affect the organization's future reality by compromising its competitive advantage. Students contribute to environmental conservation by recognizing the consequences of resource mismanagement and its long-term impact on the environment. Additionally, universities offering entrepreneurship courses should integrate environmental topics across the curriculum to foster long-term awareness.

Based on the study's findings, it can be concluded that academic training processes focused on environmental sustainability and entrepreneurship serve as pedagogical projects that encourage analysis, understanding, and engagement with global challenges and opportunities in international, national, local, and business contexts. These initiatives create spaces for interaction and participation, where potential solutions aligned with sustainable development are proposed to preserve and enhance the community's quality of life. Education also establishes a reference framework grounded in respect for individual, entrepreneurial, industrial, and sectoral diversity and autonomy, while considering ethical, political, and socio-cultural aspects to facilitate sustainable environmental management.

Universities should offer education on sustainability and entrepreneurship to promote sustainable development and inspire entrepreneurial projects. This equips students with knowledge, values, and attitudes that align with the requirements and needs of communities, facilitating the integration of diverse fields of knowledge, skills, and disciplines to enhance interdisciplinary problem-solving. Moreover, it nurtures a professional approach that aligns with the realities of communities and the current state of the environment. This, in turn, fosters the generation and dissemination of knowledge, understanding, and interaction between science, technology, and techniques within an environmental, sustainable, and entrepreneurial culture.

The study encountered several limitations related to the availability of student databases. Access to these databases was restricted, which limited the researchers' ability to gather comprehensive data for analysis. Furthermore, the participation of students in the study was relatively low due to concerns regarding potential perception bias. This limitation arose from the possibility that students might have been influenced by their perception of the study's objectives or desired outcomes, leading to biased responses or a lack of genuine engagement. Another limitation was associated with the use of self-administered questionnaires, which may not have effectively motivated students to respond consciously. This approach potentially resulted in incomplete or unreliable data, as students may have provided hurried or inaccurate responses without carefully considering their answers. Additionally, the study faced limitations due to legal constraints, particularly governed by Law 2157 of 2021 (habeas data) and the Personal Data Protection Law or Law 1581 of 2012. These laws regulate the handling and use of personal data, which imposed restrictions on the researchers' access to certain types of information and hindered the full utilization of available data sources.

Universities have the potential to explore various research areas in the future pertaining to environmental sustainability. Some prospective avenues for research could be: Universities can delve into the research and development of more efficient and eco-friendly technologies aimed at reducing environmental impacts across sectors such as energy, agriculture, industry, and transportation. Future research can focus on deepening the understanding of climate change effects, analyzing the most effective measures for mitigation and adaptation, and exploring innovative solutions to curb greenhouse gas emissions. Universities can investigate strategies to safeguard and preserve biodiversity, including the identification of ecologically significant areas, promotion of sustainable practices in natural resource utilization, and research on ecosystem restoration for degraded environments. Universities can explore inventive approaches to educate and enhance awareness regarding the significance of environmental sustainability. This can involve promoting sustainable behaviors and lifestyles through innovative educational programs and initiatives.

References

Akkuş, Y., & Çalıyurt, K. (2022). The role of sustainable entrepreneurship in UN sustainable development goals: The case of TED talks. *Sustainability, 14*(13), 8035. https://doi.org/10.3390/su14138035

Andersén, J. (2022). An attention-based view on environmental management: The influence of entrepreneurial orientation, environmental sustainability orientation, and competitive intensity on green product innovation in Swedish small manufacturing firms. *Organization & Environment, 35*(4), 627–652. https://doi.org/10.1177/10860266221101345

Banco Interamericano de Desarrollo (BID). (2018). *Sostenibilidad financiera en empresas: la clave para el éxito a largo plazo*. Accessed from https://publications.iadb.org/publications/spanish/document/Sostenibilidad-financiera-en-empresas-La-clave-para-el-exito-a-largo-plazo.pdf

Bansal, P. R. (2000). Why companies go green: A model of ecological responsiveness. *The Academy of Management Journal, 43*(4), 717–736. https://doi.org/10.5465/1556363

Barney, J. (1991). Firm resources and sustained competitive advantage. *Journal of Management, 17*(1), 99–120. https://doi.org/10.1177/014920639101700108

Biagi, B., Dettori, B., Paci, R., & Usai, S. (2021). Economic development in Sardinia: Overcoming the insularity gap. *Rivista Internazionale Di Scienze Sociali, 129*(2), 165–186. https://doi.org/10.26350/000518_000067

Bouguerra, A., Hughes, M., Cakir, M. S., & Tatoglu, E. (2023). Linking entrepreneurial orientation to environmental collaboration: A stakeholder theory and evidence from multinational companies in an emerging market. *British Journal of Management, 34*(1), 487–511. https://doi.org/10.1111/1467-8551.12590

Brones, F. A., de Carvalho, M. M., & de Zancul, E. S. (2017). Reviews, action and learning on change management for ecodesign transition. *Journal of Cleaner Production, 142*(1, SI), 8–22. https://doi.org/10.1016/j.jclepro.2016.09.009

Camon Luis, E., & Celma, D. (2020). Circular economy. A review and bibliometric analysis. *Sustainability, 12*(16). https://doi.org/10.3390/su12166381

Chen, M. (2020). Effects of psychological distance perception and psychological factors on pro-environmental behaviors in Taiwan: Application of construal level theory. *International Sociology, 35*(1), 70–89. https://doi.org/10.1177/0268580919881870

Cohen, B., & Winn, M. I. (2007). Market imperfections, opportunity and sustainable entrepreneurship. *Journal of Business Venturing, 22*(1), 29–49. https://doi.org/10.1016/j.jbusvent.2004.12.001

Comisión Europea. (2016). *Entrepreneurship competence framework: A reference framework to promote entrepreneurship education across Europe*. Accessed from https://op.europa.eu/en/publication-detail/-/publication/7a0e57a8-7e1f-11e6-b6ed-01aa75ed71a1

Cone Communications. (2015). *2015 cone communications/ebiquity global CSR study*. Accessed from https://www.conecomm.com/research-blog/2015-cone-communications-ebiquity-global-csr-study.

de Oliveira Lacerda, R. T., Klein, B. L., Fulco, J. F., Santos, G., & Bittarello, K. (2017). Innovative integration between incubated startups and universities to achieve continuous competitive advantages in dynamic enviroments. *NAVUS-Revista de gestao e tecnologia, 7*(2), 78–96. https://doi.org/10.22279/navus.2017.v7n2.p78-96.497

Dean, T. J., & McMullen, J. S. (2007). Toward a theory of sustainable entrepreneurship: Reducing environmental degradation through entrepreneurial action. *Journal of Business Venturing, 22*(1), 50–76. https://doi.org/10.1016/j.jbusvent.2005.09.003

DiMaggio, P. J., & Powell, W. W. (1983). The iron cage revisited: Institutional isomorphism and collective rationality in organizational fields. *American Sociological Review, 48*(2), 147–160. https://doi.org/10.2307/2095101

Eisenhardt, K. M. (1989). Building theories from case study research. *The Academy of Management Review, 14*(4), 532–550. https://doi.org/10.2307/258557

Elkington, J. (1998). *Cannibals with forks: The triple bottom line of 21st century business*. Capstone.

Ellen MacArthur Foundation. (2015). *Growth within: A circular economy vision for a competitive Europe*. Accessed from https://www.ellenmacarthurfoundation.org/assets/downloads/publications/Ellen-MacArthur-Foundation-Growth-Within.pdf

Errigo, A., Choi, J. K., & Kissock, K. (2022). Techno-economic-environmental impacts of industrial energy assessment: Sustainable industrial motor systems of small and medium-sized enterprises. *Sustainable Energy Technologies and Assessments, 49*, 101694. https://doi.org/10.1016/j.seta.2021.101694

Fernandez-Lopez, S., Dios-Vicente, A., & Rodeiro-Pazos, D. (2020). Working capital management and profitability on the ecologic and non-ecologic cheese sector in Spain. Revista española de estudios agrosociales y pesqueros-Reeap, Editorial Ministerio de Agricultura, Pesca y Alimentación. *Madrid (España), 255*, 15–45.

Filippin, K. K., & Dreher, L. (2004). Major hazard risk assessment for existing and new facilities. *Process Safety Progress, 23*(4), 237–243. https://doi.org/10.1002/prs.10045

Fornell, C., & Larcker, D. F. (1981). Evaluating structural equation models with unobservable variables and measurement error. *Journal of Marketing Research, 18*(1), 39–50. https://doi.org/10.1177/002224378101800104

Fraj, E., & Martínez, E. (2005). El nivel de conocimiento medioambiental como Factor moderador de la relación entre la actitud y el comportamiento ecológico. *Investigaciones Europeas de Dirección y Economía de la Empresa, 11*(1), 223–243.

Gast, J., Gundolf, K., & Cesinger, B. (2017). Doing business in a green way: A systematic review of the ecological sustainability entrepreneurship literature and future research directions. *Journal of Cleaner Production, 147*, 44–56. https://doi.org/10.1016/j.jclepro.2017.01.065

Global Entrepreneurship Monitor. (2016). *GEM 2015/2016 global report*. Accessed from https://www.gemconsortium.org/report/49912

Gregori, P., & Holzmann, P. (2022). Entrepreneurial practices and the constitution of environmental value for sustainability. *Business Strategy and the Environment, 31*(7), 3302–3317. https://doi.org/10.1002/bse.3077

Gulla, J. A., & Brasethvik, T. (2002). A model-driven ERP environment with search facilities. *Data & Knowledge Engineering, 42*(3), 327–341. https://doi.org/10.1016/S0169-023X(02)00051-4

Hall, J. K., Daneke, G. A., & Lenox, M. J. (2010). Sustainable development and entrepreneurship: Past contributions and future directions. *Journal of Business Venturing, 25*(5), 439–448. https://doi.org/10.1016/j.jbusvent.2010.01.002

Haveman, H. A., & Nedzhvetskaya, N. (2022). Community, enterprise, and self-help: The coevolution of capitalism and non-profit and for-profit businesses in Britain and Germany. In *The corporation: Rethinking the iconic form of business organization*. Emerald Publishing.

Henseler, J., Ringle, C. M., & Sarstedt, M. (2015). A new criterion for assessing discriminant validity in variance-based structural equation modeling. *Journal of the Academy of Marketing Science, 43*(1), 115–135. https://doi.org/10.1007/s11747-014-0403-8

Hernandez, M., & Munoz, P. (2022). Reformists, decouplists, and activists: A typology of ecocentric management. *Organization & Environment, 35*(2), 282–306. https://doi.org/10.1177/1086026621993204

Hockerts, K., & Wüstenhagen, R. (2010). Greening goliaths versus emerging Davids—Theorizing about the role of incumbents and new entrants in sustainable entrepreneurship. *Journal of Business Venturing, 25*(5), 481–492. https://doi.org/10.1016/j.jbusvent.2009.07.005

Iqbal, M. J., & Khizar, H. M. U. (2022). The epistemology of Entrepreneurial Responsible Orientation (ERO): Theory, conceptualization, and future research. *SAGE Open, 12*(2), 21582440221089320.

Jabeur, S. B., Ballouk, H., Arfi, W. B., & Khalfaoui, R. (2022). Machine learning-based modeling of the environmental degradation, institutional quality, and economic growth. *Environmental Modeling & Assessment, 27*(6), 953–966. https://doi.org/10.1007/s10666-021-09807-0

Jayawardhena, C., & Raluca, S. (2016). The impact of green entrepreneurship on environmental sustainability. *Journal of Business Research, 69*(5), 1723–1730.

Kaida, N., & Kaida, K. (2015). *Facilitating pro-environmental behavior: The role of pessimism and anthropocentric environmental values.* https://doi.org/10.1007/s11205-015-0943-4

Kirchherr, J., Piscicelli, L., Bour, R., & Kostense-Smit, E. (2018). Barriers to the circular economy: Evidence from the European Union. *Resources, Conservation and Recycling, 135*, 1–10. https://doi.org/10.1016/j.ecolecon.2018.04.028

Kraus, S., Burtscher, J., Niemand, T., Roig-Tierno, N., & Syrjä, P. (2017). Configurational paths to social performance in SMEs: The interplay of innovation, sustainability, resources and achievement motivation. *Sustainability, 9*(10). https://doi.org/10.3390/su9101828

Liu, W., & Chen, J. (2020). Modified two major environmental values scale for measuring Chinese children's environmental attitudes. *Environmental Education Research, 26*(1), 130–147. https://doi.org/10.1080/13504622.2019.1697431

Lv, M., Zhang, H., Georgescu, P., Li, T., & Zhang, B. (2022). Improving education for innovation and entrepreneurship in Chinese technical universities: A quest for building a sustainable framework. *Sustainability, 14*(2), 595. https://doi.org/10.3390/su14020595

Mediavilla, M. E., Medina Quintana, S., & González López, I. (2020). Diagnóstico de sensibilidad medioambiental en estudiantes universitarios. *Educación y Educadores, 23*(2), 179–197. https://doi.org/10.5294/edu.2020.23.2.2

Mupfasoni, B., Kessler, A., & Lans, T. (2018). Sustainable agricultural entrepreneurship in Burundi: Drivers and outcomes. *Journal of Small Business and Enterprise Development, 25*(1), 64–80. https://doi.org/10.1108/JSBED-03-2017-0130

Nikolaou, I. E., Tsagarakis, K. P., & Tasopoulou, K. (2018). An examination of ecopreneurs' incentives through a combination between institutional and resource-based approach. *Management of Environmental Quality: An International Journal, 29*(2), 195–215. https://doi.org/10.1108/MEQ-01-2017-0004

Omri, A., & Afi, H. (2020). How can entrepreneurship and educational capital lead to environmental sustainability? *Structural Change and Economic Dynamics, 54*, 1–10. https://doi.org/10.1016/j.strueco.2020.03.007

Pangarso, A., Sisilia, K., Setyorini, R., Peranginangin, Y., & Awirya, A. A. (2022). The long path to achieving green economy performance for micro small medium enterprise. *Journal of Innovation and Entrepreneurship, 11*(1), 1–19. https://doi.org/10.1186/s13731-022-00209-4

Parrish, B. D. (2010). Sustainability-driven entrepreneurship: Principles of organization design. *Journal of Business Venturing, 25*(5), 510–523. https://doi.org/10.1016/j.jbusvent.2009.05.005

Philip, L. D., Emir, F., & Alola, A. A. (2022). The asymmetric nexus of entrepreneurship and environmental quality in a developing economy. *International Journal of Environmental Science and Technology, 19*(8), 7625–7636. https://doi.org/10.1007/s13762-021-03670-y

Portillo, E. (2017). Supremacy of the economic power for sustainable business development. *Revista CICAG, 15*(1), 110–132. Accessed from http://ojs.urbe.edu/index.php/cicag/article/view/338/273

Rizos, V., Behrens, A., Kafyeke, T., Hirschnitz-Garbers, M., & Ioannou, A. (2015). *The circular economy: barriers and opportunities for SMEs*. Accessed from https://publications.jrc.ec.europa.eu/repository/bitstream/JRC95512/ldna27248enn.pdf

Schaltegger, S., Lüdeke-Freund, F., & Hansen, E. G. (2016). Business cases for sustainability: The role of business model innovation for corporate sustainability. *International Journal of Innovation and Sustainable Development, 10*(4), 284–309. https://doi.org/10.1504/IJISD.2012.046944

Schneiderhan, O., & Bogner, F. (2020). The relation between knowledge acquisition and environmental values within the scope of a biodiversity learning module. *Sustainability, 12*(5), 2036. https://doi.org/10.3390/su12052036

Shepherd, D. A., & Patzelt, H. (2011). The new field of sustainable entrepreneurship: Studying entrepreneurial action linking "what is to be sustained" with "what is to be developed". *Entrepreneurship Theory and Practice, 35*(1), 137–163. https://doi.org/10.1111/j.1540-6520.2010.00426.x

Soomro, B. A., Ghumro, I. A., & Shah, N. (2020). Green entrepreneurship inclination among the younger generation: An avenue towards a green economy. *Sustainable Development, 28*(4), 585–594. https://doi.org/10.1002/sd.2010

Tatsuo, K. (2010). An analysis of the eco-efficiency and economic performance of Japanese companies. *Asian Business & Management, 9*(2), 209–222. https://doi.org/10.1057/abm.2010.3

Tsai, S.-B. (2019). Green energy and applications. *Applied Sciences-Basel, 9*(15). https://doi.org/10.3390/app9153088

Universidad de Exeter. (2017). *The impact of sustainable business practices on workers' wellbeing*. Accessed from https://www.exeter.ac.uk/news/research/title_562474_en.html

Walker, A. M., Vermeulen, W. J., Simboli, A., & Raggi, A. (2021). Sustainability assessment in circular inter-firm networks: An integrated framework of industrial ecology and circular supply chain management approaches. *Journal of Cleaner Production, 286*, 125457. https://doi.org/10.1016/j.jclepro.2020.125457

Yacob, P., Wong, L. S., & Khor, S. C. (2019). An empirical investigation of green initiatives and environmental sustainability for manufacturing SMEs. *Journal of Manufacturing Technology Management, 30*(1), 2–25. https://doi.org/10.1108/JMTM-08-2017-0153

York, J. G., & Venkataraman, S. (2010). The entrepreneur–environment nexus: Uncertainty, innovation, and allocation. *Journal of Business Venturing, 25*(5), 449–463. https://doi.org/10.1016/j.jbusvent.2009.07.007

Zhang, M., & Kong, Z. (2022). A tripartite evolutionary game model of emergency supplies joint reserve among the government, enterprise and society. *Computers & Industrial Engineering, 169*, 108132. https://doi.org/10.1016/j.cie.2022.108132

Part III
Construction and Energy as CE strategies

Architecture for Deconstruction: A Methodological Approach to Eco-construction

Carlos Almeida Marques

Abstract Sustainability, construction, environment, materials, recycling, demolition, reuse, waste management, life cycle, and renewal are keywords that have become part of the architectural project. All of them reflect the process of transition in the construction industry from a concept of linear economy to a circular one.

The demolition process usually has a direct correlation with the construction sector. This circumstance has gained relevance in recent decades, in which the development of new infrastructure and the construction of new housing implied a significant increase in construction and demolition waste. While there are still barriers to overcome in design for the deconstruction and reuse of reclaimed materials and products, there are sound economic reasons for using demolition as an economic resource. Deconstruction as a methodological approach can bring benefit from the point of view of public policies, namely those that may be related to climate change mitigation by enhancing minimize construction waste, aiding the local economy, avoiding new materials and retaining the cultural value of existing materials, reduce transport cost, reduce greenhouse gas emissions and reduce demand on natural and virgin resources.

Keywords Architecture · Deconstruction · Circular economy · Life cycle · End of the loop

1 Introduction

Deconstruction as a methodological approach to both the architectural project and building production is directly related to two other concepts: the *Life Cycle Assessment,* which is defined according to International Standard ISO 14040 as a "compilation and evaluation of the inputs, outputs and the potential environmental

C. A. Marques (✉)
CIAUD-FAUL and CAPP-ISCSP-UL, Lisbon, Portugal
e-mail: camarques@caam.com.pt

C. Rego et al. (eds.), *Transitioning to a Circular Economy*, Studies on Entrepreneurship, Structural Change and Industrial Dynamics,
https://doi.org/10.1007/978-3-031-77661-8_11

impacts of a product system throughout its life cycle" (AIA, 2010, p. 12) and the concept of *Circular Economy* understood as a "systems solution framework that tackles global challenges like climate change, biodiversity loss, waste, and pollution" (Ellen MacArthur Foundation).[1]

All these three concepts are associated with two important strategies: *end-of-life management*, which takes multiple forms, including deconstruction, component and material reuse and recycling of debris, and the idea of *closing the loop*. "An industrial ecosystem mimics a natural ecosystem through an interacting web of inputs, processes and wastes, which close the loop by turning waste back into resources" (Morgan & Stevenson, 2005, p. 11).

As a result of the strong growth in construction activities, a number of environmental impacts of great importance are produced, both from the perspective of the consumption of natural resources—those that are inert and others like water, for example—or by the effect of producing a large volume of waste—cement, metals, plastics, etc.—resulting from demolition processes.

According to the Global Material Resources Outlook to 2060 report (OECD, 2019), the global consumption of raw materials will almost double by 2060 due to "an increase in total materials use between 2017 and 2060 of 78 Gt" (2019, p. 18), as the world economy grows and living standards increase, exacerbating the environmental overloading we are experiencing today. Construction materials are set to dominate resource consumption in fast-growing developing economies, with building material-related emissions projected to increase by 3.5 to 4.6 Gt CO2_eq/year by 2060 (UN, 2022, p. 72).

The OECD report (2019) also underlines the significant environmental consequences brought by the large-scale use of construction materials. As an example, the projected environmental impacts of concrete show that they have an effect on most indicators: acidification, climate change, cumulative energy demand, eutrophication, freshwater ecotoxicity, human toxicity, land use, and photochemical oxidation (OECD, 2019, p. 196). This report estimates "the environmental impacts of concrete production further increase to 2060 (…). In particular, GHG emissions related to concrete production double to 2060" (OECD, 2019, p. 197) and are as large as the emissions of seven select materials, most of them intensively used as construction materials like aluminum, copper, iron, lead, manganese, nickel, and zinc.[2]

As the building sector will continue to grow to meet citizens' needs for housing and workplaces, this growth must be in alignment with the Paris Agreement and other international strategies to accelerate the transition to net-zero carbon like the Human Settlements Pathway of the Marrakech Partnership for Global Climate Action, which has explicitly called for the following targets:

---

[1] Ellen MacArthur Foundation, Circular economy introduction . https://ellenmacarthurfoundation.org/topics/circular-economy-introduction/overview (Accessed 30-03-2023).

[2] Source: CML's CMLCA model, based on the central baseline scenario of the OECD ENV-Linkages model.

- By 2030, the built environment should halve its emissions whereby 100 percent of new buildings must be net-zero carbon in operation.
- By 2050, all new and existing assets must be net zero across the whole life cycle, including operational and embodied emissions (MPGCA, 2021, p. 9).

The latest assessment report from the Intergovernmental Panel on Climate Change (IPCC, 2022) for the mitigation working group (AR6) sent a clear message that the building and construction industry offers significant global mitigation potential for reaching the Paris Agreement by improving existing buildings' efficiency integrating renewables in buildings, and decarbonizing production of building materials (IPCC, 2022).

In the process of finding methods to deal with the historical failures and disruptive pressures of the construction industry, some organizations have established programs to incentivize best practices or work within the current limitations of the industry, while others have been trying to radically restructure and reengineer the industry (Razkenaria et al., 2020).

The demolition process usually has a direct correlation with the construction sector. This circumstance has gained relevance in recent decades, in which the development of new infrastructure and the construction of new housing implied a significant increase in construction and demolition waste. This reality is most evident when it comes to the rehabilitation of existing buildings, mainly housing buildings.

For this reason, waste management has been the target of various policies for sustainability, with the aim of achieving adequate management of materials resulting from construction and demolition processes, aiming to protect, preserve and improve the quality of the environment; to protect human health; to ensure prudent, efficient, and rational use of natural resources; to reduce pressure on the regenerative capacity of ecosystems; to promote the principles of the circular economy; to strengthen the use of renewable energy; to increase energy efficiency; to reduce dependence on imported resources; to provide new economic opportunities; and to foster competitiveness in the construction industry.

This shift in economic paradigms has, as a corollary, the recognition that a developed country is not the one that generates the most waste but the one that generates less waste, or at least that best treats it. The application of these paradigms in the field of construction has been favoring the adoption of another approach for the architectural project, based on the concept of *architectural deconstruction,* which is the subject of this paper.

2 The Economics of Construction: Waste Generation

There is a strong link between economic development and materials use, as materials are an important input for all production processes. Additionally, a particularly strong link between economic development, investment, construction activity and demand for construction materials can also be established (OECD, 2019).

At the end-of-life management, the waste from construction, renovation, and demolition activities (typically labeled as construction and demolition C&D), generated in the processes of constructing new buildings and renovating or demolishing existing buildings, become a loss or a cost as much as an economic opportunity.

Waste generation and treatment in the European Union (EU) and several non-member countries is a relevant economic issue. Waste is defined as "any substance or object which the holder discards or intends or is required to discard" (EU, 2008, Directive 2008/98/EC Article 3(1)). Potentially it represents an enormous loss of resources in the form of both materials and energy. In addition, the management and disposal of waste can have serious environmental impacts. A landfill, for example, takes up land space and may cause air, water, and soil pollution, while incineration may result in emissions of air pollutants.

Therefore, EU waste management policies aim to reduce the environmental and health impacts of waste and to improve the EU's resource efficiency. The long-term goal of these policies is to reduce the amount of waste generated and, when waste generation is inevitable, to promote it as a resource and achieve higher levels of recycling and the safe disposal of waste.

Simultaneous, the Directive establishes the legislative framework for the handling of waste in the Community, introducing an approach that takes into account the whole life cycle of products and materials and not only the waste phase, and focuses on reducing the environmental impacts of waste generation and waste management, thereby strengthening the economic value of waste (Directive 2008/98/EC).

The waste as an economic value has a strong correlation with the concept of so-called "urban mining," "in which *urban* means the area inside city borders and *mining* is understood as the extraction of secondary metal resources from obsolete, and in that sense accessible, reservoirs situated in these areas" (Krook & Baas, 2013, p. 2). Cities hold vast amounts of materials that can be reused, and doing so can tackle the over-exploitation of scarce natural resources. To decrease the use of fossil fuels and raw materials, more effort can be directed towards recovering waste materials as well as valuable mineral products and reusing them in building construction. This circular approach makes the most of materials such as concrete, bricks, steel reinforcements, roofing materials, copper pipes and aluminum. It is a key component in making cities carbon neutral. Urban mining has advantages over primary mining: materials are already in the city where they are likely to be needed again and the environmental impact, particularly in land use, is clearly lower than mining for natural resources.

In fact, the built environment has a significant impact on many sectors of the economy, on local jobs and the quality of life. It requires vast amounts of resources and accounts for about 50% of all extracted material (EU, COM (2020) 98 final, p. 11). Globally, approximately 100 billion tonnes of construction, renovation and demolition waste is generated annually (OECD, 2019, p. 72).

In the EU, the production of construction and demolition waste accounted for of the total waste generated by economic activities and households in 2022, with very different figures among its members. France, Luxembourg, Malta, the Netherlands and Austria stand out all of them with figures equal to or greater than 64.0% of the total waste produced and, right below, Denmark and Germany, with more than 55.0%. In the case of Portugal, for the same year, the production of construction and demolition waste corresponded to 19.7% of the total, which we can consider to be a low waste generation compared to the majority of EU members (Table 1). Although the data refers to 2022, construction activities rebounded back to pre-pandemic levels in most major economies in 2021, thus increasing both raw materials consumption and waste production (UN, 2022).

The share of different economic activities and of households in total waste generation in 2022 is presented in Figure 1. In the EU, construction contributed 38.0% of the total waste generation, followed by mining and quarrying (23.0%), manufacturing (10.0%), wastewater (10.0%) and households (9.0%); the remaining 10.0% was waste generated from other economic activities, mainly services (5.0%) and energy (3.0%). Numbers show by far the relevance of construction among all other economic activities.

A significant part of the waste from mining and quarrying and from construction and demolition is classified as major mineral waste. Figure 2 shows that of the total waste generated in the EU in 2022, 64.0% (3.2 tonnes/inhabitant) was major mineral waste. The relative sahre of major mineral waste generated varied considerably between EU countries. Several of the Member States with particularly high levels of waste generated per inhabitant reported very high shares of waste from mining and quarrying activities, such as Finland, Romania and Sweden, and/or construction and demolition activities such as Luxemburg and Bulgaria; in these countries major mineral waste accounted for between 82.0% and 91.0% of all waste generated (Eurostat, 2022).

In a further development of our research, it will be of interest to evaluate how much of this volume of construction waste is reprocessed into products, materials or substances, whether for the original purpose or other purposes. Knowledge and guidelines have been created to develop policies and strategies to reduce the amount of waste generated, promoting this as a resource and achieving higher levels of reuse or recycling when waste generation is unavoidable within the safe disposal process.

3 An Architecture for Deconstruction

The conventional demolition of buildings carried out, for instance, by pulling down a building with a backhoe often leads to the mixing of various materials and contamination of non-hazardous components. Hence, advanced approaches aim at the deconstruction or selective dismantling of buildings, where a building is disassembled into various parts (Schultmann & Rentz, 2002, p. 1).

Taking this into consideration, it is important to highlight the difference between *conventional demolition* and *deconstruction*: (a) conventional demolition is associated with the partial or total destruction of the construction, not considering the possible reuse of the resulting materials or components; (b) in turn, deconstruction

Table 1 Waste generation by economic activities and households, 2022 (% share of total waste)

	Mining and quarrying	Manufacturing	Energy	Wastewater	Construction and demolition	Other economics activities	Households
EU	22.7	10.3	3.0	10.5	38.4	6.1	9.0
Belgium	0.0	12.6	0.2	24.7	41.4	13.4	7.7
Bulgaria	72.4	5.5	10.0		1.6	3.5	2.6
Czechia	0.3	12.5	1.0	13.5	43.8	17.9	11.0
Denmark	0.1	5.6	4.0	7.7	56.1	10.0	16.6
Germany	1.3	14.4	2.1	13.1	55.1	4.7	9.4
Estonia	11.5	17.2	50.2	3.5	9.0	6.0	2.6
Ireland	8.8	31.3	1.6	13.4	18.5	14.3	12.0
Greece (1)	11.6	14.5	5.8	13.5	33.5	7.8	13.4
Spain	7.0	12.0	0.8	19.6	30.6	9.9	20.0
France	0.1	5.3	0.2	7.3	71.6	5.3	10.0
Croatia	10.0	7.6	1.4	20.1	24.8	18.8	17.3
Italy	0.8	14.9	1.2	21.5	42.5	3.8	15.3
Cyprus	5.4	15.0	0.1	12.5	28.0	26.3	12.7
Latvia	0.1	12.3	1.1	36.3	12.5	9.9	28.0
Lithuania	0.4	19.2	3.8	23.1	9.9	18.7	24.9
Luxemburg	8.2	6.0	0.2	4.5	75.3	3.2	2.6
Hungary	0.2	16.0	18.3	4.0	37.4	6.2	17.9
Malta	1.5	0.9	0.0	6.4	77.7	6.1	7.4
Netherlands	0.1	10.8	0.5	7.7	64.7	9.5	6.8
Austria	0.0	7.9	0.6	3.9	76.6	4.9	6.1
Poland	36.5	14.8	7.9	13.9	12.3	6.6	8.0
Portugal	0.2	15.7	1.1	20.5	19.7	15.6	27.2
Romania	85.2	4.1	3.4	1.8	0.5	2.1	2.8
Slovenia	0.1	12.9	4.3	3.4	6.6	67.0	5.7

	Mining and quarrying	Manufacturing	Energy	Wastewater	Construction and demolition	Other economics activities	Households
Slovakia	1.2	27.0	6.7	9.4	9.8	26.5	19.4
Finland	76.4	9.4	0.9	1.0	9.6	1.2	1.5
Sweden (2)	77.3		1.2	5.2	8.3		2.6
Iceland (3)	0.0	24.2	0.0	2.0	3.6	31.0	39.2
Liechtenstein	0.0	0.6	0.0	0.3	91.8	0.1	7.3
Norway	1.6	13.9	2.0	8.7	39.3	18.7	15.9
Montenegro	21.0	1.8	26.6	0.7	18.4	11.2	20.2
North Macedonia	26.1	25.8	0.5	11.5	3.4	6.1	26.6
Serbia	93.2	0.6	3.7	0.5	0.4	0.3	1.3
Türkiye	24.1	20.3	25.4	0.3	0.0	5.3	24.6
Bosnia and Herzegovina	0.2	35.6	47.2	0.0	1.6	0.5	15.0
Kosovo (4)	0.3	10.2	84.6	0.4	2.5	2.0	0.0

Source: Eurostat (online data code: env_wasgen)
(1) Provisional data. (2) Confidential data on waste generated by manufacturing. (3) 2020 data. (4) This designation is without prejudice to possitions on status and is in line with UNSCR 1244/1999 and the ICJ Opinion on the Kosovo Declaration of Independence.

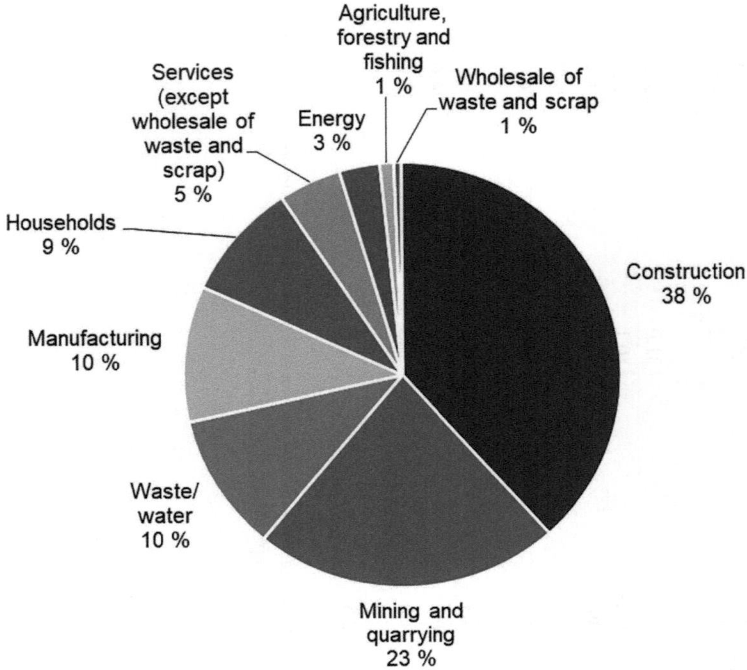

Fig. 1 Waste generation by economic activities and households, EU, 2022 (% share of total waste). Source: Eurostat (online data code: env_wasgen)

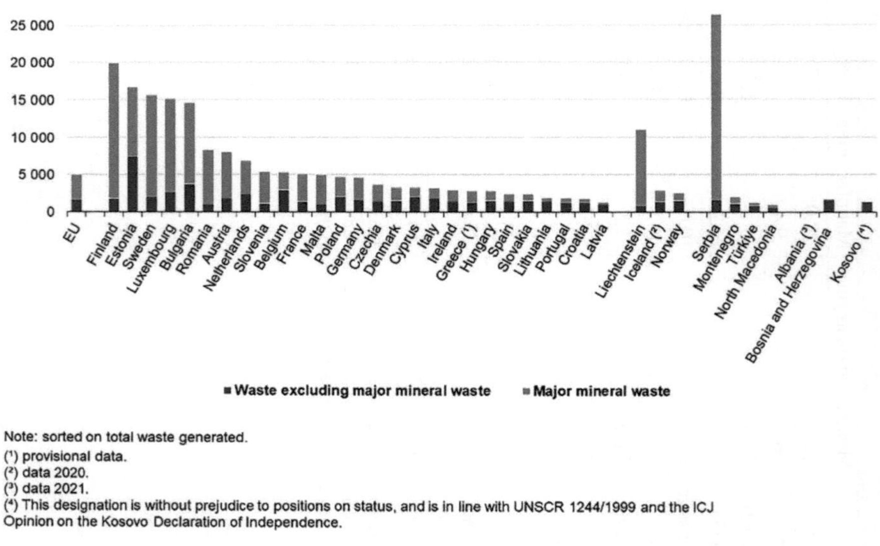

Fig. 2 Waste generation, 2022(kg per capita). Source: Eurostat (online data code: env_wasgen)

or selective demolition aims for the construction to be carefully dismantled in order to enable maximum recovery of materials and components, promoting its subsequent reuse or recycling.

Chris Morgan and Fionn Stevenson (2005) define deconstruction "as the dismantling of a building in such a manner that its component parts can be re-used" (p. 7). Deconstruction involves the careful disassembly of buildings in order to maximize recovered materials' reuse and recycling (Chini & Schultman, 2002; Nakajima & Russell, 2014). Although deconstruction takes longer than demolition, research indicates that deconstruction and reuse of materials and components can offer higher environmental and sometimes economic benefits than demolition and recycling (Geyer & Jackson, 2004).

An architecture strategy for deconstruction involves thinking of building conservation as the management of a physical asset (ISO 55000 Standards of the International Organization for Standardization) that involves all phases of the asset's life cycle, which is defined as the period from its conception to the end of its useful life. For this purpose, it is very useful to consider the "life cycle costing" as a "tool for assessing the total cost performance of an asset over time, including the acquisition, operating, maintenance, and disposal costs" (Langdon, 2007, p. 3).

4 Designing Architecture for Deconstruction

Designing architecture for deconstruction (DAD) at the start of a project enables one building, at the end of its useful lifespan, to be the resource for the next and helps "close the loop" for resource use. DAD can close the waste loop in two ways—firstly, by reusing existing construction elements and, secondly, by encouraging the designed elements to be reused easily and locally. Ideally, DAD should be contained as far as possible within a given regional area to minimize transportation and maximize the local economy (Morgan & Stevenson, 2005, p. 11).

Typically, the process of deconstruction and reuse of materials is more expensive because it lacks the economies of scale enjoyed by new construction. The design of new structures to facilitate the economic recovery of structural components for reuse is important to achieve such benefits (Gorgolewski, 2008).

Design is a crucial stage in the life cycle of a building, shaping innumerable aspects of its construction, performance and end-of-life. All material efficiency strategies can be facilitated, or in some cases, can only be achieved through appropriate design (Allwood et al., 2012; Carruth et al., 2011). According to Philip Crowther, DAD may, in the short term, add economic and possibly environmental costs. Still, on the much larger scale of the life cycle of resources, the long-term benefits are potentially much greater. (Crowther, 2002).

Deconstruction architectural design also approaches *Baukultur*'s concept, which "as an aspect of cultural identity and diversity, holistically embraces every human activity that changes the built environment, including every built and designed asset that is embedded in and relates to the natural environment. Thus, Baukultur refers to both detailed construction methods and large-scale transformations and developments, embracing traditional and local building skills as well as innovative techniques" (Davos Declaration, 2018).

5 Methodology

From a methodological perspective, architecture for deconstruction needs to concern itself with a holistic view of the goals above referred. From an operative point of view, it means to conceive the architectural object with its *buildability* assessment base, which is to say "the extent to which the design of a building facilitates its construction, subject to the overall requirements" (CIRIA, 1983, p. 26). The main references to an architecture for deconstruction aim to design for *adaptability*, *disassembly* and *reuse*, of which the concept we can resume as follows:

- *Adaptability* can be clustered in three strategies—"avoid" the use of high-volume, carbon-intensive materials, and replace them with low-carbon and circular alternatives; "shift" towards low-carbon and bio-based materials and use local, low-embodied-carbon materials; and improve construction with green infrastructures and biomaterial systems (UN, 2020). Other strategies aiming to design buildings to be adaptable will tend to extend their service lives, thus minimizing the energy and resources required over that period. Increasing the longevity of existing and new building stock and reusing existing components whenever possible are urgent priorities. Yet the lifespan of buildings and infrastructure is not solely determined by physical durability but also by social, cultural and economic factors (Cao et al., 2021). (p. 75) "The golden rule is to anticipate change, and to design buildings that make such changes easy to achieve" (Morgan & Stevenson, 2005, p. 24).
- *Disassembly* on a technical level is related to framing, prefabrication and modularity. Modularity can refer to modular construction, where large components of a building are produced off-site and then assembled into an entire building. The term is also sometimes used to refer to buildings designed for adaptation and repurposing for changing use. Designing a building for disassembly means conceiving a construction made of components. The strategy can involve the design of the component itself or of the building, in a way that facilitates the recovery of the component for reuse, providing material efficiency benefits in construction at both the component and whole building level. Design-for-disassembly and modular construction can enhance longevity while enabling dismantling at the end of life to retain the building elements' value and potential for reuse (Keena & Dyson, 2017).
- *Reuse* of building components is a strategy that avoids the impacts of primary materials production. Building design is an important factor in ensuring component reuse. Such design includes ensuring easy disassembly of components and standardization of joints (UN, 2020, p. 41). Reuse has a direct correspondence

with lifetime extension, which can allow additional savings of approx. 0.1 Gt CO2e in cumulative O2 emissions, mainly thanks to reuse. Lifetime extension reduces material consumption; it prolongs the use of older and less efficient buildings (UN, 2020, p. 48).

By designing for adaptability, disassembly and reuse, design practitioners are finding new opportunities early in the design process to reduce environmental impacts, conserve resources, and reduce costs (EPA, 2023).

According to Chris Morgan and Fionn Stevenson (2005), in practical terms, design for the deconstruction means to adopt a set of procedures, in particular, the following key principles:

- Know the place—nothing can replace "local knowledge" when projecting to a certain place.
- Minimize waste production by increasing the number of times a material can be reused.
- Use of natural materials or biodegradable composite materials.
- Reduction of costs with reused or recycled materials.
- Use of materials whose production or recycling is least harmful to the environment as a result of the pollution associated with its manufacture.
- Avoid excessive transport of materials by ensuring that the buildings are adapted to the prefabrication resources and systems of the site.
- Implementation of environmental comfort systems through the use of constructive techniques of passive energy recovery.
- Use of equipment with low energy consumption, associated with clean, renewable energy production systems.
- Prior planning of the deconstruction process in order to avoid the destruction of reusable building elements.
- Look for maximum flexibility in the spatial configuration and structure of the building seeking to adapt it to different usage patterns.
- Incorporate in the design of the work the participation of the client and the entire project team: architects, engineers, builders, and suppliers.
- Audit contractors and ensure that the initial briefing and training for demolition and deconstruction have taken place.
- The adoption of construction systems that allow easy removal of all its components.
- Add all alterations to drawings and specifications so there is an integrated set of "as built" drawings for maintenance and deconstruction purposes.

Designing for circularity is gaining traction across education and practice (UN, 2020, p. 78). Although there is a big difference between the "construction" and "product manufacture" worlds, lessons learned from other sectors can help improve construction practice and be adapted to the *buildability* we had associated with the architectural object.

The above concepts and key principles are to be understood as guidelines/suggestions for conducting and reinforcing the decisions made by architects, promoters, builders or politicians by providing them with a scientific justification for those decisions. The state of research about deconstruction is a work in progress needed to be reviewed to find answers that will improve and demonstrate the "green-ness" of the project as part of the decarbonization process.

6 Results

Eco-edification and the architecture for deconstruction are not simple procedures, both from the conceptual and technical-constructive point of view, since its main characteristic is to seek an efficient link between the construction process and the future deconstruction activity, both with a view to the useful purpose of each building. However, for some years now, public and private initiatives have been taking shape in this direction of construction that is more adapted to the preservation of the environment.

The Gaspar Frutuoso School, built in 2016 at Ribeira Grande, Azores (Marques, 2017), can be presented as a result of this new environmental policy. From the construction point of view, the building was conceived to combine the concept of deconstruction with a bio-constructive process where each ecosystem formation follows the circular economy general principles.

To this end, it was necessary to incorporate in the building project the participation of the entire team: client, technicians, suppliers and builders, whose work had two main objectives: (a) adoption of construction systems that allow the easy removal and reuse of its components; and (b) maximum flexibility in the spatial configuration and structure of the building by seeking to adapt it to different patterns of use. Both objectives can be exemplified by the 45,000 m^2 of movable walls, made of recycled composite material panels consisting of a mixture of wood particles and compressed dry cement, which divide the all-internal spaces with 15,000 m^2 built area (Figure 3).

The use of removable walls throughout almost all the interior spaces will provide easy access to maintain the thousands of meters of electrical cables and infrastructures, ventilation and water and sanitation facilities.

In their carrying structure, the use of steel pillars and beams and reinforced concrete, all of them reusable or recyclable, allowed lightening of the entire construction process while ensuring the conditions of robustness and resistance to use and local climatic factors.

The exterior walls are ventilated facades built with prefabricated concrete panels and compact phenolic resin panels, thus providing a better adaptation to local climatic conditions and greater design flexibility.

The desired result was to obtain a systematized and modulated building in the various design scales so the building was able to adapt to the school program's evolutionary transformations while presenting a clear soundness of construction. The building construction is firmly focused on the deconstructive architecture concepts of adaptability, disassembly and reuse, which, in conjunction with other eco-construction approaches, aims to result in a sustainable design.

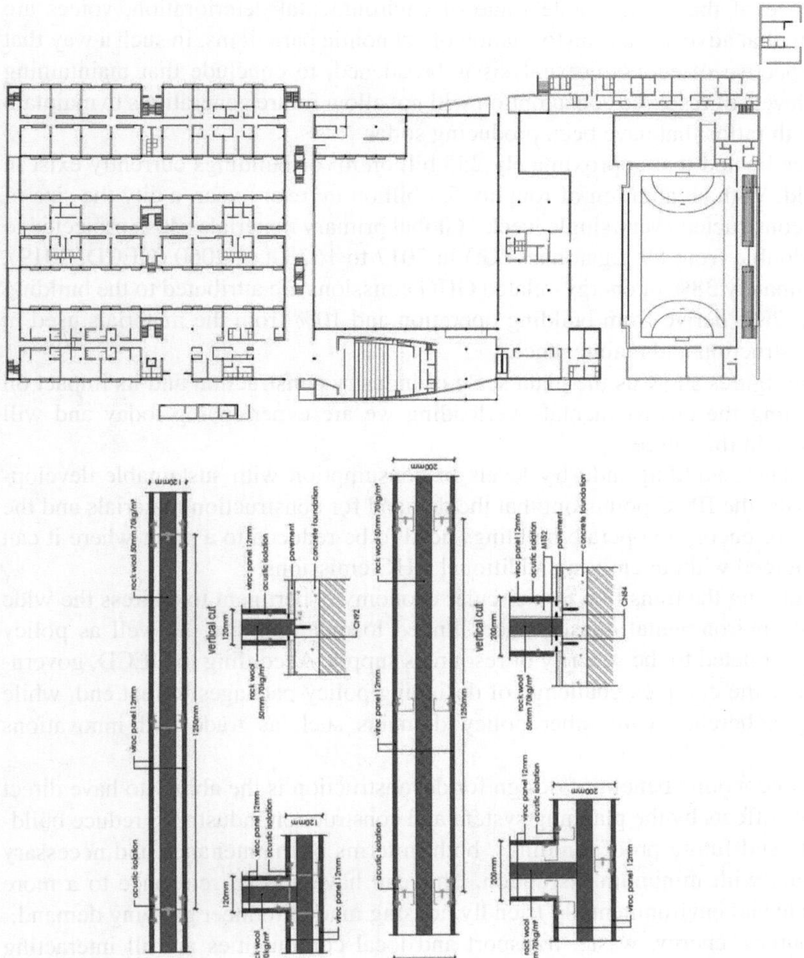

Fig. 3 Gaspar Frutuoso School, detail of walls and ground floor plan, 2016. Source: Author's elaboration

7 Discussion and Conclusions

In the face of the recognizable stage of environmental deterioration, voices are emerging that advocate a transformation of economic paradigms, in such a way that the perspective of economic analysis is broadened, to conclude that maintaining current levels of resource consumption will not allow future generations to maintain the growth ratios that have been producing so far.

It is estimated that approximately 255 billion m^2 of buildings currently exist in the world. With an addition of roughly 5.5 billion m^2 every year, a city the size of Paris is constructed every single week.[3] Global primary materials use is projected to almost double from 89 gigatonnes (Gt) in 2017 to 167 Gt in 2060 (OECD, 2019). Approximately 38% of energy-related GHG emissions are attributed to the building industry, 28% derive from building operation and 10% from the materials used in their construction and maintenance.[4]

Those figures show us the giant scale of industry construction and its impact on exacerbating the environmental overloading we are experiencing today and will experience in the future.

To balance building industry levels of consumption with sustainable development goals, the IPCC points out that the demand for construction materials and the demand for energy to operate buildings need to be reduced to a point where it can all be sourced without emitting additional GHG emissions.

Stimulating the transition to a circular economy is pertinent to address the wide range of environmental consequences linked to materials use, as well as policy objectives related to the security of resources supply. According to OECD, governments face the complex challenge of designing policy packages to that end, while ensuring coherence with other policy domains such as trade and innovations policies.

A key economic benefit of design for deconstruction is the ability to have direct economic effects by the planning system and construction industry to reduce building costs and future-proof buildings, both in terms of maintenance and necessary upgrading, with minimum disruption. This can have special relevance to a more affordable and environmentally friendly housing market to meet growing demand.

Resources, energy, waste, transport and local communities are all interacting aspects of a regional approach to design for deconstruction. The proximity and self-sufficiency principles require waste to be dealt with as close as possible to where it is produced. Minimizing transportation is a significant part of a successful architecture for deconstruction.

We can conclude by saying that for any given set of objectives, the best practicable environmental option, like deconstruction, identifies strategies that provide

---

[3] Global ABC/IEA/UNEP (2017), Global Alliance for Buildings and Construction: 2017 Global Report for Buildings and Construction. https://www.worldgbc.org/sites/default/files/UNEP%20188_GABC_en%20%28web%29.pdf. (Accessed 29-03-2023).

[4] Global ABC/IEA/UNEP (2020), Global Alliance for Buildings and Construction: 2020 Global Report for Buildings and Construction. https://globalabc.org/sites/default/files/inline-files/2020%20Buildings%20GSR_FULL%20REPORT.pdf. (Accessed 29-03-2023).

the most benefits or the least damage to the environment as a whole, at an acceptable cost, in the long short terms, thus evaluating environmental, social and economic consequences.

References

AIA. (2010). *AIA guide to building life cycle assessment in practice*. American Institute of Architects. Accessed March 20, 2023. Accessed from https://content.aia.org/sites/default/files/2016-04/Building-Life-Cycle-Assessment-Guide.pdf

Allwood, J. M., Cullen, J. M., Carruth, M. A., Cooper, D. R., McBrien, M., Milford, R. L., Moynihan, M. C., & Patel, A. C. (2012). *Sustainable materials: With both eyes open*. UIT Cambridge.

Cao, Z., Masanet, E., Tiwari, A., & Akolawala, S. (2021). *Decarbonizing concrete: Deep decarbonization pathways for the cement and concrete cycle in the United States, India, and China*. Industrial Sustainability Analysis Laboratory, Northwestern University.

Carruth, M. A., Allwood, J. M., & Moynihan, M. C. (2011). The technical potential for reducing metal requirements through lightweight product design. *Resources, Conservation and Recycling, 57*(C), 48–60. https://doi.org/10.1016/j.resconrec.2011.09.018

Chini, A., & Schultman, F. (2002). *Design for deconstruction and materials reuse, CIB Publication 272. Proceedings of the CIB Task Group 39 – Deconstruction Meeting, 9 April 2002*.

CIRIA. (1983). *Buildability: An assessment*. Construction Industry Research and Information Association.

Crowther, P. (2002). Design for buildability and the deconstruction consequences. In A. Chini, & F. Schultman (Eds.), *Design for deconstruction and materials reuse. CIB Publication 272, Proceedings of the CIB Task Group 39 – Deconstruction Meeting, 9 April 2002*.

Davos Declaration. (2018). *Towards a European vision of high-quality Baukultur, Conference of Ministers of Culture 20–22 January 2018*. Accessed from https://baukultur%2D%2Dproduction%2D%2Dstorage.s3.amazonaws.com/baukultur/2022-06-09-075742%2D%2Dcontext-document-en.pdf

EPA. (2023). Designing for adaptability, disassembly and reuse. United States Environmental Protection Agency (EPA). Accessed from https://www.epa.gov/smm/best-practices-reducing-reusing-and-recycling-construction-and-demolition-materials#design

European Union. (2020). *A new circular economy action plan for a cleaner and more competitive Europe, Communication from the Commission to the European Parliament*. The Council, The European Economic and Social Committee and the Committee of the Regions.

EUROSTAT. (2022). *European statistics*. Accessed from https://ec.europa.eu/eurostat/statistics-explained/index.php?title=Waste_statistics#Total_waste_generation

Geyer, R. & Jackson, T. (2004). Supply Loops and Their Constraints: The Industrial Ecology of Recycling and Reuse. In California Management Review, Vol. 46, No.2 Winter 2004, *University of California, Berkeley*, 54–73. https://doi.org/10.2307/41166210

Gorgolewski, M. (2008). Designing with reused building components: Some challenges. *Building Research and Information, 36*, 175–188.

IPCC. (2022). *Synthesis report of the IPCC Sixth Assessment Report (AR6), longer report*. Accessed from https://www.ipcc.ch/ar6-syr/

Keena, N., & Dyson, A. (2017). Qualifying the quantitative in the construction of built ecologies. In *Embodied energy and design* (pp. 196–205). Columbia University.

Krook, J., & Baas, L. (2013). Getting serious about mining the technosphere: a review of recent landfill mining and urban mining research. *Journal of Cleaner Production*. https://doi.org/10.1016/j.jclepro.2013.04.043

Langdon, D. (2007). *Life cycle costing (LCC) as a contribution to sustainable construction: Guidance on the use of the LCC methodology and its application in public procurement, guidance document*. European Commission.

Marques, C. (2017). *Gaspar Frutuoso School – Architecture as a pedagogic space.* Editora Caleidoscópio.

Morgan, C., & Stevenson, F. (2005). *Design for Deconstruction.* SEDA.

MPGCA. (2021). *Human settlements action Table, Marrakech Partnership for Global Climate Action Pathway co-led by GlobalABC.* United Nations.

Nakajima, S. & Russell, M. (2014). Barriers for Deconstruction and Reuse/Recycling of Construction Materials. *In CIB Publication, 397.* Working Commission W115 - Construction Materials Stewardship. Accessed from https://www.researchgate.net/publication/264671518_Barriers_for_Deconstruction_and_ReuseRecycling_of_Construction_Materials_in_Norway

OECD. (2019). *Global material resources Outlook to 2060: Economic drivers and environmental consequences OECD.* OECD. https://doi.org/10.1787/978924307452-en

Razkenaria, M., Fennerc, A., Shojaeib, A., Hakimc, H., & Kibertc, C. (2020). Perceptions of offsite construction in the United States: An investigation of current practices. *Journal of Building Engineering, 29.* https://doi.org/10.1016/j.jobe.2019.101138

Schultmann, F., & Rentz, O. (2002). Resource-constraint project scheduling for deconstruction projects. In A. Chini & F. Schultmann (Eds.), *Design for deconstruction and materials reuse. CIB Publication 272, Proceedings of the CIB Task Group 39. Deconstruction Meeting*.

UN. (2020). *Resource efficiency and climate change. Material efficiency strategies for a low-carbon future, A report of the International Resource Panel.* United Nations. Accessed from https://www.resourcepanel.org/reports/resource-efficiency-and-climate-change

UN. (2022). *Global status report for buildings and construction.* United Nations Environment Programme.

Circularity of Household-Generated Construction and Demolition Waste: Management Principles for Green Transition in Latvia

Natālija Cudečka-Puriņa, Inese Mavļutova, Rudīte Vesere, Sanita Osipova, Dzintra Atstāja, Māris Purviņš, Jekaterina Kuzmina, and Jānis Butkevičs

Abstract The chapter provides a holistic view of the construction waste problem, revealing the lack of a unified approach toward waste sorting and further reuse possibilities. The goal of the study is to investigate the principles of circular economy in building construction and demolition waste management and to find possible solutions for their development in household-generated waste management for green transition in Latvia.

The current research presents the study of principles of circularity of household-generated waste based on a systematic literature review; data analysis gained out of the survey to draw practical suggestions for professionals towards sustainable development in the short-term and long-term future. The above-mentioned systematic analysis of the literature is grounded in text analytics; the best practices from Austria and Scandinavia were explored. The collected data on construction waste in Latvia were analyzed by applying statistical methods. The research results revealed a significant increase in building construction and demolition waste and the lack of circularity of them. The authors conclude that it is necessary to use best practices on how to apply circularity in building construction and demolition waste management and how to develop the cooperation links between local building authorities and householders using digital solutions for the green transition.

Keywords Construction and demolition waste · Circular economy · Waste management · Digital forum · Latvia

---

N. Cudečka-Puriņa · I. Mavļutova · R. Vesere · S. Osipova · D. Atstāja (✉) · M. Purviņš · J. Kuzmina · J. Butkevičs
Riga, Latvia
e-mail: natalija.cudecka-purina@ba.lv; inese.mavlutova@ba.lv; sanita.osipova@lu.lv; dzintra.atstaja@ba.lv; maris.purvins@ba.lv; jekaterina.kuzmina@ba.lv; janis.butkevics@ba.lv

C. Rego et al. (eds.), *Transitioning to a Circular Economy*, Studies on Entrepreneurship, Structural Change and Industrial Dynamics,
https://doi.org/10.1007/978-3-031-77661-8_12

1 Introduction

The academic community has already made extensive contributions to the fields of sustainability, transitions to sustainability, circular economy, etc. The latest research emphasizes the importance and sustainability of the development of circularity, which has a positive effect in increasing the productivity of direct input materials, water, and energy, reducing the emission impact, and saving economic investment, as described by Chen et al. (2022), as well as reducing the environmental load rate. Technology also keeps advancing, offering innovative solutions for various methods and opportunities for profit creation. However, there has been little progress in the circularity status in all countries. The authors assume that it is justified by the diversity of information regarding sustainability issues and adopted solutions worldwide. Moreover, there is a great lack of cooperation between European Union member states and the city level to combat similar challenges. Even within cities, interventions often assume stakeholder cooperation rather than ensuring it. At the same time, as highlighted by Patricio et al. (2022), we are facing the Fourth Industrial Revolution (Industry 4.0), which, among other things, relies on the availability of large amounts of data and on digitalization, as well as development of different digital platforms and digital models for resource tracing (Cudecka-Purina et al., 2024). The usage of digital technologies will become crucial in this green transition (Uvarova et al., 2021). This chapter aims to respond to such challenges by generating scientific results that can be easily transformed into real solutions, mapping solutions to problems for which they would be most suitable (or compatible), and exploring behavioral nudges that are required to support them. The Authors are concerned that wise assessment of the construction and demolition waste management principles will undoubtedly not only positively impact the decrease of primary resource consumption and increase the circularity for green transition through digital solutions (Mikhno et al., 2023; Atstaja et al., 2022).

Still, a lot of issues have been identified that all together form the existing research GAP and highlight scientific excellence.

In the twenty-first century, the demand for sustainability is no longer just a problem highlighted by individual scientists, researchers, and practitioners. There is a call for the whole society to find an immediate solution by changing the culture of production and consumption. Bringing the principle of sustainability to life has become a public demand for the state. Sustainability has become a priority on the agenda of fundamental human rights and national and international organizations. In Latvia, sustainability is embedded in the constitution (Satversme) as a principle of state activity and a basic human right to live in a clean environment (Constitution of the Republic of Latvia, 1922). The authors believe government support at national, regional, and local levels is essential to strengthening the values of sustainable consumption (Rutitis et al., 2022). The authors agree with Esteban Cantillo and Quesada (2022) that the municipality does not have selective collection routes, which effectively contribute to the separation of waste at the source and thus avoid the disposal of recyclable waste in the landfill.

Waste management is seen as a sector with a huge potential to recover the resources from the waste flow and divert them from landfills to either promote reuse, recycling, or recovery. Applying this approach, it is possible to:

- Decrease initial waste amount
- Implement principles of Circular Economy (CE) into practice
- Decrease consumption of primary resources
- Prolong the use of resources by bringing them back to the economic cycle

All this highlights the importance of current research as in particular construction and demolition waste (C&D) has a high potential for reuse, if managed appropriately. The types of C&D waste being generated daily globally vary by region specifics but, in most cases, are common for the following materials: concrete, glass, cement, bricks, wood, plastics, etc. (please see Fig. 1 for more details). This chapter will also address the problems that households are facing when doing small-scale construction or renovation works, and the aim of the study is to reveal potential solutions for this problem by diverting waste from disposal and optimizing household expenditures, as management of C&D waste is fairly an expensive matter.

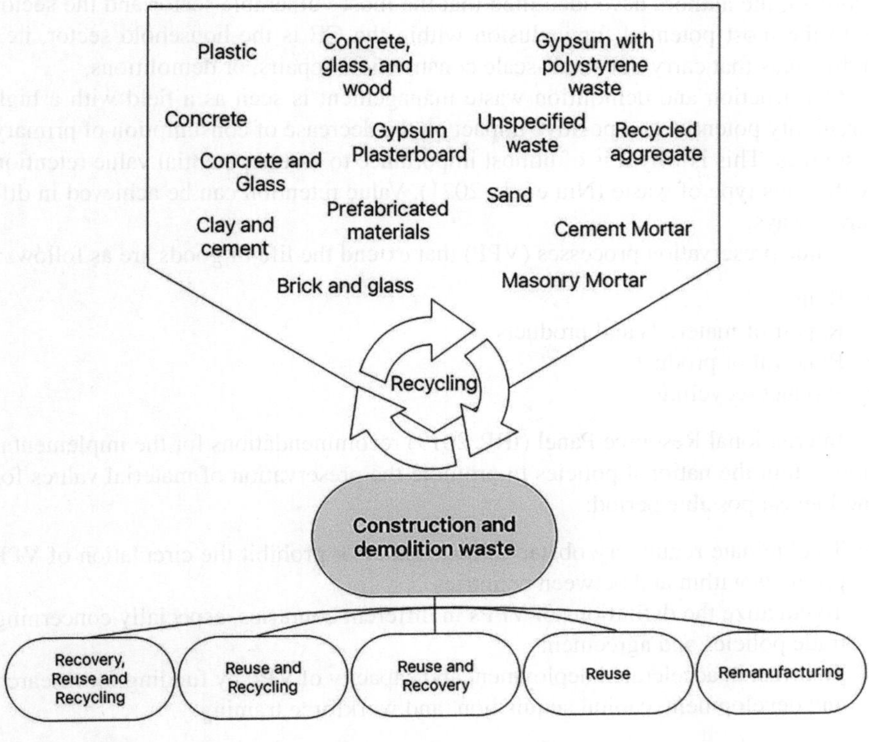

Fig. 1 Circularity of the construction and demolition waste (Adapted from Oluleye et al. (2022))

2 Characteristics of Buildings Construction and Demolition Waste and Value Retention

Construction and demolition waste, also sometimes called building construction and demolition waste (BCDW), reveals an extensive list of waste sub-types that have huge potential for reuse under a circular economy. The building industry is accountable for significant environmental impacts due to the usage of energy or other resources, as well as the production of waste. CE, a new paradigm, can substantially amend the sustainability of the building sector. Main BCDW types and possible directions of their circularity are clearly depicted in Fig. 1:

As Fig. 1 shows the main directions of building construction and demolition waste circularity are recovery, reuse, recycling, and remanufacturing. An important distinction between reuse and recycling is the new use or application of waste in its original form and the practice that converts waste into new products with some physicochemical reprocessing. Recycling requires more energy than reuse and potentially new material inputs (Lizárraga-Mendiola et al., 2022). Obviously, building construction and demolition waste is mostly generated by the construction industry, but it is the management of household construction waste, which is also important and is the least discussed issue. Taking into consideration the existing problem, the authors have identified that the most vulnerable sector and the sector with the most potential for inclusion within the CE is the household sector, i.e., individuals that carry out small-scale construction, repairs, or demolitions.

Construction and demolition waste management is seen as a field with a high circularity potential and positive impact on the decrease of consumption of primary resources. This is why it is of utmost importance to assess potential value retention within this type of waste (Niu et al., 2021). Value retention can be achieved in different ways.

Value preservation processes (VPP) that extend the life of goods are as follows:

- Reuse
- Repair of materials and products
- Renewal of products
- Product recycling

International Resource Panel (IRP, 2019) recommendations for the implementation within the national policies to promote the preservation of material values for the longest possible period:

- To eliminate regulatory obstacles that hinder or prohibit the circulation of VPP products within and between countries.
- To equalize the definitions of VPPs in different countries, especially concerning trade policies and agreements.
- To invest in accelerated deployment and capacity of VPP by funding for research and development, capital acquisition, and workforce training.

- To implement stakeholder education and awareness campaigns to promote the adoption of VPP products and strengthen the business case for VPP manufacturers.
- For non-industrialized countries to focus in the short term on "closing the [circulation] cycle," identifying access and regulatory barriers to VPPs that may inadvertently exist due to related policy priorities (e.g., consumer protection, anti-dumping and domestic policy, trade). In the long term, focus on expanding VPPs' production capacity through knowledge and technology transfer and training programs to increase the supply of skilled labor (Leder et al., 2020; Morseletto, 2020).

Material efficiency strategies can significantly reduce emissions in residential buildings' material cycle. IRP simulations (IRP, 2019; OECD, 2018) for resource consumption in 2050 show that, with the implementation of policies promoting the circular economy, the emission reduction for the built environment in the G7 countries can decrease from 80–100% to 35–40%.

The best material efficiency strategies for the built environment, ranked by emissions reduction potential according to IRP modeling results, will be as follows:

1. Improved reuse of building components and end-of-life recycling of materials, as well as improved productivity in manufacturing:
   - Prohibitions on waste disposal in landfills
   - Promotion of production of modules, panels, and building systems
   - Promotion of Building Information Modeling (BIM)
   - Reduction of subsidization of raw materials
2. Substitution of materials:
   - Revision of building and fire safety regulations for timber and timber frame structures
   - Standards that allow cement with clinker substitutes
   - Revision of building codes to prevent long-term effects of materials

Along with the return of primary resources used in construction into secondary circulation, the waste landfilled in landfills and the consumption of primary resources decrease (European Commission, 2018).

To reduce the amount of construction and demolition waste or construction waste, promoting its collection, the use of recycled materials, and the reflection and identification of resource consumption, the EU put forward several policy initiatives in various stages of implementation (EU, 2012, 2014).

Construction waste management, including sorting, is mandatory for all EU countries. However, each national state sets its priorities in the overall process of construction waste management and circulation. The consumption patterns and preferences in Eastern Europe and especially in Latvia are significantly different from those in Western European countries for several reasons (Mavlutova et al., 2021). The data on the amount of construction waste collected and sent for sorting or processing in Latvia show that the EU's target of construction waste circulation is already significantly exceeded (no less than 70% of the collected construction waste must end up in secondary circulation). In Latvia, the proportion of sorted, recycled, and recovered construction waste in the period from 2014 to 2020 has been in the range of 91–97%); the priority issue to be solved is the most efficient possible return of raw materials into circulation and the most effective use of materials in construction, or transfer for reuse, including other sectors. However, at the same time, it should be noted that the amount of construction waste generated by residents is only sometimes adequately and efficiently collected, managed, and reused (Norouzi et al., 2021; Yu et al., 2022).

Analyzing the current situation, the authors believe that the development of a unified digital platform and industrial symbiosis forum might promote the transition to CE within the society as well as with the current trend of increasing costs for waste management, including BCDW management, general inflation, and increase of household expenditures, this forum might allow the society to optimize their expenditures and provide support for those who cannot afford to purchase new construction materials due to their economic situation. In addition, the major contribution within CE is the demand for new resources and new construction materials that can be decreased. Involvement of residents helps to get a wider range of ideas about the necessary multifunctional solutions for the needs; this, in turn, requires much greater engagement from the community and representatives of local communities to participate in and self-manage various areas transformation processes (Mikelsone et al., 2021).

Getting an accurate picture of the volume (including volume forecast) and composition of materials and components generated as construction waste is essential. Accordingly, for the study of the existing regulations and good practices, Austria was chosen, where since 2015, the procedure for the classification, use, and accounting of recycled construction waste has been established, and Finland, which in its National Strategic Program for the Promotion of the Circular Economy has defined construction as an essential industry in the circular economy, and is also implementing several large-scale projects in the fields of construction and real estate to introduce probity in these industries and test it in practice. In one of the projects—the Nordic network for sustainable construction—Finland is involved with four other Scandinavian countries: Sweden, Denmark, Iceland, and Norway, which we will briefly cover in this article.

When turning to Latvia, according to the information provided in the Waste management plan for 2021–2028, the annual generation of construction and demolition waste in Latvia currently accounts for ~383,000 tons. In 2019, 70% of the construction waste that was collected was a mixture of construction waste and inert materials, while 20% was solely made up of inert materials (VARAM, 2021). Recycling, including using excavated areas for engineering purposes, is the primary waste management method for construction waste.

According to Latvian legislation, construction and demolition waste is waste generated by construction because of construction or demolition (Saeima, 2012). Construction waste must be disposed of by the operator who has received the appropriate permits for waste collection, transportation, and recovery. Each shipment of construction waste by legal persons must be registered in the transportation accounting system. In addition, to be able to use recycled waste in the construction site, it must be foreseen within the construction projects in advance. One more important aspect is that backfilling is not waste regeneration itself, although materials that have been regenerated can be used to fill the spaces. Although the system is well established, the field of construction and demolition waste faces a range of different problems. Some of the most common ones are listed below:

- The builder cannot present documents to prove that it has handled construction waste to the waste management company.
- The construction waste is handed over to the company that did not receive a waste transportation permit or has received it.
- The waste management company posseses necessary waste treatment permit, but has not submitted financial security, and cannot operate.
- Construction projects foresee the use of construction waste, but the construction project does not provide the conditions for the quality of the materials used.
- Transportation of construction waste is not registered in the construction waste transportation accounting system.
- Construction waste has been dumped in forests and meadows or any other illegal areas that do not foresee such activity.

Latvia does not collect separate statistics on the collection of construction and demolition waste that is generated by industry and the household sector. However, it is assumed by the authors that within the household sector there might be a higher probability of lack of information on appropriate and more cost-efficient management of this type of waste with minimal harm to the environment. To assess the best potential solutions that can be applied to Latvia, the best practices from a sample of European countries have been analyzed. The authors chose the following countries: Austria, Finland, and Denmark.

3 Best Practices from Austria and Scandinavia

In Austria, since 2015, the Ordinance on the recycling of construction materials (the Ordinance) has been in force, which introduced the classification of recycled construction waste according to the technical quality of the obtained product (Rechtsinformationssystem des Bundes, 2013). The arguments of the Austrian Federal Ministry of Climate Protection, Environment, Energy, Mobility, Innovation, and Technology for determining the national policy for construction waste management when deciding on a special regulation for construction waste recycling were:

- Various dangerous substances in construction waste (asbestos, tar, etc.)
- Types of construction waste unsuitable for processing due to the uneven quality of recycling aggregates

- Low costs of primary resources and high costs of disposal in landfills create an excellent incentive for low-quality recycling
- Limited range of applications for some recycled materials, the surplus of recycled materials on the market
- No end-of-waste criteria, no product status of recyclable materials
- Legal uncertainty due to lack of precise legal regulation (Honic et al., 2019; Çimen, 2023)

Below the authors provide a short summary of the initiatives of local governments and industry representatives with respect to circular construction.

Municipalities have introduced specific initiatives for properly managing construction materials and construction waste based on their initiative, as well as by educational and scientific institutions and the private sector—real estate developers and managers. The Vienna city municipality has launched the "Do Tank Circular City Wien 2020-203" project—a circular economy in the built environment, including various activities such as workshops. However, there are regulations and standards regarding life cycle and recycling-oriented planning and conservation of construction resources (Minixhofer et al., 2022).

Another related project is the "Digital Urban Mining Platform," creating a platform that will primarily be used to capture raw materials and resources built into the city with the help of technical means. The system is based on the digital scanning of buildings and resource modeling in BIM.

Digital, Internet-based platforms have been created in the private sector, which also actively participates in the circular economy system, offering materials recovered during demolition or interior change or their disassembly and assembly services.

Some examples of platforms:

- Material Nomaden https://www.materialnomaden.at/
- BauKarussell https://www.baukarussell.at/
- Material nomads, Re:Store: https://www.restore.or.at/store/

Like most EU countries, Austria has a relatively high uniformity of recoverable materials. This is related to the requirements of the Building Materials Regulation regarding the CE marking, declaration of essential technical characteristics, absence of maintenance and use instructions, as well as the provision of guarantees to protect consumer rights and ensure the safety of users. So, for example, none of the platforms offer used doors or windows, a standard building structure obtained due to dismantling. The Austrian Sustainable Real Estate Industry Association has a circular economy working group that works on the topic of the circular economy from different points of view; for example, procurement requirements are discussed, and guidelines are compiled on what should be part of the legal procurement framework to promote the circular economy (Priess et al., 2017).

Finland's experience is analyzed, considering that Finland is a solid global influencer and a provider of sustainable solutions in the international market. Finland's approach to promoting the circulation of construction waste and construction materials was significantly updated and strengthened in 2022 when the Finnish government adopted the Circular Economy Strategic Program. Construction is an important sector for the transition to a circular economy. Finland's vision and goals for the circular economy in 2023 are:

- Sustainable products and services are the mainstream of the economy, and the sharing economy is an everyday activity.
- Future-friendly choices and decisions that strengthen the welfare of society.
- "More for less": the use of natural resources is sustainable, and materials are kept in circulation for longer and in a safe way.
- Achievements based on innovations and digital solutions or a breakthrough in the circular economy with intelligent regulation and responsible investments, with the participation of companies and consumers (Government of Finland, 2021).

Regarding the construction sector, the basic principle set out in the strategy is that "the construction and real estate sectors are based in the circular economy in the future." Regarding materials (resources), similar to elsewhere in Europe, both industries make up a significant part of Finland's steel, concrete, wood, and plastic consumption. Finland's policy framework aims to use these materials efficiently and keep them in circulation, thereby significantly reducing the emissions of manufacturing new products. Circular economy's business models, such as rental and reuse, offer the opportunity to rationalize the use of assets and reduce the number of empty properties (Hosseinian et al., 2021; Lazarevic et al., 2022). Regarding the built environment, Finland's vision is that cleanliness is the defining criterion throughout the life cycle. Finland's view of the life cycle stages of the construction and real estate industry covers the design, construction, renovation, demolition, and manufacturing of construction products. In each stage, Finland has defined the basic principles of circularity in the circular economy strategy. Finland's 2035 goals for sustainable construction:

- The use of existing properties should be significantly more efficient.
- Must achieve a significant reduction of CO_2 emissions caused by the real estate and construction industry in connection with the use of natural resources (Lazarevic et al., 2022).
- The existing building fund should be evaluated as a resource that can be transformed, developed, and qualitatively improved.
- Buildings are resource banks, and CO_2 attractors; the amount of CO_2 attracted must be indicated in digital material passports.
- Digital solutions, modular construction, and industrial production of building products will be industrial solutions that will increase the industry's productivity. They will contribute to the circular economy.
- Space is a resource; its optimal use should be modeled with digital tools and constructed with new profit models.

Denmark's experience in promoting the circulation of construction waste and construction materials during the work was also evaluated. In the middle of 2020,

Denmark decided to introduce Deutsche Gesellschaft für Nachhaltiges Bau (DGNB) certification requirements for construction plans in its territory. In addition to this, in 2021, a decision was made to apply the DGNB framework to the construction industry. In 2022, the municipality of Copenhagen issued a dedicated circular economy manual (Byrggeri Kobenhavn, 2022). The generally defined requirements of the municipality apply to groups of processes and materials with the most significant impact:

- Resource mapping
- Inspections of compliance with the rules for sorting construction waste at selected construction sites where demolition is carried out

The municipality implements 14 pilot projects for system development and applicability evaluation. One of the principles to be implemented in the future of the circular economy in Copenhagen municipality projects is material passports, which will also make available information about the material in its subsequent life cycles. According to Ramsheva et al. (2020), by shifting from a focus on product to system, it is possible to create greater value for the industry, society, and the environment. Concentrating on products or materials may result in savings on input material costs, as well as an overall decrease in primary resource consumption. However, adopting a service perspective, along with consistent monitoring of performance for potential improvements and updates, can extend the lifespan of buildings and slow down the need for their end-of-use.

Another principle is from the development stage to general practice—which is related to the transfer of knowledge from pilot projects to general requirements and provides an opportunity to facilitate the process; for example, within the Icelandic Nordic project, the pilot project "Vaxtarhus and a clean neighborhood" is implemented. According to Nordic Innovation (2023), a workshop series initiated by the program has been conducted thrice between 2019 and 2022, with over 140 companies taking part. The series has effectively heightened awareness and knowledge of circular business models while also providing valuable insights into the challenges companies face during transformation, as well as their current maturity levels.

In general, following the lessons learned in Austria over the past 5 years, it is essential to pay the most attention to the groups of construction and demolition wastes or potential materials with the largest volume and economic value. Invest more resources in initiatives with a higher level of potential for recycling materials and adding value (use by combustion, energy extraction is the end of life cycle).

To plan the further use of construction and demolition wastes or their processing products, as well as by-products from production processes, building material manufacturers or service providers must be able to predict the availability of such resources. Finland's experience with the creation of a register of such materials would facilitate the availability and exchange of potentially commercially valuable information.

4 Methodology and Results

This study was conducted in 2022 and the qualitative approach was used to examine the current situation and possible solutions. The questionnaire was developed from the extant literature to conclude the picture of the subject under study.

Figure 2 presents a stepwise design of the research to address the research problems outlined above.

To carry out the research, the authors, being engaged in LIFE Waste to Resources IP project, have developed a survey, which has been performed by Research Center SKDS and the results of the survey have been handed back to the researchers to be able to process and analyze. This country-wide survey for Latvia covered permanent residents of Latvia between the ages of 18 and 75. The survey's sample size is 2005 respondents, of whom 67% have been directly involved in the construction, repair, improvement, and/or demolition of their own or family real estate. The survey was structured in a way to gain comprehension of the overall knowledge level of the society on the construction and demolition waste subject, as well as to identify the main gaps in the management of this particular waste stream. Overall the survey consisted of 65 questions. The following section of the research will provide an analysis of the obtained results of the survey and later form a set of conclusions and recommendations (Figs. 3 and 4).

The authors would like to highlight that according to the results of a survey of the population of Latvia, slightly more than 2/3 of respondents (67%) have carried out repair or construction work in their household in the last 5 years, which resulted in discarded repair and/or construction waste. From the respondents, the distribution in detail was as follows:

- Currently doing—5%
- Have carried out in the last year—25%
- Have carried out 1–2 years ago—16%
- Have carried out 3–5 years ago—21%)

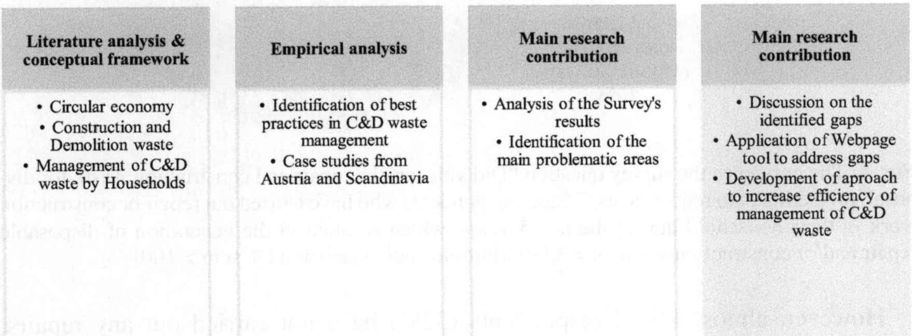

Fig. 2 A step-by-step overview of the research design of this study

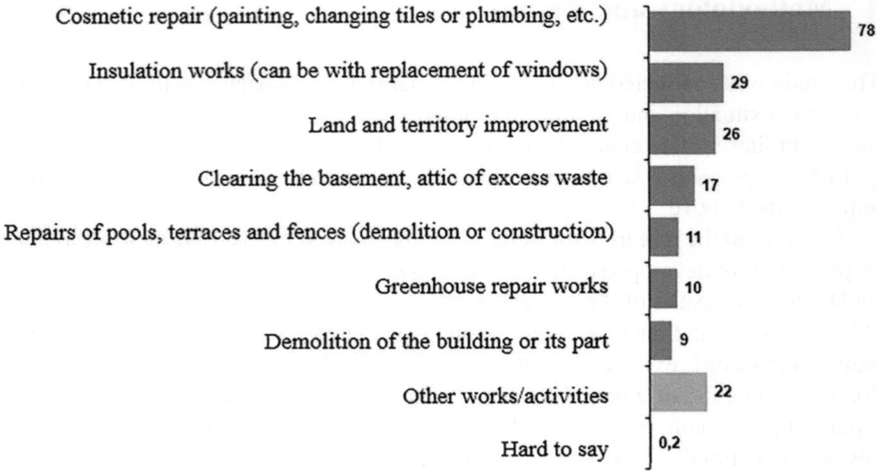

Fig. 3 Type of repair or construction work, answer to the survey question "What type of repair or construction work did you/your household do?" Base: respondents who have carried out repair or construction work in their household during the last 5 years, which resulted in the generation of disposable repair and/or construction waste, $n = 1350$ Multiple choice question (% sum > 100)

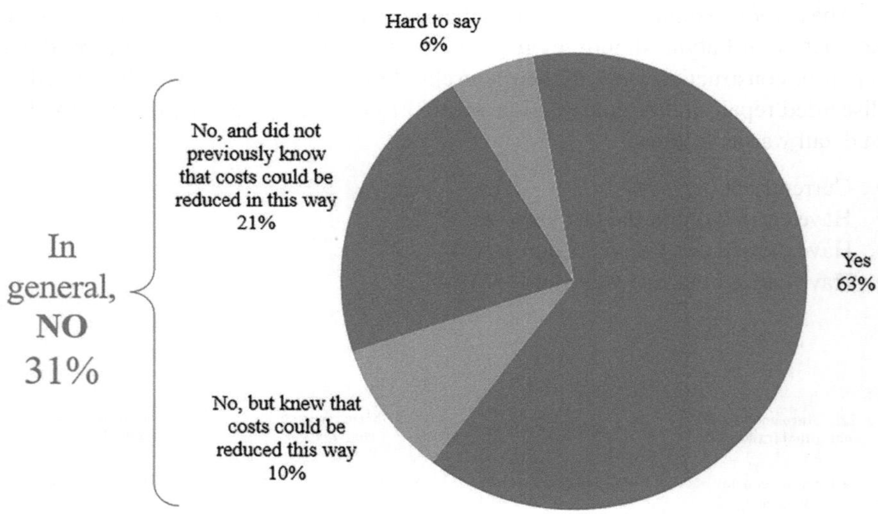

Fig. 4 The answer to the survey question "Did you sort the repair and construction waste for disposal, for example, to reduce costs?" Base: respondents who have carried out repair or construction work in their household during the last 5 years, which resulted in the generation of disposable repair and/or construction waste, $n = 1350$ Multiple choice question (% sum > 100)

However, almost 1/3 of respondents (32%) have not carried out any repairs, resulting in repair or construction waste. The results show that it is quite common to do some small renovation or construction work in a timely manner in Latvian society.

When analyzing the type of repair works performed, it can be summarized that the most common type of repair or construction work performed is cosmetic repair, such as painting, changing tiles or plumbing, etc. (78% of respondents did this). Other jobs or activities are mentioned relatively less frequently. And of course, for this type of activity, the households do not need to receive any approval from the building authority institution, which makes the process less bureaucratic and complex; still on the other hand it assumes that the society is aware of all information publicly available on the management of construction and demolition waste. To support the above-mentioned, most respondents (73%) who have carried out repair or construction work in their household in the last 5 years, which resulted in the generation of discarded repair and/or construction waste, did not carry out such work that required approval from the building authority. 17% of respondents made an agreement with the construction board for those works that required it, but 3% of respondents indicated that the works were only partially coordinated with the building board—not all works that required approval were agreed. 7% of respondents stated that they did not know whether the works performed required approval from the building board.

The figure below provides a comprehensive visualization of the approach that the citizens chose for performing repair and construction works at their individual real estate. In addition, this figure represents the scale of the construction or repair works carried out, showing that in most cases the responses covered really small-scale works that do not require certain approval, and this also means that these particular households did not have a chance to receive more explicit information on the management of construction and demolition waste that could be received by the ones who undergo the official process due to the scale of the works. Meaning that the households either must be educated enough to know where to look for required information or the municipalities must have this information provided to the inhabitants to make sure that they discard construction and demolition waste in the most resource-efficient manner while still complying with all applicable regulations (Fig. 5).

The survey carried out by the authors also addressed the question of waste generation. From the analysis of the responses, it has been concluded that more than half of the respondents (57%) indicate that they generated up to 1 cubic meter or 10 large bags of waste when carrying out repair or construction work. More than 1 cubic meter of waste was generated by 29% of respondents:

- 1 to 4 cubic meters—20%
- More than 4 cubic meters—9%

It can be observed that the amount of waste was smaller for those respondents who carried out repairs in the apartment.

The next section of the survey also covered the discharge of construction and demolition waste, which is important to be assessed and to reveal tipping points, if any. According to the survey's results, most often, respondents get rid of repair and

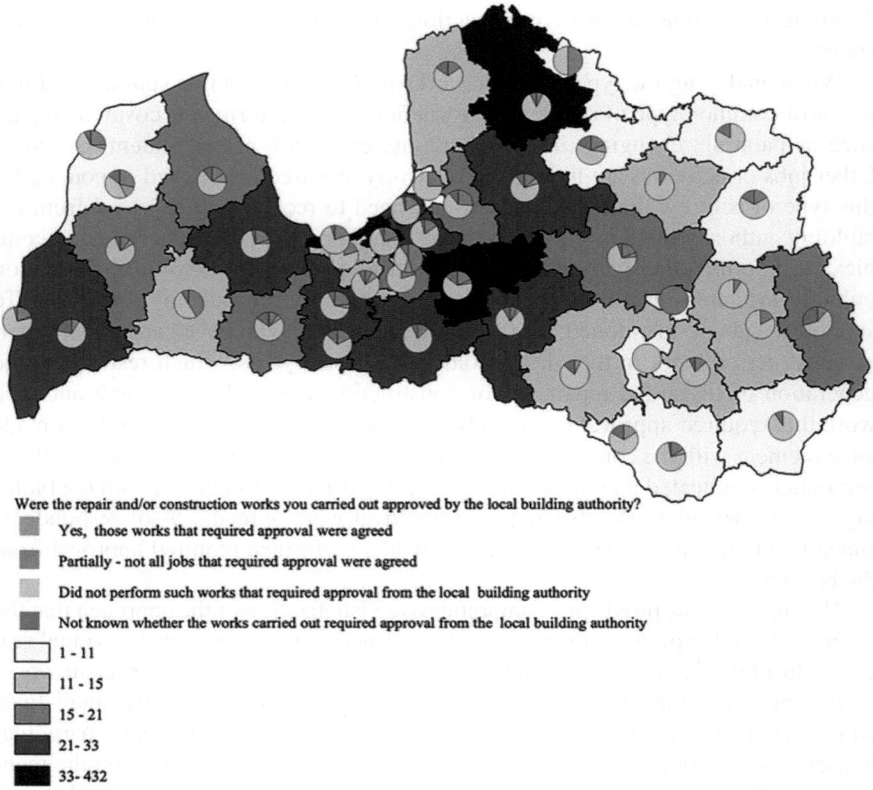

Fig. 5 Response distribution for the question "Were the repair and/or construction works you carried out approved by the local building authority"

construction waste by throwing it into the common household waste container (34%) or burning it (29%). These results undoubtedly reveal a significant gap in access to information and the level of overall education in society; for instance, it is forbidden to discharge construction and demolition waste into common household waste containers for unsorted waste, not to mention waste incineration—which is forbidden on the legislative level.

Respondents also tend to get rid of repair and construction waste by taking it to the landfill with a special container for construction waste (19%), taking it to the landfill themselves (16%), as well as by contracting a company or private person that they found on the Internet or by other people recommendations to transport the waste to the landfills (16%). It is worth noting that there are also responses that can be biased, i.e., 14% of the respondents got rid of the repair and construction waste by using it for strengthening the road, and 10% used it for filling low (wet) places. It is quite important to highlight that these activities, if performed on a larger scale and not in the real estate owned by the same person, require specific permits.

Respondents also got rid of repair and construction waste by taking it to the landfill with special construction waste bags (6%), selling (4%), burying it (3%), and throwing it in a forest, quarry, ditch, or similar places (1%). 17% of respondents stated that the waste (or part of it) is still stored with them. From the whole survey, this was the main question, revealing also illegal activities, like burying and throwing in the forest. The authors conclude that these actions are caused by a range of reasons, such as:

- Lack of information on the management of construction and demolition waste
- High cost of construction and demolition waste management

In fact, it can be concluded that both problems have quite clear and straightforward solutions. Lack of information can be tackled by:

- Revising the information on the municipality web resources
- Enhancing cooperation between municipalities and waste management companies to provide more educational materials, communication with society, etc.
- Informing the society on waste sorting activities with respect to construction and demolition waste as well (as generally waste sorting is associated with municipal waste solely), to decrease the volume of the waste discharged and thus using the waste sorting points or stations for sorted waste, which is collected free of charge

Although, according to the results of the survey, this covers only a small proportion of the respondents, this result cannot be neglected as it has a direct impact on environmental pollution.

A positive aspect is that a little less than 2/3 of respondents (63%) have sorted the repair and construction waste for disposal, for example, to reduce costs. However, almost 1/3 of respondents (31%) have not done so in general. It is worth noting that out of these 31%, 10% did not sort waste but knew that this way could reduce costs, and 21% out of the respondents clearly stated that they previously did not know that this way could reduce costs.

It can be observed that repair and construction waste was sorted more often by those respondents who carried out repairs in a private house and summer house/garden house, as well as respondents living outside of Riga.

The most convenient way for respondents to obtain information about the disposal of repair/construction waste is on their municipality's website or by calling the municipality's hotline (48%). Likewise, respondents would gladly obtain such information by following means:

- Contacting the company that takes care of the respondent's household waste removal (34%); finding the most advantageous offer for them in internet advertisements, for example, private advertisement platform ss.com (27%)
- In printed informational materials (21%)
- From friends or acquaintances (17%)

Another topic that has been addressed within the survey is secondary use of materials. Here, 41% of the respondents believe that there is a generally high possibility that the leftover materials after repair or construction would be offered to others on a special portal (rather high—26%, very high—15%). However, 1/3 of respondents (33%) indicate that the likelihood of offering leftover repair/construction materials to others is generally low or none (rather low—13%, very low—11%, none—9%). It has been assumed that such distribution of the results is explained by the limited availability of information on potential reuse and by the lack of legislative support or explanations, i.e., on municipal or waste management company resources, providing information on what are the legal and permitted actions that individual can do with construction and demolition waste to foster reuse.

One positive aspect that has been revealed by the survey is that almost half of all respondents (48%) generally know places in their neighborhood where people (individuals) can hand over various repair and construction waste (clearly know where—22%, roughly guess where it could be—26%). However, 45% of the respondents indicated that they do not know if and where people in the neighborhood can hand over various repair and construction waste. 7% of respondents indicated that they know that there is no such place in their neighborhood to hand over repair and construction waste. Those living in Riga and those respondents who have not carried out repair or construction work in the last 5 years more often stated that they do not know if and where people in the neighborhood can hand over various repair and construction waste.

Having identified the main outcomes of the survey's results to be a lack of comprehensive and available information to the public, the authors considered using qualitative research based on text information analysis and applied their own developed program in order to scan municipal Websites with the aim of evaluating the level of available information.

To find out the information available to citizens on how to get rid of construction debris, the information on municipalities' websites was analyzed using the program developed by the authors. The approach used for analysis consists of a two-stage approach. In the first stage, the project is created—specifying the search keywords and web addresses where they will be explored. In this case, they are "Circular economy," "Construction waste," "Construction and demolition waste," "Bulky waste," and "Bulky and construction waste." In the second stage, the user can manage the project to correct the keywords and addresses in which they will be searched. Within this approbation, 6 local government websites were chosen by random selection. Keywords were searched in the Latvian language to find the information available to residents on the municipality's website (Fig. 6).

The search tool has some usage restrictions:

- The tool searches directly for the pre-defined keywords or combinations of words and does not consider the variant with their inflections. To improve search results, it is recommended to do it without word endings.
- The tool searches standard HTML websites. If the site contains non-standard links (e.g., they are created with JavaScript or similar, or JavaScript is used to load the content), then the tool may only find some of the information you are looking for.

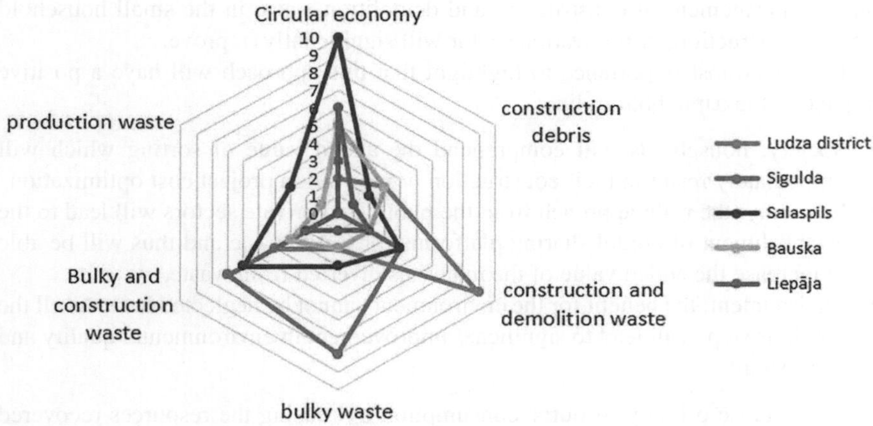

Fig. 6 The result of the scanning of websites of Latvian municipalities, March 2023 (created by authors, based on LIFE Waste to Resources IP (2023a, 2023b))

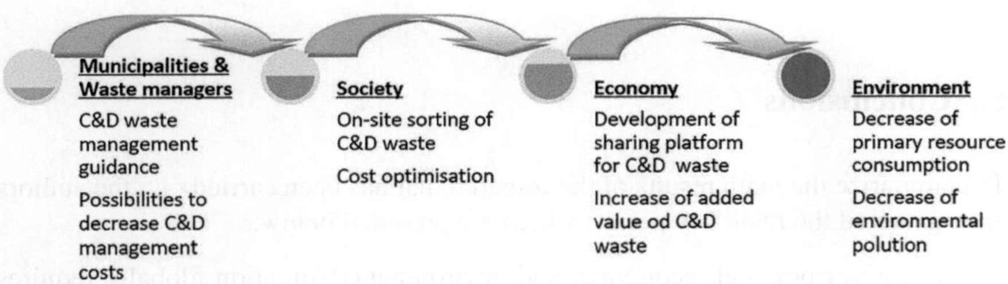

Fig. 7 Approach to increase the efficiency of management of construction and demolition waste

- The tool searches standard HTML websites. The information may not be found if the site contains PDF, DOC, or any other documents with the necessary information (Kuzmina et al., 2023).

Based on the research carried out within this chapter, the authors have developed an approach to solve the outstanding issues that have been identified in Latvia, based primarily on the survey of the society and the gap assessment of existing information and guidance on effective management of construction and demolition waste (Fig. 7).

The authors foresee that this approach can be offered to regulatory institutions. This approach considers that the main precondition for effective and efficient management of construction and demolition waste is providing precise, clear, and comprehensive information on what is construction and demolition waste, how can it be managed (i.e., that it cannot be discarded together with other household waste), how can a household reduce costs by sorting out some of the materials from the total flow. As soon as this information is provided, it is foreseen that the overall situation on the management of construction and demolition waste in the small household repair, construction, or renovation sector will significantly improve.

It is of utmost importance to highlight that this approach will have a positive impact on the triple bottom line:

- Society: households will comprehend the added value of sorting which will immediately result in their construction or renovation project cost optimization.
- Economy: the right approach from the public and private sectors will lead to the establishment of digital sharing platforms for C&D waste and thus will be able to increase the added value of the materials diverted from waste.
- Environment: the benefit for the environment cannot be neglected here. As all the previous steps will lead to significant improvement in environmental quality and ecosystem:
  - Decrease primary resource consumption by reusing the resources recovered from C&D flow
  - Decrease illegal activities, and pollution of the environment as management of C&D waste will become more affordable for households

5 Conclusions

To summarize the main results of the research that has been carried out, the authors have gathered the main key points, which are provided below:

- The current political, economic, and environmental situation globally requires immediate and radical actions to sustain all three bottom lines as well as to secure sustainability for the next generations. One of the most important directions is the implementation of the principles of waste circulation of construction and demolition of buildings because construction waste is an important waste stream that can be turned into a resource. To do this as efficiently as possible, it is necessary to significantly improve the appropriate management of construction waste generated in households and the citizens' awareness of the activities to be performed.
- To motivate households to sort construction and demolition waste at its source and hand it over for reuse and/or recycling, households must have a clearly defined state of play and appropriate infrastructure for the collection of this waste type at the municipal level. The collection system must be comprehensible, simple, and clear. At the same time, the waste collection and management fee should be comprehensible to society.
- As the experience that has been assessed within current research shows, effective results in the beneficial use of household construction and demolition waste can be achieved by creating and developing public platforms and exchange points so that residents can transfer for further use construction waste that can be usefully

used, including construction materials left over from repair, construction and demolition works.

- To find out the information available to society on how to get rid of construction and demolition waste, a program developed by the authors was used—scanning of municipal Websites. The method can be approved and applied also to evaluate the information available to waste producers about the collection system of other waste streams. The successful creation and development of a separate waste collection system depend on the information available to society.
- The authors have developed an approach to increase the efficiency of management of construction and demolition waste which has a beneficial impact on the triple bottom line—society, economy, and environment—offering concrete and practical steps to reach this qualitative improvement.
- In order to achieve a better result in the sorting of construction waste generated in households, shared waste collection, and the arrival of this waste for reuse and recycling, it is necessary to increase the responsibility of not only municipalities and waste managers for informing the population about the management of this waste, but such responsibility should also be determined for the house managers, who are in the closest contact with the population.

The authors also have identified the directions for further research, which would be most important to link in a more coherent manner the household and commercial sectors with respect to the management of construction and demolition waste. The second direction of the research can be a more detailed analysis of international practice to develop a sharing economy-based digital platform and to analyze possibilities of aligning and merging existing control and accounting waste tools with the sharing platform, as well as going beyond—assessing the compatibility of the platforms developed in other countries.

References

Atstāja, D., Koval, V., Purviņš, M., Butkevičs, J., & Mikhno, I. (2022). Construction waste management for improving resource efficiency in the reconstruction of war-destroyed objects. *Economics Ecology Socium, 6*(2).

Byrggeri Kobenhavn. (2022). *Circular economy handbook.* Accessed from Cirkulær økonomi | Byggeri København (kk.dk)

Chen, X., Dong, M., Zhang, L., Luan, X., Cui, X., & Cui, Z. (2022). Comprehensive evaluation of environmental and economic benefits of industrial symbiosis in industrial parks. *Journal of Cleaner Production, 354*, 131635.

Çimen, Ö. (2023). Development of a circular building lifecycle framework: Inception to circulation. *Results in Engineering, 17*, 100861. https://doi.org/10.1016/j.rineng.2022.100861

Cudecka-Purina, N., Kuzmina, J., Butkevics, J., Olena, A., Ivanov, O., & Atstaja, D. (2024). A Comprehensive Review on Construction and Demolition Waste Management Practices and Assessment of This Waste Flow for Future Valorization via Energy Recovery and Industrial Symbiosis. *Energies, 17*, 5506. https://doi.org/10.3390/en17215506

Esteban Cantillo, O. J., & Quesada, B. (2022). Solid waste characterization and management in a highly vulnerable tropical city. *Sustainability, 14*, 16339. https://doi.org/10.3390/su142416339

European Commission. (2012). *Communication from the commission to the European Parliament and the council strategy for the sustainable competitiveness of the construction sector and its enterprises, COM/2012/0433 final.*

European Commission. (2014). *Communication from the commission to the European Parliament, the council, the European economic and social committee and the committee of the regions on resource efficiency opportunities in the building sector, COM/2014/0445 final.*

European Commission. (2018). *Guidelines for the waste audits before demolition and renovation works of buildings.*

Government of Finland. (2021). *A new direction. Circular Economy Strategic Program Proposal 13.01.2021*, p. 33. Accessed from Uusi suunta: Ehdotus kiertotalouden strategiseksi ohjelmaksi - Valto (valtioneuvosto.fi)

Honic, M., Kovacic, I., & Rechberger, H. (2019). Improving the recycling potential of buildings through material passports (MP): An Austrian case study. *Journal of Cleaner Production, 217*, 787–797. https://doi.org/10.1016/j.jclepro.2019.01.212

Hosseinian, A., Ylä-Mella, J., & Pongrácz, E. (2021). Current status of circular economy research in Finland. *Resources, 10*, 40. https://doi.org/10.3390/resources10050040

IRP. (2019). *Global resources outlook 2019: Natural resources for the future we want. Report of the international resource panel.* United Nations Environment Programme. ISBN: 978-92-807-3741-7.

Kuzmina, J., Atstaja, D., Purvins, M., Baakashvili, G., & Chkareuli, V. (2023). In search of sustainability and financial returns: The case of ESG energy funds. *Sustainability, 15*(3), 2716.

Lazarevic, D., Salo, H., & Kautto, P. (2022). Circular economy policies and their transformative outcomes: The transformative intent of Finland's strategic policy programme. *Journal of Cleaner Production, 379*(part 2), 134892. https://doi.org/10.1016/j.jclepro.2022.134892

Leder, N., Kumar, M., & Sanchez Rodrigues, V. (2020). Influential factors for value creation within the circular economy: Framework for waste valorisation. *Resources, Conservation and Recycling, 158*, 104804. https://doi.org/10.1016/j.resconrec.2020.104804

LIFE Waste to Resources IP. (2023a). *Ziņojums par pārstrādāto būvmateriālu aprites sistēmas attīstību mājsaimniecībās un privātajā sektorā.* Accessed from Ziņojums par pārstrādāto būvmateriālu aprites sistēmas attīstību mājsaimniecībās un privātajā sektorā.pdf(Shared) - Adobe cloud storage

LIFE Waste to Resources IP. (2023b). Accessed from LIFE 3.0 - LIFE Project Public Page (europa.eu)

Lizárraga-Mendiola, L., López-León, L. D., & VázquezRodríguez, G. A. (2022). Municipal solid waste as a substitute for virgin materials in the construction industry: A review. *Sustainability, 14*, 16343. https://doi.org/10.3390/su142416343

Mavlutova, I., Kuzmina, J., Uvarova, I., Atstaja, D., Lesinskis, K., Mikelsone, E., & Brizga, J. (2021). Does car sharing contribute to urban sustainability from user-motivation perspectives? *Sustainability, 13*, 10588. https://doi.org/10.3390/su131910588

Mikelsone, E., Atstaja, D., Koval, V., Uvarova, I., Mavlutova, I., & Kuzmina, J. (2021). Exploring sustainable urban transformation concepts for economic development. *Journal Studies of Applied Economics (Estudios de Economía Aplicada), 39*(5). https://doi.org/10.25115/eea.v39i5.5209

Mikhno, I., Ihnatenko, N., Cherniaiev, O., Vynogradnya, V., Atstaja, D., & Koval, V. (2023). Construction waste recycling in the circular economy model. In *IOP Conference Series: Earth and Environmental Science* (Vol. 1126, No. 1, p. 012003). IOP Publishing.

Minixhofer, P., Scharf, B., Hafner, S., Weiss, O., Henöckl, C., Greiner, M., Room, T., & Stangl, R. (2022). Towards the circular soil concept: Optimization of engineered soils for green infrastructure application. *Sustainability, 14*, 905. https://doi.org/10.3390/su14020905

Morseletto, P. (2020). Targets for a circular economy. *Resources, Conservation and Recycling, 153*, 104553. https://doi.org/10.1016/j.resconrec.2019.104553

Niu, Y., Rasi, K., Hughes, M., Halme, M., & Fink, G. (2021). Prolonging life cycles of construction materials and combating climate change by cascading: The case of reusing timber in Finland. *Resources, Conservation and Recycling, 170*, 105555.

Nordic Innovations. (2023). *Circular business models in the Nordic manufacturing industry — Current status and development*. Accessed from FULLTEXT01.pdf (diva-portal.org)

Norouzi, M., Chàfer, M., Cabeza, L. F., Jiménez, L., & Boer, D. (2021). Circular economy in the building and construction sector: A scientific evolution analysis. *Journal of Building Engineering, 44*, 102704.

OECD. (2018). *The macroeconomics of the circular economy transition: A critical review of modelling approaches. Environment Working Paper No. 130, ENV/WKP(2018)4.*

Oluleye, B. I., Chan, D. W., Saka, A. B., & Olawumi, T. O. (2022). Circular economy research on building construction and demolition waste: A global review of current trends and future research directions. *Journal of Cleaner Production, 131927.*

Patricio, J., Kalmykova, Y., Rosado, L., Cohen, J., Westin, A., & Gil, J. (2022). Method for identifying industrial symbiosis opportunities. *Resources, Conservation and Recycling, 185*, https://doi.org/10.1016/j.resconrec.2022.106437

Priess, P., Rajnoha, R., Losert, S., Vogel, S., & Teufel, H. (2017). Sustainable real estate development and its implications on investment: Statistical relations on the case from Austria. *Journal of Security and Sustainability Issues, 6*(3), 419–434. https://doi.org/10.9770/jssi.2017.6.3(8)

Ramsheva, Y. K., Moalem, R. M., & Milios, L. (2020). Realizing a circular concrete industry in Denmark through an integrated product, service and system perspective. *Sustainability, 12*, 9423. https://doi.org/10.3390/su12229423

Rechtsinformationssystem des Bundes. (2013). *Order of the Federal Minister of Agriculture, Forestry, Environment and Water Management on obligations during construction or demolition activities, sorting and processing of waste generated during construction or demolition activities, production and disposal of recycled building materials*. Accessed from https://www.ris.bka.gv.at/GeltendeFassung.wxe?Abfrage=Bundesnormen&Gesetzesnummer=20009212

Rutitis, D., Smoca, A., Uvarova, I., Brizga, J., Atstaja, D., & Mavlutova, I. (2022). Sustainable value chain of industrial biocomposite consumption: Influence of COVID-19 and consumer behavior. *Energies, 15*(2), 466. https://doi.org/10.3390/en15020466

Saeima. (2012). *Waste management law*. Accessed from Atkritumu apsaimniekošanas likums (likumi.lv)Atkritumu apsaimniekošanas likums (likumi.lv)

Satversmes Sapulce. (1922). *Constitution of the Republic of Latvia*. Accessed from https://likumi.lv/ta/id/57980-latvijas-republikas-satversme

Uvarova, I., Mavlutova, I., & Atstaja, D. (2021). Development of the green entrepreneurial mindset through modern entrepreneurship education. *Journal IOP Conference Series: Earth and Environmental Science, 628*, 012034. https://doi.org/10.1088/issn.1755-1315. ISSN: 1755-1315, Print ISSN: 1755-1307, https://iopscience.iop.org/issue/1755-1315/628/1 (SCOPUS) https://iopscience.iop.org/article/10.1088/1755-1315/628/1/012034/pdf

VARAM. (2021). *State waste management plan for 2021-2028*. Accessed from Atkritumu apsaimniekošanas valsts plāns 2021.-2028.gadam. | Vides aizsardzības un reģionālās attīstības ministrija (varam.gov.lv)

Yu, Y., Yazan, D. M., Junjan, V., & Iacob, M.-E. (2022). Circular economy in the construction industry: A review of decision support tools based on information & communication technologies. *Journal of Cleaner Production, 349*, 131335. https://doi.org/10.1016/j.jclepro.2022.131335

Impact of Home Energy Efficiency Labeling Investment on Rent Housing Decision of Students in Spain: A Choice Experiment Approach

Raúl González González, Eva Crespo-Cebada, Carlos Díaz-Caro, and Ángel-Sabino Mirón Sanguino

Abstract The real estate sector is one of the economic sectors which, from the initial production phase to the consumption of the products and services it offers, makes a negative contribution to climate change. In particular, the housing market in Spain presents multiple opportunities to improve in terms of energy efficiency and shape a path towards a green and circular economy. The energy efficiency certificate is one of the instruments that allow improvements in this regard. The purpose of this chapter is to assess students' preferences about the rental housing market and energy efficiency certificate. For this purpose, we propose the use of choice experiments as a tool to quantify these preferences. The study sample consists of a set of 254 students from the region of Extremadura, obtained from a simple random sample of students who rented a flat during their university studies. The main results show that the energy certificate is a relevant attribute in the choice of the apartment to rent; however, when the market segmentation is carried out, different groups are observed.

Keywords Real estate · Choice experiment · Housing decision · Investment · Energy certification

---

R. González González · C. Díaz-Caro (✉) · Á.-S. Mirón Sanguino
Cáceres, Spain
e-mail: rgonzalewm@alumnos.unex.es; carlosdc@unex.es; asmiron@unex.es

E. Crespo-Cebada
Badajoz, Spain
e-mail: ecreceb@unex.es

C. Rego et al. (eds.), *Transitioning to a Circular Economy*, Studies on Entrepreneurship, Structural Change and Industrial Dynamics,
https://doi.org/10.1007/978-3-031-77661-8_13

1 Introduction

The real estate sector is one of the economic sectors that, from the initial production phase to the consumption of the products and services it offers, makes a negative contribution to climate change. Specifically, both the use of products that are necessary for the development of housing, as well as the use of energy resources to carry out the use of housing, are key factors to improve and be more efficient in terms of sustainable development and even the achievement of the Sustainable Development Goals (SDGs).

In particular, the housing market in Spain presents multiple opportunities to improve in terms of energy efficiency and shape a path towards a green and circular economy. Thus, both the European Union and the national level have created and defined a system for classifying housing by means of labels according to their degree of energy efficiency and, therefore, better use of resources for the use of real estate services (Brounen & Kok, 2011).

The energy performance certificate or energy certificate is an official document developed by a competent expert that provides objective information on the energy performance of a property. This certification assesses the energy performance of the dwelling by calculating the annual energy consumption necessary to meet the energy demand of a building under normal occupancy and operating conditions (including the production of hot water, heating, lighting, cooling and ventilation). The energy efficiency rating scale is a seven-letter scale ranging from A (most energy efficient) to G (least energy efficient).

This labeling is part of the measures aimed at improving efficiency in energy consumption and promoting sustainability by reducing emissions into the environment, as well as reducing energy use, which, in turn, is one of the main lines of action of the Kyoto Protocol international pact (protocol of the United Nations Framework Convention on Climate Change). Therefore, this policy can be considered an instrument that enables and enhances the transition to a greener and more circular economy, contributing not only to economic savings but also in environmental terms (Kwak et al., 2010).

According to Royal Decree 235/2013, the energy efficiency certificate has been mandatory in Spain since the beginning of June 2013 for all homes intended for rent or sale from that date onwards. In fact, this classification has led to significant investments to improve the property with the dual objective of complying with the legal requirements for sale or rental and increasing the value of the asset in response to new business models that require greater knowledge to adapt the homes to efficiency improvements that were not previously demanded. In turn, this new business system will require the start of entrepreneurship that will foster the creation of companies with a greater technological component, given that it requires the specialization of entrepreneurs towards the efficiency improvements. These business models will also fall within the scope of the green and circular economy, contributing to the reduction of emissions in the real estate sector.

In turn, student housing rental is a niche market that generates significant profits in different European countries such as Germany (Miessner, 2021) or the United Kingdom (Rugg et al., 2002). However, this segment has not been analyzed and studied in depth in the academic field (Zróbek-Rózańska & Szulc, 2018). Thus,

Spain is also a country where young students choose to rent housing while studying at university and, therefore, presents a market gap that can be exploited and improved both in economic and environmental terms.

Knowledge of preferences about the main characteristics of rental flats has recently been addressed in the literature (Franke & Nadler, 2019; Heinzle et al., 2013; Kim & Lee, 2022; Kwak et al., 2010; Lee et al., 2018). However, research on the youth segment of the rental market presents an opportunity for further investigation given the decision-making power of this group.

According to the type of data available, research on residential preferences often uses one of two approaches: revealed preference (RP) data, based on actual market behavior; or stated preference (SP) data, based on subjects' responses to hypothetical scenarios. In addition to these data-driven analysis methods Anas and Arnott (1991) introduced a dynamic simulation model of the housing market with taste heterogeneity. Using PR data to assess residential choice behavior, spatial correlations between alternatives unavoidably emerge, i.e., individuals may prefer to consider two dwellings that are both similar in some attributes as being quite similar (McFadden, 1973), and a positive relationship emerges between similar attributes and spatial location proximity of dwellings.

Hence, PR choice data were not suitable for use in this study, as it looked at the nature of general residential choice behavior rather than predicting a specific residential market situation. If detailed data were available not only on residential choice but also on hedonic prices, it could be feasible to analyze preference heterogeneity using a combination of both. RP data are very trustworthy because they are indeed based on the actual choices of individuals in the market but are also relatively unreliable (Louviere et al., 2000). For all these reasons, it has been decided to carry out an analysis based on the SP to analyze the preferences in the residential market and the effect of energy certification on the Spanish student population.

The aim of this chapter is to analyze students' preferences about the rental housing market. For this purpose, we propose the use of choice experiments as a tool to quantify these preferences. This approach has been widely used in the measurement of preferences in the housing market (Collins & Curtis, 2018; Jiang & Chen, 2016; Kim & Lee, 2022; Kwak et al., 2010). Furthermore, we aim to segment the sample by applying the latent class technique to quantify unobservable heterogeneity. The study sample consists of a set of 254 students from the region of Extremadura obtained from a simple random sample of students who rented a flat during their university studies.

1.1 Literature Review

In this section we will try to provide a review of the literature that we will later use in the discussion section. A literature review will be carried out on the use of choice experiments for the analysis of decisions and choices in Real Estate, the impact of energy certification on the choice of place of residence and the impact of real estate eco-labeling.

First of all, the work of Marmolejo-Duarte and Ruiz-Lineros (2013) analyzes the usefulness of choice experiment models to carry out decision analysis in the field of long-term consumer products, such as those related to Real Estate. These authors implement a choice experiment to analyze the possibilities of rehabilitation in an industrial area in Catalonia. In addition, they use latent class models to perform market segmentation. The results show the usefulness of the application of choice models in this sector.

For their part, Ibraimovic and Hess (2018) also carry out a model based on a choice experiment with the application of latent classes to analyze the choice in the location, given the importance that they contemplate the preferences in the choices, such as the location of the residential property. Thus, they show the importance of latent class models for analyzing unobservable heterogeneity in residential choice and more specifically between different ethnic groups in Switzerland. The main results show how ethnic attributes shape the latent classes while the origin of the person is the best predictor of group membership.

Another paper (Hoshino, 2011) that also addresses the heterogeneity of preferences in residential choices shows how relevant the measurement of preferences, and in particular heterogeneous preferences in Japan, is for effective and efficient planning and development. The methodological approach consists of a mixed logit model with different attributes such as income, floor area, building age, distance to transportation, etc. The main results show a high degree of preference heterogeneity in many attributes.

In turn, Heinzle et al. (2013) analyze with a choice experiment to what extent the features related to green energy certification are included in the preferences of potential real estate investors in Singapore. The results show that investors would be willing to pay more for a property with a green certificate, with variations depending on the levels of the green certification program, increasing according to the green requirements of the program. These results highlight the innovation in the venture on the possibilities that this type of certification presents for the green economy and its high demand by investors.

The energy efficiency levels of rental apartments tend to be lower compared to owner-occupied properties; however, an improvement in the same can increase the income and profits of owners who dedicate it to renting. Collins and Curtis (2018) analyze the willingness to pay Irish tenants based on the improvement in the scales of energy efficiency certification, although they show that there is an overvaluation of energy labels given the costs incurred.

For Korea, it is worth mentioning the work of Kim and Lee (2022), who try to analyze consumer preferences on the energy efficiency of housing in the real estate market, also applying choice models. The results obtained show a small preference for this attribute, with only 20% of consumers considering the energy certificate. In this sense, the authors indicate that there could be an improvement in the information program to achieve the objective and be more and better valued by consumers. In this line of improvement of energy certificates, the work of Ramos et al. (2015) also applies to the residential sector.

Also for Korea Lee et al. (2018) analyze the disclosure of energy efficiency information for the reduction of CO_2 emissions, through a consumer residential choice experiment in South Korea. The results show that there is a significant impact on residential choice is obtained. Therefore, the communication of the formation of such a certificate by landlords contributes to better decision-making by buyers/landlords, in addition to the possible premium obtained for its improvement.

Franke and Nadler (2019) analyze the German market with various methodologies, including the choice experiment for 206 tenants. The results obtained show a divergence between landlords who attach greater importance to the energy label than tenants. The proposals for improvement to increase its valuation focus on the following aspects: regulation and transparency and compliance, environmental awareness and communication of economic incentives.

Continuing with heterogeneity, Gibler and Tyvimaa (2014) show the importance of segmentation, in their case by cluster analysis, of residents' demand with respect to housing, given the diverse population in Finland. These results are of great importance for builders and entrepreneurship and the circular economy given the non-standard demand in housing, presenting multiple opportunities for innovation to meet demand and improve business profits.

Students as tenants of real estate are an undervalued and understudied aspect of the market. To date, there is a widespread view that students' decisions in the real estate market and their reach are mostly confined to the realm of academic institutions. However, as they expose Zróbek-Rózańska and Szulc (2018) the case of Poland, the student market presents a relevant niche market to consider in business aspects.

In Germany Miessner (2021) also conducted a study on the student population and showed how benefits are obtained by applying rental strategies for that population. Also Rugg et al. (2002) highlight the importance of the student population in the UK and how demand and preferences can vary, indicating the need to understand the dynamics of rental market development for this niche market.

2 Methodology

The database to carry out the objective of this work has been obtained from a questionnaire carried out to a sample of students who have rented an apartment as a place of residence to pursue their higher education studies. The final sample consists of 254 individuals collected during the year 2021 through different ways (online, social networks, emails, face-to-face, etc.). The sample obtained was compiled by means of simple random sampling without being able to count on objective representativeness given that the characteristics of the students who rent an apartment in Spain are unknown.

Simple random sampling is distinguished by the fact that each element of the population has the same probability of being selected to form part of the sample.

Table 1 Statistic descriptive of the sample

Variable	Mean
Age	21.66
Gender (woman)	62.99%
Income less 900€/month	24.80%
Between 900 and 1500 €/month	42.68%
Between 1501 and 2500 €/month	12.20%
More than 2501 €/ month	20.33%

Source: Own compilation

This type of sampling is distinguished by its simplicity and easy comprehension (Seden Magazine, 2006).

The elaboration of this questionnaire has been carried out through Google Forms, a type of online tool increasingly used in research for its advantages in flexibility, speed in the collection of information and lower cost than with traditional surveys (Kayser et al., 2013; Koutsimanis et al., 2012; Ortiz et al., 2020; Sama et al., 2018).

Table 1 shows the different variables chosen to describe the sample, as well as their respective means.

It should be noted, first, that the average age of the people who responded to the survey is 22 years old. Most of them are young students who have had to look for an apartment every year to carry out their careers or work in the most comfortable way possible. Thus, in this way, they must choose a home from among many, according to their preferences and personal tastes, and this is what we have tried to capture in this work. Another characteristic to highlight is the average income of the sample, which ranges between 900 and 1500 euros per month.

2.1 Choice Experiment

Choice experiments are based on the theory of random utility (McFadden, 1973). This theory proposes that the utility of a good is decomposed into an observable component that is a function of a vector of attributes and characteristics of the individual and an unobservable component known as error.

Given certain assumptions for the error distribution, the probability of choosing an option can be expressed as a logistic distribution (McFadden, 1973). The attributes vary at different levels, which combine to generate specific scenarios (alternatives) that are selected from a universe of possible scenarios (Hensher et al., 2005; Louviere et al., 2000). These scenarios are presented to participants in a series of choice sets, where each set contains two or more scenarios.

By repeating the sets and systematically varying the levels of the attributes, the researcher can infer which attributes influence the choice of a given scenario and evaluate transactions between attributes.

Table 2 below breaks down the attributes chosen to make comparisons for the choice experiment. Each attribute is represented by its different levels and the coding used to conduct the experiment and obtain the results.

Table 2 Attributes and levels of the choice set

Attribute	Levels	Code
Square meter	60	M1
	90	M2
	120	M3
	150	M4
House condition	Bad	Bad
	Regular	Reg
	Good	Good
Location	Center	Cent
	Residential	RESD
	Surrounding	SURND
Rent cost	300	Price
	400	
	500	
	600	
Energy certificate	A	A
	B	B
	C	C
	D	D
	E	E
	F	F
	G	G

Accounting for the 4 attributes previously mentioned, given the different levels in each case, a total of 6480 (3*3*3*3*3*3*3*3 = 81) possible combinations of possible plausible scenarios can be set up. Given the high number of resulting comparisons, due to economic and time constraints, it was decided to implement a factorial design (Hole, 2015). This approach returned a total number of 16 alternatives, which implied that each respondent was faced with a set of 8 options. This kind of design practice is commonly used in choice experiments (Louviere et al., 2000). An example choice set is presented in Table 3.

2.2 Econometric Model

To analyze the preferences on apartment renting and more specifically on the energy certificate, this study has used the conditional logit model and latent classes to study the heterogeneity of the responses obtained.

This model comes from the random utility models (McFadden, 1973), which assume that the utility function of each individual is the sum of two terms, a deterministic part that can be described as a function of the factors influencing the utility of individuals and another random, unobserved, and which is considered stochastic.

Table 3 Choice set example

Comparison 1	A	B	C
Square meters	60	60	No option
House condition	Regular	Good	
Location	Center	Residential	
Rent cost	400	400	
Certificate	D	C	
Selected option	()	()	()

Source: Own compilation

Following Revelt and Train (1998), we assume a sample of N individuals with the option of choosing J alternatives on T occasions so that the utility of an individual n derived from the choice of alternative j on occasion t is as follows:

$$U_{njt} = \beta_n^{'} x_{njt} + \varepsilon_{njt} \tag{1}$$

$\beta_n^{'}$: vector of coefficients specific to each individual.

x_{njt}: is the vector of observable attributes of individual n and alternative j at choice occasion t,

ε_{njt}: independent and identically distributed extreme value.

The probability of responding n by choosing alternative i at choice t is given by the following expression:

$$L_{nit}\left(\beta_n\right) = \frac{\exp\left(\beta_n^{'} x_{nit}\right)}{\sum_{j=1}^{J} \exp\left(\beta_n^{'} x_{njt}\right)} \tag{2}$$

Expression 2 is the conditional logit formula (McFadden, 1973). In this paper we will use the simulation approach (Train, 2009), in which the log likelihood is given by Eq. (3):

$$\text{SLL}\left(\theta\right) = \sum_{n=1}^{N} \ln \left\{ \frac{1}{R} \sum_{r=1}^{R} S_n\left(\beta'\right) \right\} \tag{3}$$

where R is the number of repetitions and β' is the rth draw from $f(\theta)$.

In order to identify unobservable heterogeneity and groups, latent classes are applied, and their estimation is carried out on the basis of the following:

$$\text{SLL}\left(\theta\right) = \sum_{n=1}^{N} \ln \left\{ \sum_{q=1}^{Q} H_{nq} \prod_{t=1R}^{T} \prod_{t=1R}^{J} \left[\frac{\exp\left(x_{njt}^{'} \beta_n^{r}\right)}{\sum_{j=1}^{J} \exp\left(x_{njt}^{'} \beta_n^{r}\right)} \right]^{y_{njt}} \right\} \tag{4}$$

where H_{nq} is the probability of belonging to a certain class and is obtained from:

$$H_{nq} = \frac{\exp\left(z_n^t \gamma_q\right)}{\sum_{q=1}^{Q} \exp\left(z_n^t \gamma_q\right)} \tag{5}$$

Therefore, the functional form of the U_{njt} derived from individual n for alternative j in choice set t can be defined as follows:

$$\begin{aligned} U_{njt} = {} & \beta_0 \text{ASC} + \beta_1 M2_{njt} + \beta_2 M3_{njt} + \beta_3 M4\_2_{njt} + \beta_4 \text{REG}\_3_{njt} + \beta_5 \text{GOOD}_{njt} + \\ & \beta_6 \text{RESD}_{njt} + \beta_7 \text{SURND}_{njt} + \beta_8 A_{njt} + \beta_9 B_{njt} + \beta_{10} C_{njt} + \beta_{11} D_{njt} + \beta_{12} E_{njt} + \\ & \beta_{13} F_{njt} + \beta_{14} \text{PRICE}_{njt} + \varepsilon_{njt} \end{aligned} \tag{6}$$

The variables taken as a basis were 60 meters ($M1$), bad condition (Bad), center (CENT) and level G of the energy certificate (G).

By introducing the price coefficient in the choice experiment, it is possible to calculate the marginal rate of substitution between a non-price coefficient and the price attribute, i.e. the willingness to pay, and it is calculated:

$$\text{WTP}_K = -\left(\beta_K / \beta_{\text{Price}}\right) \tag{7}$$

WTP indicates how much we would be willing to pay in monetary terms for each increase in the level of attribute K provided by the rent of the property.

3 Results

The results obtained after applying the econometric logit regression to the model previously described in Eq. (6) are shown below. Table 4 shows the results of the choice experiment for the overall sample.

The results obtained cannot be interpreted in quantitative terms, only in terms of their negative or positive influence on the utility function, that is, on the probability of choosing a particular property with these characteristics. Thus, a positive coefficient will indicate that this attribute has a positive utility or that the apartment containing this attribute will have a higher probability of being chosen compared to the apartment that does not have this attribute. The P-value will show us the significance of the coefficient, whether it is statistically different from zero. The Z-value will indicate the relative importance of the level of each attribute, i.e., the higher the Z-value in absolute terms, the greater the relative importance of the coefficient in the choice.

As can be seen, the attribute square meters present positive coefficients in the three levels that have been divided, although only coefficient M4 is statistically significant. In other words, there is a greater probability of choosing a larger apartment compared to a smaller apartment. On the other hand, as for the condition of the

Table 4 Results of choice experiment

Variable	Coefficient (Std. Err.)	*Z*-value	*P*-value
M2	0.0070 (0.1997)	0.04	0.972
M3	0.1418 (0.1961)	0.72	0.470
M4	0.3745 (0.1578)	2.37	0.018
REG	11.776 (0.1725)	6.82	0.000
GOOD	16.755 (0.1614)	10.38	0.000
RESD	−0.1219 (0.1212)	−1.01	0.314
SURND	−0.6914 (0.1109)	−6.23	0.000
PRICE	0.0009 (0.0001)	6.09	0.000
A	0.7980 (0.1792)	4.45	0.000
B	0.5536 (0.1613)	3.43	0.001
C	0.6472 (0.1313)	4.93	0.000
D	0.2515 (0.1253)	2.01	0.045
E	0.4735 (0.1418)	3.34	0.001
F	0.3202 (0.1415)	2.26	0.024
ASC	11.649 (0.2606)	4.47	0.000

Source: Own compilation

dwelling, both coefficients are positive and statistically significant, which seems to indicate that a better state of repair of the dwelling has a positive influence on the choice.

On the other hand, the location of the dwelling presents negative coefficients (RESD and SURND) compared to the downtown location. Only the surrounding is statistically significant, which shows that the central area is an attribute that influences apartment selection to a greater extent than other locations.

The price turns out to be positive and statistically significant. Energy certification through its different grades (A–G) shows that all coefficients are statistically significant and positive, which shows that certification has a positive effect on the rental of the dwelling.

On the other hand, given the relative importance of the coefficient given by the *Z*-value, it can be observed that the highest value in positive terms of the *Z*-value is found in the coefficient of good housing condition, followed by fair. What is indicated is that housing condition is the first attribute in terms of relative importance for the selection of a student apartment rental. These coefficients are followed by location, price, and energy certificate in the negative, with size in the last place.

Table 5 presents the results after applying the latent class model. Three fit indicators are shown: Akaike's Information Criterion (AIC), Corrected Akaike's Information Criterion (CAIC) and Bayesian Information Criterion (BIC). The number of classes chosen will be determined by the smallest value of these three criteria.

Table 6 shows the latent class model. Each latent class is a group of individuals (a series of tenants). The sample of 254 people has been divided into 3 groups, since, as shown in Table 6, the indicators take the smallest value for class 3.

Table 6 shows the probability of belonging to each class; specifically there is a 20.5% probability of belonging to the first class, a 45.3% probability of belonging to the second class and a 34.2% probability of belonging to the third class.

Table 5 Latent class model

Classes	LLF	NPARAM	AIC	CAIC	BIC
2	−2343.784	31	4749.568	4890.226	4859.226
3	−2235.924	47	4565.848	4779.103	4732.103
4	−2214.285	63	4554.577	4840.423	4777.423
5	−2173.057	79	4504.113	4862.563	4783.563

Source: Own elaboration

Table 6 Three latent classes model

Variable	CLASS 1	CLASS 2	CLASS 3
ASC	3.101	0.350	10.399
PRICE	−0.005	0.001	0.001
M2	−0.024	0.006	0.424
M3	−2.400	0.258	1.755
M4	1.778	0.320	0.291
REG	5.331	0.406	7.421
GOOD	10.228	0.471	8.399
RESD	0.172	0.429	0.970
SURND	2.436	0.129	−2.578
A	19.423	0.555	1.156
B	25.063	−0.167	3.619
C	−0.389	0.231	1.748
D	7.206	−0.386	3.453
E	25.102	−0.130	3.481
F	20.378	−0.212	2.634
CLASS SHARE	0.205	0.453	0.342

Source: Own elaboration

Table 7 shows the results for each of the three classes that have been obtained.

In class 1, we can observe that all variables are not statistically significant, only the B level of green certification. The value of *Z* determines the relative importance of the variables from highest to lowest. In this case, the level of energy certification has the highest relative importance. The level of energy efficiency positively influences the final decision of users. Class 1 has been characterized because they have given the same importance to the condition of the house as to the level of energy efficiency. Therefore, for this class, the energy efficiency certificate can be a relevant attribute in the rental market.

In the table above for class 2, the price and the condition of the house (good) have a strong and positive influence on the rental decision, as do the different levels of energy certification (A and D). In this case, the value of *Z* indicates that the relative importance lies in the price, the location of the property (RESD) and the condition of the property (good). This class seeks to rent a house at a good price and in good condition.

Finally, for class 3 has positive and significant coefficients for price and a good and fair state of repair, in addition to the energy certificate, making it the only class with all significant coefficients. On the other hand, it presents negative and significant coefficients only for properties located in the surrounding area. In terms of relative importance, the location of the property is in first place, followed by the state of conservation and the level of the energy certificate.

It has not been possible to calculate the willingness to pay for each attribute according to Eq. (7) due to the result obtained in the price for which a positive coefficient of the same has been obtained.

Table 7 Table of results for the estimation of latent classes

	Class 1			Class 2			Class 3		
	Coefficient	*Z*-value	*P*-value	Coefficient	*Z*-value	*P*-value	Coefficient	*Z*-value	*P*-value
ASC	17.822	0.02	0.984	0.3505	1.04	0.297	103.999	3.46	0.001
PRICE	−0.0056	−0.41	0.682	0.0007	2.93	0.003	0.0125	1.95	0.051
M2	−21.797	−0.02	0.984	0.0064	0.02	0.982	0.4230	0.46	0.649
M3	−40.460	−0.06	0.950	0.2586	0.94	0.346	17.547	1.37	0.172
M4	0.6995	0.01	0.990	0.3198	1.45	0.147	0.2910	0.79	0.427
REG	48.638	0.11	0.916	0.4060	1.78	0.075	74.224	3.42	0.001
GOOD	106.118	0.37	0.715	0.4714	2.37	0.018	84.014	4.20	0.000
RESD	−0.6657	−0.01	0.988	0.4295	2.82	0.005	0.9689	0.90	0.367
SURND	22.333	0.06	0.954	0.1291	0.86	0.391	−25.784	−4.93	0.000
A	198.460	0.33	0.739	0.5549	1.82	0.069	11.562	1.77	0.077
B	279.248	0.00	0.000	−0.1672	−0.67	0.504	36.183	4.31	0.000
C	0.6909	0.01	0.990	0.2316	1.33	0.185	17.468	3.45	0.001
D	83.885	0.45	0.653	−0.3854	−2.15	0.031	34.524	3.42	0.001
E	268.947	0.49	0.623	−0.1296	−0.63	0.530	34.794	2.80	0.005
F	207.168	0.26	0.797	−0.2117	−0.90	0.369	26.341	4.14	0.000

Source: Own elaboration

4 Discussion

The aim of this work has been to analyze the preferences of youth tenants on real estate and how the level of energy certification affects their choices as a tool to contribute to a more neutral economy in emissions and with lower impacts on climate change. In this sense, the energy certification of housing is a challenge to contribute to these objectives, especially if it is valued positively as seen in the work. It is worth highlighting the importance of the student tenant sector in the Spanish real estate market and how it is both an opportunity and a niche market as in other countries (Miessner, 2021; Rugg et al., 2002; Zróbek-Rózañska & Szulc, 2018).

First, it is worth noting the preferences of the overall sample, which have allowed us to identify the drivers that tenants consider when making their choices. First, the size attribute is not notable in general terms, since its coefficient is not significant except in the case of a large apartment compared to a smaller apartment. This may be explained by the fact that tenants are looking for rooms and not apartment size, with the overall size of the apartment being indifferent. In this case, the results are not in line with previous literature dealing with the size in square meters of the apartment (Heinzle et al., 2013; Hoshino, 2011; Jiang & Chen, 2016). A possible explanation for the differences in the results may be motivated by several characteristics. Firstly, the sample analyzed are students; secondly, the geographical area and different preferences are analyzed; and finally, the possible effect of the characteristics of the real estate market in the countries are studied.

The second attribute analyzed refers to the state of repair of the property, i.e., whether the property is in poor, fair, or good condition. This attribute is the one that respondents attach the greatest relative importance to, possibly because they are going to live in the property for a period and want to find a sufficiently pleasant level to rent the property. In addition to the fact that the real estate market in Spain for students has a wide variety of housing conditions, characterized by regular levels on many occasions, which leads or may explain why students want to rent an apartment in good condition. The results are in line with other studies that directly or indirectly address the condition of the housing, either by the characteristics of the housing or by the number of years that the housing has been in good condition (Gibler & Tyvimaa, 2014; Heinzle et al., 2013; Hoshino, 2011; Jiang & Chen, 2016).

The third attribute, the location of the property, is also a factor that is considered when selecting a property, especially if the property is not located outside the city and therefore students show a clear preference for a central location or at least in the city, since they are indifferent to renting an apartment in the center of the city than in the outskirts of the city. This can be explained by the tendency of students to live in the city and the city centers where most of the leisure and cultural services are located, although it is true that most of the universities in Spain are on the outskirts of the cities. Once again, the results coincide with previous studies that analyze the location of housing, either by location near transport links or by the location of the university (Hoshino, 2011), proximity to the main streets of the city (Jiang & Chen,

2016) or by an approach similar to the one followed in this work (Heinzle et al., 2013).

Furthermore, the coefficient of price turns out to be statistically significant and positive, which shows a positive preference for price, which contradicts what economic theory establishes for the case of the price of any normal type of good. However, housing is not a normal good, at least in certain segments, and, in addition, the situation occurs in which, in this study, when surveying students, the figures of the buyer and payer surely differ, since although the students have their preferences, the financial action of payment is carried out by the parents who finance the degree and the expenses involved. Therefore, this coefficient may be positive and be contrary to the results obtained in the literature on housing (Gibler & Tyvimaa, 2014; Heinzle et al., 2013; Hoshino, 2011; Jiang & Chen, 2016; Liao et al., 2015; Matsunaka et al., 2013). However, there are several papers in the literature that show positive coefficient results for choice experiments (Gao et al., 2019; Richetin et al., 2022).

Finally, the energy certificate attribute turns out to be statistically significant and positive in all its coefficients, which indicates that tenants show a positive preference for the fact that the dwelling obtains the corresponding energy certificate, and, in addition, they value a more efficient level of the certificate to a greater extent compared to lower levels. This factor is attractive to tenants who, on the one hand, find many dwellings with very low energy certificates and, on the other hand, a higher energy certificate leads to greater energy savings for the tenant in terms of electricity bills, among others, generating a higher level of well-being for the tenant. The literature is not conclusive regarding the impact of the energy certificate on choice and preferences, with some studies showing a negative impact (Kim & Lee, 2022; Ramos et al., 2015) while other studies show positive results of energy certification (Franke & Nadler, 2019; Heinzle et al., 2013; Lee et al., 2018). However, most of the studies agree that communication and knowledge of the certificates are relevant for an adequate contribution to the elections.

The analysis applying latent classes has allowed segmenting tenants to consider heterogeneity and differentiated behavior. This effect is supported by the literature in which a certain degree of heterogeneity in preferences is observed (Gibler & Tyvimaa, 2014; Ibraimovic & Hess, 2018; Marmolejo-Duarte & Ruiz-Lineros, 2013). Specifically, a few latent classes have been obtained based on the estimated coefficients.

With respect to the probability of belonging to the classes, there is a difference between the second class and the first and third classes. Therefore, there is a higher probability of belonging to the second class than there is of belonging to the first or third class. The first group can be defined as an indifferent group or class, since most of the coefficients are not significant and it is also the group with the lowest probability of the three; these indifferent tenants are characterized by being satisfied with any of the levels offered without showing a clear preference among any of them.

The second group or class, which is the most likely, is therefore the one that the largest number of tenants choose to have. This class is characterized by being more concerned about the location and condition of the housing and being more indifferent to the energy certification; therefore, these tenants can be considered as traditional or conservative, in which the most standardized attributes in the housing market are the ones that influence them to a greater extent.

Finally, the third group or class, which is the second in terms of probability or size, is the group most similar to the results of the overall sample, presenting positive preferences both for the most standardized aspects of housing and for the energy certificate, which is placed as one of the most important attributes, in addition to presenting significant and positive coefficients for all of them. The preferences of this group seem to indicate that they could be tenants more aware of climate change and green and circular economy. This analysis allows us to observe the heterogeneity that is not directly observable and segment the tenants into different groups to know their preferences and establish different strategies and not only consider a single market strategy. In addition, it is observed that the energy certificate still has a long way to go due to the lack of interest of the respondents; only one group has a positive preference for it. In this sense, the development of knowledge and publicity about this type of instrument to combat climate change and move towards a greener and more circular economy should be a political and business objective in order to increase the benefits of the company and develop this market niche.

5 Conclusions

Effective and efficient planning and development of residential environments requires clarifying the nature of residential preferences. The objective of this work was to investigate the preferences of student-age tenants in the rental market and, more specifically, how the energy efficiency certificate influences their choice of property. Residential preferences are likely to be heterogeneous, so standard econometric models that assume a single type of preference are not optimal. In this study we employed experimental joint choice methods with a latent class approach, which allowed us to deal with heterogeneity among respondents. Our results revealed significant heterogeneity with respect to some residential attributes.

The present application contributes to opening new avenues for quantitative research in real estate design. Real estate design, although the relevant attributes to be evaluated depend on the specific project being evaluated and the socio-political context in which it is developed, so choice experiment could be used in a wide range of real estate developments.

This work allows us to obtain different strategies in the Spanish real estate rental market. Firstly, we highlight the incorporation and exploitation of the student market niche as a way of increasing income, as has been shown in other countries both at the European and international levels. The second of the strategies deals with how to address the issue of energy certification in the rental market, for which much remains to be done even though this is a key instrument in the sector for a greener and more circular development. Thus, the undertaking of new formulas for the certificate so that it is known and people become more aware of this factor is relevant for the real estate market. Finally, certification currently exists at relatively low levels, both in terms of the number of certifications and the results that are produced and valued; therefore, an improvement and rehabilitation of housing along these lines will contribute significantly to global improvement and climate change.

However, the work presents certain limitations that could well be considered for future lines of research. Firstly, the consideration of a broader study over time will result in obtaining more consistent information over time, especially given the changes experienced after the pandemic. Second, the development of variables that attempt to explain the clusters obtained through a survey in which more information about the tenants would be obtained. Finally, the development and implementation of behavioral economics techniques in the experiment would help to understand in greater depth the choices made.

References

Anas, A., & Arnott, R. J. (1991). Dynamic housing market equilibrium with taste heterogeneity, idiosyncratic perfect foresight, and stock conversions. *Journal of Housing Economics, 1*(1), 2–32.

Brounen, D., & Kok, N. (2011). On the economics of energy labels in the housing market. *Journal of Environmental Economics and Management, 62*(2), 166–179. https://doi.org/10.1016/j.jeem.2010.11.006

Collins, M., & Curtis, J. (2018). Rental tenants' willingness-to-pay for improved energy efficiency and payback periods for landlords. *Energy Efficiency, 11*(8), 2033–2056. https://doi.org/10.1007/s12053-018-9668-y

Franke, M., & Nadler, C. (2019). Energy efficiency in the German residential housing market: Its influence on tenants and owners. *Energy Policy, 128*(C), 879–890. https://doi.org/10.1016/j.enpol.2019.01.052

Gao, Z., Yu, X., Li, C., & McFadden, B. R. (2019). The interaction between country of origin and genetically modified orange juice in urban China. *Food Quality and Preference, 71*, 475–484. https://doi.org/10.1016/j.foodqual.2018.03.016

Gibler, K. M., & Tyvimaa, T. (2014). The potential for consumer segmentation in the Finnish housing market. *Journal of Consumer Affairs, 48*(2), 351–379. https://doi.org/10.1111/joca.12037

Heinzle, S. L., Boey Ying Yip, A., & Low Yu Xing, M. (2013). The influence of green building certification schemes on real estate investor behaviour: Evidence from Singapore. *Urban Studies, 50*(10), 1970–1987. https://doi.org/10.1177/0042098013477693

Hensher, D., Rose, J., & Greene, W. (2005). *Applied choice analysis: A primer*. Cambridge University Press.

Hole, A. (2015). *DCREATE: Stata module to create efficient designs for discrete choice experiments*. Boston College Department of Economics.

Hoshino, T. (2011). Estimation and analysis of preference heterogeneity in residential choice behaviour. *Urban Studies, 48*(2), 363–382. https://doi.org/10.1177/0042098010363498

Ibraimovic, T., & Hess, S. (2018). A latent class model of residential choice behaviour and ethnic segregation preferences. *Housing Studies, 33*(4), 544–564. https://doi.org/10.1080/02673037.2017.1373749

Jiang, H., & Chen, S. (2016). Dwelling unit choice in a condominium complex: Analysis of willingness to pay and preference heterogeneity. *Urban Studies, 53*(11), 2273–2292. https://doi.org/10.1177/0042098015593023

Kayser, M., Nitzko, S., & Spiller, A. (2013). Analysis of differences in meat consumption patterns. *International Food and Agribusiness Management Review, 16*(2), 43–56.

Kim, J., & Lee, J. (2022). Impact of home energy efficiency labeling program on housing decisions in South Korea: A choice experiment approach. *Journal of Housing and the Built Environment, 37*(1), 483–504. https://doi.org/10.1007/s10901-021-09854-9

Koutsimanis, G., Getter, K., Behe, B., Harte, J., & Almenar, E. (2012). Influences of packaging attributes on consumer purchase decisions for fresh produce. *Appetite, 59*(2), 270–280. https://doi.org/10.1016/j.appet.2012.05.012

Kwak, S. Y., Yoo, S. H., & Kwak, S. J. (2010). Valuing energy-saving measures in residential buildings: A choice experiment study. *Energy Policy, 38*(1), 673–677. https://doi.org/10.1016/j.enpol.2009.09.022

Lee, H., Lee, M., & Lim, S. (2018). Do consumers care about the energy efficiency of buildings? Understanding residential choice based on energy performance certificates. *Sustainability (Switzerland), 10*(11). https://doi.org/10.3390/su10114297

Liao, F. H., Farber, S., & Ewing, R. (2015). Compact development and preference heterogeneity in residential location choice behaviour: A latent class analysis. *Urban Studies, 52*(2), 314–337. https://doi.org/10.1177/0042098014527138

Louviere, J. J., Hensher, D. A., & Swait, J. D. (2000). *Stated choice methods: Analysis and applications*. Cambridge University Press. https://doi.org/10.1017/CBO9780511753831.008

Marmolejo-Duarte, C., & Ruiz-Lineros, M. (2013). Using choice-based-experiments to support real estate design decisions. *Journal of European Real Estate Research, 6*(1), 63–89. https://doi.org/10.1108/17539261311312979

Matsunaka, R., Malaithm, S., Oba, T., Yoon, J., & Nakagawa, D. (2013). An analysis of residential location choice behavior in Bangkok metropolitan region: An application of discrete choice models for the ranking of alternatives. *Journal of the Eastern Asia Society for Transportation Studies, 10*, 989–1008.

McFadden, D. (1973). *Conditional logit analysis of qualitative choice behavoir*.

Miessner, M. (2021). Studentification in Germany: How investors generate profits from student tenants in Goettingen and the impacts on urban segregation. *European Urban and Regional Studies, 28*(2), 133–154. https://doi.org/10.1177/0969776420934850

Ortiz, A., Tejerina, D., Díaz-Caro, C., Elghannam, A., García-Torres, S., Mesías, F. J., Carmona, J. T., & Crespo-Cebada, E. (2020). Is packaging affecting consumers' preferences for meat products? A study of modified atmosphere packaging and vacuum packaging in Iberian dry-cured ham. *Journal of Sensory Studies*. https://doi.org/10.1111/joss.12575

Ramos, A., Gago, A., Labandeira, X., & Linares, P. (2015). The role of information for energy efficiency in the residential sector. *Energy Economics, 52*, S17–S29. https://doi.org/10.1016/j.eneco.2015.08.022

Revelt, D., & Train, K. (1998). *Mixed logit with repeated choices: Households' choices of appliance efficiency level* (Vol. 80). Accessed from https://about.jstor.org/terms

Richetin, J., Caputo, V., Demartini, E., Conner, M., & Perugini, M. (2022). Organic food labels bias food healthiness perceptions: Estimating healthiness equivalence using a Discrete Choice Experiment. *Appetite, 172*, 105970. https://doi.org/10.1016/J.APPET.2022.105970

Rugg, J., Rhodes, D., & Jones, A. (2002). Studying a Niche Market: UK students and the private rented sector. *Housing Studies, 17*(2), 289–303. https://doi.org/10.1080/02673030220123234

Sama, C., Crespo-Cebada, E., Díaz-Caro, C., Escribano, M., & Mesías, F. J. (2018). Consumer preferences for foodstuffs produced in a socio-environmentally responsible manner: A threat to fair trade producers? *Ecological Economics*. https://doi.org/10.1016/j.ecolecon.2018.04.031

Train, K. E. (2009). *Discrete choice methods with simulation* (2nd ed.). Cambridge University Press. https://doi.org/10.1017/CBO9780511805271

Zróbek-Rózañska, A., & Szulc, L. (2018). Over a million student tenants in Poland. Analysis of preferences. *Real Estate Management and Valuation, 26*(2), 104–113. https://doi.org/10.2478/remav-2018-0020

Photovoltaic Module Waste and the Circular Economy: A Compressive Review

Donaji Jiménez-Islas, Miriam Edith Pérez-Romero, José Álvarez-García, and María de la Cruz del Río-Rama

Abstract The need for comfort in society demands more energy every day; various industries were created recently to supply energy, including that associated with photovoltaic technology. The usable life of photovoltaic modules is 20 to 30 years; the disposal of materials at the end of their use constitutes a source of contamination and value if treated correctly. With the intention of reducing the environmental impact, researchers and technologists have focused on studying and developing treatment alternatives. The aim of the present chapter was to explore the technologies associated with the disposal of photovoltaic module waste and its relationship with the circular economy. In addition, it is proposed to explain the evolution of scientific publications associated with the field of study through bibliometric indicators such as Publications and citations, countries, characteristics of the main authors, sources, network analysis of keywords in the previous 10 years using the database of Scopus. The main technologies associated with the treatment of photovoltaic

D. Jiménez-Islas
Renewable Energy Division, Higher Technological Institute of Huichapan, TecNM (Tecnológico Nacional de México campus Huichapan), Huichapan, Hidalgo, Mexico
e-mail: djimenez@iteshu.edu.mx

M. E. Pérez-Romero
Business Management Division, Higher Technological Institute of Huichapan, TecNM (Tecnológico Nacional de México campus Huichapan), Huichapan, Mexico

Sustainable Territorial Development (R015), International Doctoral School of Universidad de Extremadura (UNEX), Badajoz, Extremadura, Spain
e-mail: mperezgp@alumnos.unex.es; meperez@iteshu.edu.mx

J. Álvarez-García (✉)
Financial Economics and Accounting Department, University of Extremadura, Cáceres, Extremadura, Spain
e-mail: pepealvarez@unex.es

M. del Río-Rama
Business Management and Marketing Department, University of Vigo, Ourense, Galicia, Spain
e-mail: delrio@uvigo.es

C. Rego et al. (eds.), *Transitioning to a Circular Economy*, Studies on Entrepreneurship, Structural Change and Industrial Dynamics,
https://doi.org/10.1007/978-3-031-77661-8_14

panel waste will be described based on the profile of keywords in recent years. From 2018 to 2022 there is an exponential growth in publications related to the topic of PVm waste. The scientific community is busy establishing mechanisms to treat PVm that reach the end of their useful life, which is caused by the environmental impact represented by the projections of panels that will go into disuse. Technologists, researchers, students and companies will benefit from the content of this chapter, and it constitutes a decision-making tool to establish policies and laws to address the described problems.

Keywords Solar cell · Circular economy · Recycle · EoL · Waste

1 Introduction

The depletion of fossil fuels as a source of energy and global warming are factors that drive the implementation of renewable energies; the conversion of solar radiation into electricity is an alternative renewable energy source to fossil fuels and represents an important step in the global energy transition process (Oteng et al., 2021). The photovoltaic industry has been expanding in different demographic regions of the world because photovoltaic technology has possibly the greatest potential of the available technologies to replace fossil fuels (Farrell et al., 2020). Solar radiation is the main source of energy that is directly or indirectly harnessed by most renewable energy systems (Rabaia et al., 2022). Photovoltaic modules use solar light energy to generate electricity, do not generate noise during operation, do not emit toxic gases and their market is expanding (Kang et al., 2012).

The implementation of photovoltaic systems has increased most recently due to the need for energy for the development of industrial activities and the comfort of society. The source of alternative energy with the most rapid growth is photovoltaic power generation, having grown at a high and consistent 37% annual rate during the 1990s (Deng et al., 2019). With an average 25-year life cycle for photovoltaic modules (PVm), few attentions are placed on the problem associated with the generation of waste, actors such as the lack of regulation and the fact that the cost of recycling exceeds the landfill option, only 10% of photovoltaic modules are treated or recycled in the world (Prasad et al., 2022).

PVm that were installed in the 1990s are nearing their End-of-life (EoL) and therefore require a disposal process (Nain & Kumar, 2022). Moreover, a substantial amount of installed crystalline-Si Pvm have not yet attained the EoL, and PV waste flows from breakages and early losses have been documented (Ganesan & Valderrama, 2022).

Disposal and management practices of PVm, when they reach the EoL, are unknown. Among the reasons for contributing to the low development of waste management are: (a) the absence in most countries of specific laws for PVm waste, (b) lack of incentive programs to encourage recycling, (c) lack of collection and handling centers for deteriorated PVm or in EoL conditions, (d) the high cost of recycling for different generations of PVm, and (e) scarce research and development for the management of PVm (Nain & Kumar, 2020).

Using appropriate recovery and recycling procedures, in addition to implementing the necessary technology, is required to treat wastes resulting from the decommissioning of the world's energy generation facilities (Sica et al., 2018). In order to determine the type of PVm waste treatment, it is necessary to take into consideration the generation to which they belong, the first of which is composed of crystalline silicon (c-Si), the second-generation cells are thin-film technologies and the third-generation dye-sensitized solar cells (DSSC) and organic-photovoltaic cells (OPV).

Disposal of waste in landfills is an economical option; however, recycling is environmentally friendly and reduces the exploitation of virgin material for the production of new PVm (Deng et al., 2019). It is difficult to visualize the recycling process only with environmental and social benefits to be accepted as an option within the circular economy model, and it is necessary to guarantee economic profitability through mechanisms such as economic incentives. Also, this industry's fundamental difficulty continues to be the lack of research attention on decommissioning, dismantling, and recycling outdated modules, which has impeded the development of recycling technologies (Jain et al., 2022).

Japan, the United States, and European countries are conducting research on the recycling of PVm; their efforts are focused on the first generation of photovoltaic panels (Chowdhury et al., 2020). Some authors explore a combination of treatment methods, including electro-hydraulic fragmentation for the recovery of copper, silver, lead, tin and aluminum (Nevala et al., 2019); polymer delamination treatments classified as chemical, thermal and mechanical (Ganesan & Valderrama, 2022); laser irradiation (Li et al., 2022); as well as methods of treatment and final disposal of PVm through stabilization using the Portland cement in a safe manner (Pavlopoulos et al., 2022).

In the analysis of previous works, recycling is considered the main way to reduce the disposal of PVm in landfills; however, there are studies that identify the Circular Economy (CE) model as a model of integrated treatment of PVm waste, making evident the need to analyze, through bibliometric indicators, the current status and the main research in the field of PVm publications under the CE model. The principles of CE promote the optimal use of all resources through recycling, reuse, repair, remanufacturing and refurbishing (Khoaele et al., 2023).

Understanding how publications have evolved is crucial to determine the relevance of the topic and its development over time. For this purpose, we use bibliometric indicators that allow us to condense the information on the field of study of the PVm-EC. This chapter aims to analyze the evolution of scientific investigation on the topic of the circular economy of PVm in the "End-of-life" (EoL) stage. For this purpose, we will use the Scopus database and implement a bibliometric analysis. The period was 2012 to 2022. We also describe the PVm generations, their management and the end of cycle of live.

2 Data and Methods

The first step in the construction of this study was to formulate the research questions: what is the current status of publications and technologies in the circular economy of PVm waste? What technologies have been published for the EoL management of PVm?

Bibliometrics is the tool used to perform the analysis of publications within the field of study related to the CE in EoL of PVm. The aim is to identify the evolution of publications and determine their impact through bibliometric indicators. In this step, using the Scopus database, the search is conducted to obtain literature data. The search criteria were combined with the Boolean operators (AND, OR), Fig. 1, step 1.

In the second step, the initial search criteria found documents that did not correspond to the objective of the study, so a second search criterion was generated considering the word "End-of-life" and "EoL," resulting in the following: (TITLE+ABSTRACT+KEYWORDS ("PV") OR TITLE+ABSTRACT+KEYWORDS ("Photovoltaic panel*") OR TIT TITLE+ABSTRACT+KEYWORDS ("solar panel") OR TITLE+ABSTRACT+KEYWORDS ("photovoltaic cells") OR TITLE+ABSTRACT+KEYWORDS ("PV panels") AND TITLE+ABSTRACT+KEYWORDS ("circular economy") AND TITLE-ABS-KEY ("End-of-life")) AND PUBLICATION-YEAR 2011 AND PUBLICATION-YEAR 2023.

In order to extract the precise phrase and any variations in the single or plural forms of the words, the symbols ("“and *) were employed in the search.

In this study, we only consider publications after 2012 due to the fact that we did not find publications in previous years related to the topic of study, this search was then limited from 2012 to 2022, and search was performed on 18th February 2023. One document was excluded from the study because it was not related to the scope of the present work.

The data were imported from the Scopus database to Microsoft Excel. Downloading data in CVS format was necessary to perform the mapping through the VOSviewer software.

The data on the evolution of scientific publications and citations per year were used to determine the specific rate of growth through the Gompertz model, which was placed in Excel and optimized through the solver function (Jiménez-Islas et al., 2022), in addition the fitting level is determined through the determination coefficient (R^2).

$$P(y) = ae^{-be^{-\mu t}} \text{ Gompertz equation}$$

$$\left(P_{(y)\text{Scopus}} - P_{(y)\text{model}}\right)^2 \text{ normalization}$$

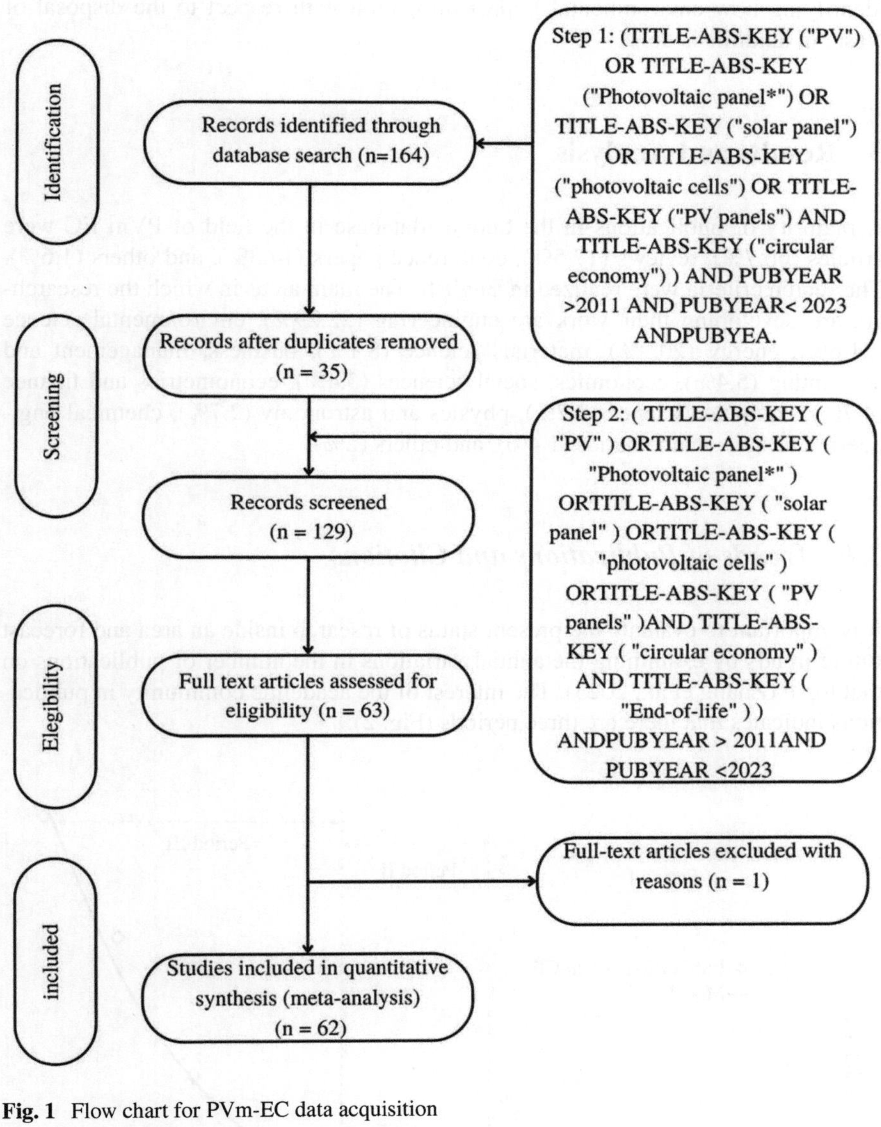

Fig. 1 Flow chart for PVm-EC data acquisition

"$P_{(y)}$" displays publications by annually, "μ" the specific growth rate of publications, "a" the asymptotic publications, "b" an integration constant, "t" is time, $P_{(y)\,Scopus}$ documents or publications reported on Scopus database and $P_{(y)model}$ estimated publications by the application of the model.

In addition to the bibliometric analysis, an analysis of the PVm that reaches the EoL and the management and recycling mechanisms was carried out through linear and circular economy models published in scientific articles, with the aim of identifying how environmental impact mitigation with respect to the disposal of waste in landfills is done.

3 Results and Analysis

A majority of publications in the Scopus database in the field of PVm EC were articles (66.7%), reviews (17.5%), conference papers (14.3%), and others (1.6%). The search criteria were realized in English. The main areas in which the researchers are developing their work are engineering (22.29%), environmental science (21.6%), energy (20.9%), materials science (8.1%), business, management and accounting (5.4%), economics, social sciences (5.4%), econometrics and finance (4.7%), computer science (2.7%), physics and astronomy (2.7%), chemical engineering (1.4%), mathematics (1.4%), and others (2%).

3.1 Trends of Publications and Citations

It is important to evaluate the present status of research inside an area and forecast future trends by examining the annual variations in the number of publications on that topic (Zhang et al., 2023). The interest of the academic community in publications indicates that there are three periods (Fig. 2):

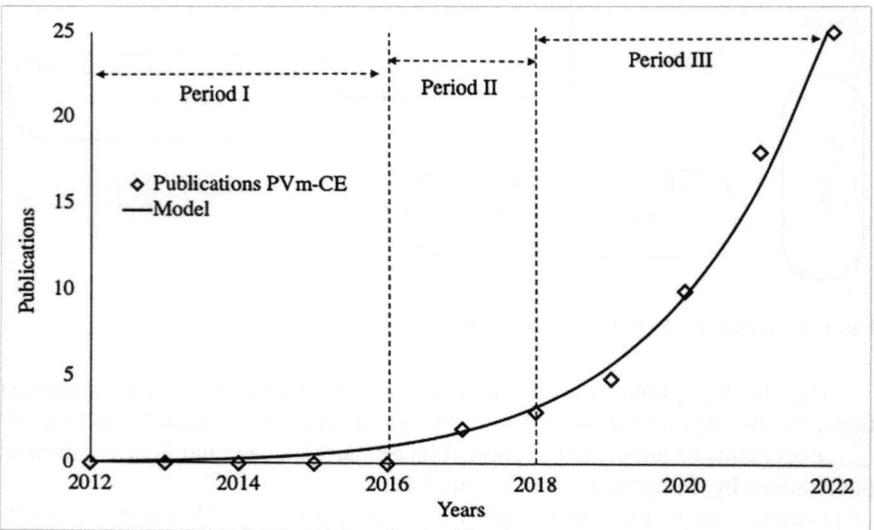

Fig. 2 Evolution of the publications in the circular economy of PV panels. R^2 = 0.9916, $\mu = 0.050\ y^{-1}$

- Period I (2012–2016) without publications.
- Period II (2017–2018) then begins a second stage of publications from 2017–2018 with a growth of 7.94%, in this stage, the publication entitled "*Management of end-of-life photovoltaic panels as a step towards a circular economy,*" analyzes the technological aspects and life-cycle stages of the most commercial panels (Sica et al., 2018). Another publication entitled "*Policies and measures for sustainable management of solar panel end-of-life in Italy,*" a review of the technical-economic and environmental conditions of photovoltaic energy production and its waste management when reaching its EoL. Its conclusion includes the need for a circularity perspective and stakeholder involvement to achieve a longer life cycle of photovoltaic energy. In addition, three publications related to PVm's circular economy have been published at conventions and conferences.
- Period III (2019–2022), the third stage with 92.06% of the publications, Fig. 3. The PV technology, namely solar energy, is currently receiving attention on a global scale (Oteng et al., 2021), the specific growth rate (μ) of publications relating to the circular economy of solar panels was calculated using the Gompertz equation. The results show a $\mu = 0.05\ y^{-1}$ with a multiple correlation coefficient of 0.9916 (R^2). The 2019 (Stage III) publications relate to PVm management, reuse, recycling and treatment issues. The 2020 publications are related to CE, supply chain and materials. The 2021 publications are related to simulation, remanufacturing, life-cycle assessment and e-waste. In 2022, publications cover recycling, PV in the circular economy and EoL management. The evolution of the themes of recycling and circular economy is also related to the keywords found in the documents. The behavior of period III was similar to that

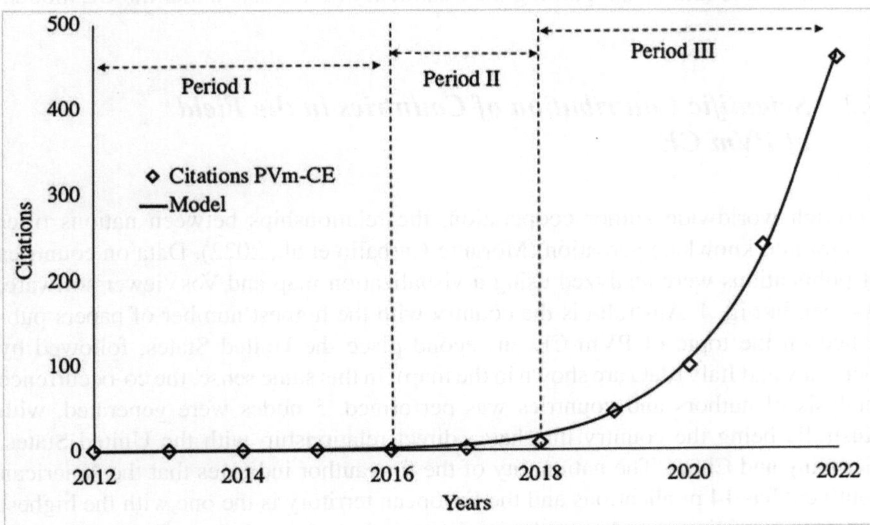

Fig. 3 Evolution of the citations in the circular economy of PVm. $R^2 = 0.99$, $\mu = 0.010\ y^{-1}$

reported for PVm and post-consumption waste management, with exponential growth from 2012 to 2017 (Salim et al., 2019).

With the growth of scientific publications on the subject of PVm CE, the number of citations related to the field of study has increased exponentially, due to international environmental policies. The 10-year period considered in the search criteria is based on the argument that publications that are 1–3 years old generate fewer citations than publications older than 5 years (Yafetto, 2022). The low number of citations is shown in periods I and II of Fig. 3 and corresponds to similar periods of the number of documents published as shown in Fig. 2. The top five countries contributing the most citations to PVm EC are Australia, Germany, the United States, Italy and the United Kingdom. The journals with the most citations are "Resources, Conservation and Recycling" and "Sustainability."

The top 5 documents with the most citations (97) are "Drivers, barriers and enablers to end-of-life management of solar photovoltaic and battery energy storage systems: A systematic literature review" (Salim et al., 2019), "Research and development priorities for silicon photovoltaic *module recycling to support a circular economy*" (Heath et al., 2020) with 95 citations, "Management of end-of-life photovoltaic panels as a step towards a circular economy" with 93 citations, "Technical challenges and opportunities in realizing a circular economy for waste photovoltaic modules" (Farrell et al., 2020) with 67 citations and "Towards a circular supply chain for PV modules: Review of today's challenges in PV recycling, refurbishment and re-certification" (Tsanakas et al., 2020) with 45 citations.

Due to the exponential behavior in the number of citations and international policies and agreements, it can be interpreted that the number of publications and citations will increase in the following years, together with the need for technologies, methods and regulations regarding the treatment of PV waste under the CE model.

3.2 Scientific Contribution of Countries in the Field of PVm CE

Through worldwide author cooperation, the relationships between nations offer insight into knowledge creation (Morante-Carballo et al., 2022). Data on countries of publications were analyzed using a visualization map and VosViewer software. As seen in Fig. 4, Australia is the country with the highest number of papers published on the topic of PVm-CE, in second place the United States, followed by Germany and Italy (data are shown in the map). In this same sense, the co-occurrence analysis of authors and countries was performed, 5 nodes were generated, with Australia being the country that has a direct relationship with the United States, Germany and China. The nationality of the first author indicates that the American continent has 14 publications and the European territory is the one with the highest number of publications due to the implementation of regulations for the management of photovoltaic panel waste upon reaching its EoL. The "National Renewable

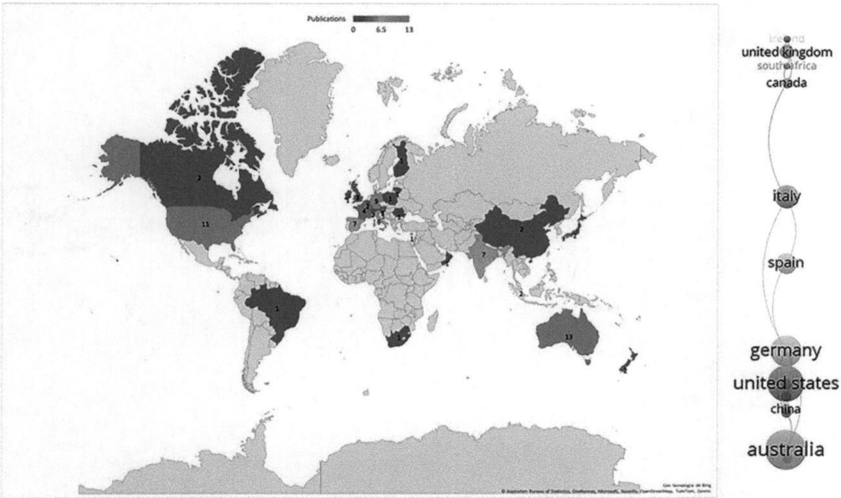

Fig. 4 Map of countries of PVm-CE publications and net-map de collaboration between countries

Energy Laboratory," "Colorado School of Mines," "HZDR—Helmholtz-Zentrum Dresden-Rossendorf," "Helmholtz-Zentrum Berlin für Materialien und Energie HZB" "Macquarie University," "Technical University of Munich," "UNSW Sydney," "Indian Institute of Technology Roorkee," "Griffith University," and "Macquarie Graduate School of Management" are the institutions ranked in the top 10 of publications on the subject under study.

3.3 The Main Authors

In the VOSviewer software, an analysis of the most influential co-authorships in the field of study was performed; 229 authors were analyzed (Fig. 5), of which Barnes T with affiliation of National Renewable Energy Laboratory, Golden, United States. The identified collaborative works are 4: "Research and development priorities for silicon photovoltaic module recycling to support a circular economy" (Heath et al., 2020), "PV in the circular economy, a dynamic framework analyzing technology evolution and reliability impacts" (Ovaitt et al., 2022), "PV Evolution in the light of Circular Economy" (Ovaitt et al., 2021) and "Quantifying Energy flows in PV Circularity Processes" (Mirletz et al., 2022). The cluster of Barnes T shows the author's node and its collaboration with other nodes.

In the same way, Wambach K with 2 documents ("Research and development priorities for silicon photovoltaic module recycling to support a circular economy" (Heath et al., 2020) and "Recycling and Reuse potential of NICE PV-Modules" (Einhaus et al., 2018) from the cluster collaborates with Barnes T, Kempe M,

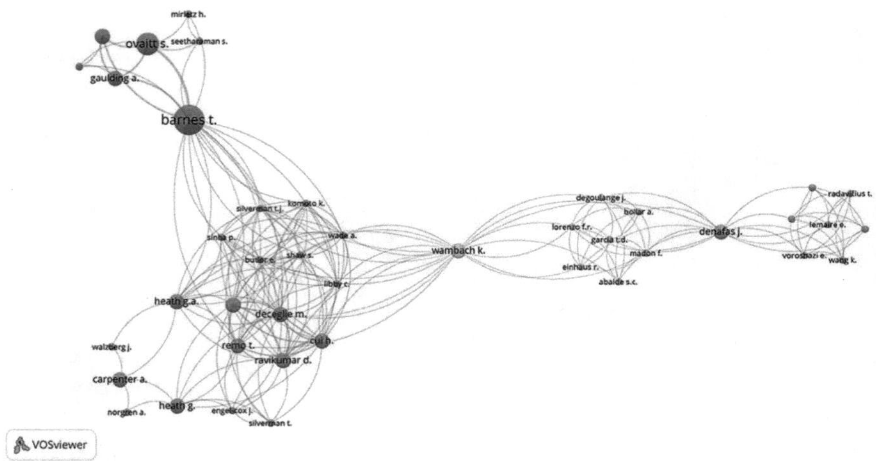

Fig. 5 Map of authors more influential

Ravikumar D, Cui H. In addition to the cluster of Heah G is a contributor with Deceglie M and Wamback K clusters.

3.4 *Keywords Co-occurrency Analysis*

The 746 keywords were divided into four clusters marked. Figure 6 shows the mapping of the co-occurrence of keywords analysis. The Keyword co-occurrence frequency is displayed through circles with a specific size, and the connection between two circles symbolizes the simultaneous appearance of words. The prominent cluster is labeled as "circular economy" and recycling; the second cluster is represented as "photovoltaic cells."

The third cluster is named "photovoltaic system" and the fourth cluster is named life cycle. There is a great connection between the words "circular economy," "recycling," and "silicon" in the cluster, which respond to the material recovery needs of the PMm with the EoL. The "Photovoltaic cells" cluster maintains a strong connection between the words PV (cell and panels) with "supply chains," "end-of-life," and "end-of-live," as well as maintaining a general connection with the "circular economy" cluster. It is evident that the authors develop research focused on the recycling of the panels; however, it is observed that they are associated with the words "life cycle," "environmental impact," and "sustainable development," which indicates the diversification of publications that consider integral treatments.

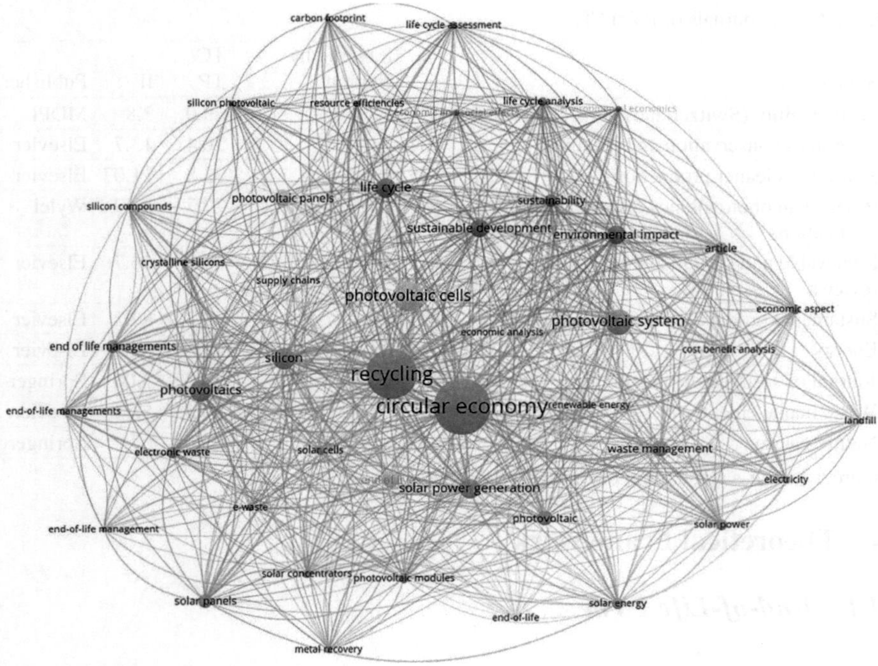

Fig. 6 Keywords co-occurrency map of PVm-EC from 2012 to 2022

3.5 *Top Journals of PVm CE*

The five most prominent journals with the most articles published are led by "*Sustainability*" with 7, followed by "*Resources Conservation and Recycling*" with 5, "*Journal of Cleaner Production*" with 4, "*Progress in Photovoltaics Research and Applications*" with 3 and "*Renewable and Sustainable Energy Reviews*" also with 3. As the impact of the publications is determined through the number of citations, Table 1 shows the ranking by citation, in which it is observed that there are three journals with more than 100 citations, in the first place *Renewable and Sustainable Energy Reviews* with 158 citations, followed by the *Journal of Cleaner Production* with 135 citations and *Resources Conservation and Recycling* with 102.

When authors publish in important journals, they seek to contribute their findings to some field of study; also as part of their work, they obtain citations through the contributions of other authors. Therefore they seek to maintain the novelty of the research in journals that have global reach; in this sense. the most prominent publishers have high citation rates and rigorous arbitration processes. Table 1 shows the impact factor (IF) of the journals in which topics related to PVm were published under the concept of CE, making evident the importance of the subject in the journals and in the scientific and technological community.

Table 1 Top journals of PVm CE

Source title	TP	TC	Ranking by citation	TC/TP	IF	Publisher
Sustainability (Switzerland)	7	63	7	9.0	3.8	MDPI
Resources conservation and recycling	5	102	3	20.4	13.7	Elsevier
Journal of cleaner production	4	135	2	33.8	11.07	Elsevier
Progress in photovoltaics research and applications	3	68	5	22.7	8.49	Wylei
Renewable and sustainable energy reviews	3	158	1	52.7	16.7	Elsevier
Sustainable production and consumption	3	67	6	22.3	8.92	Elsevier
Energy	2	44	8	22.0	8.8	Elsevier
Journal of sustainable metallurgy	2	42	9	21.0	3.06	Springer
Waste management	2	41	10	20.5	8.81	Elsevier
Nature energy	1	95	4	95.0	67	Springer

Source: Author's elaboration using Scopus database

4 Theoretical Framework

4.1 End-of-Life PVm

The development of photovoltaic technologies has generated a decrease in costs, which has caused significant growth in the European market. According to the International Renewable Energy Agency (IRENA) about 45,000 ton of PV waste was generated in 2016 (Ansanelli et al., 2021). By 2030, eight million tons of waste are projected to be generated and 60–80 million tons by 2050 (Sim et al., 2023). Lithium, tin, cadmium, and other hazardous materials included in PVm, and battery storage systems have the potential to harm the environment and human health if they are not properly disposed of after reaching the EoL (Salim et al., 2019). To determine the type of PVm waste treatment, it is necessary to know the method of manufacture and the type of components. The diversity of materials used in the PVm structure depends on its technological maturity and is classified into generations, Fig. 7. Three generations are defined: G1-crystalline-silicon cells-, G2-Thin-film cells and G3-dry cells (Czajkowski et al., 2023).

During the operation of the PVm its energy efficiency can decrease due to environmental factors of the place where it is installed, including humidity, temperature and panel soiling; the degradation rate is estimated at 0.5–0.8% per year (Seo et al., 2021). Some PVm are considered early-stage wastes for reasons such as optical failures, power loss, glass breakage, manufacturing defects, connector defects, structural damage during transportation and delamination. Also, in the first years of energy production, PVm show degradation in the anti-reflective ethylene vinyl acetate coating, and 40% of panel failures are due to microscopic (Chowdhury et al., 2020).

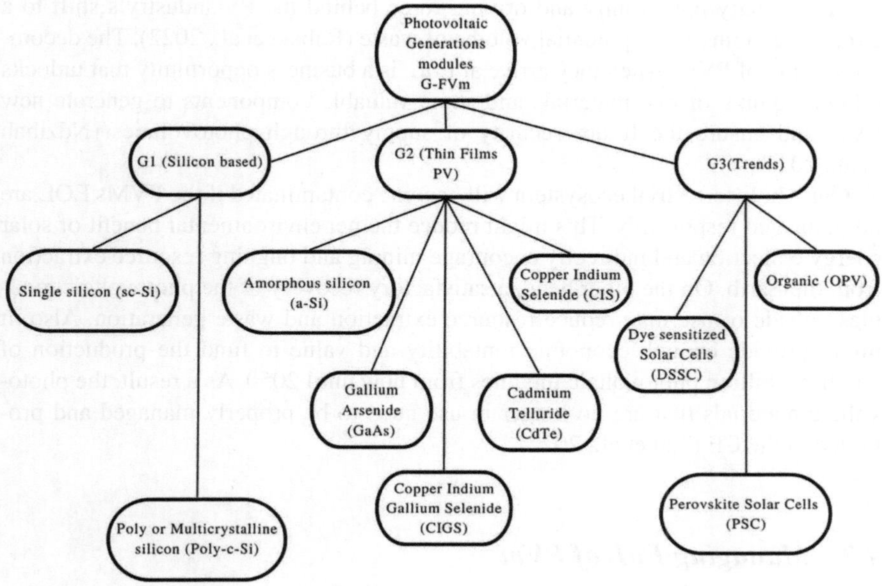

Fig. 7 Generations of PVm technologies

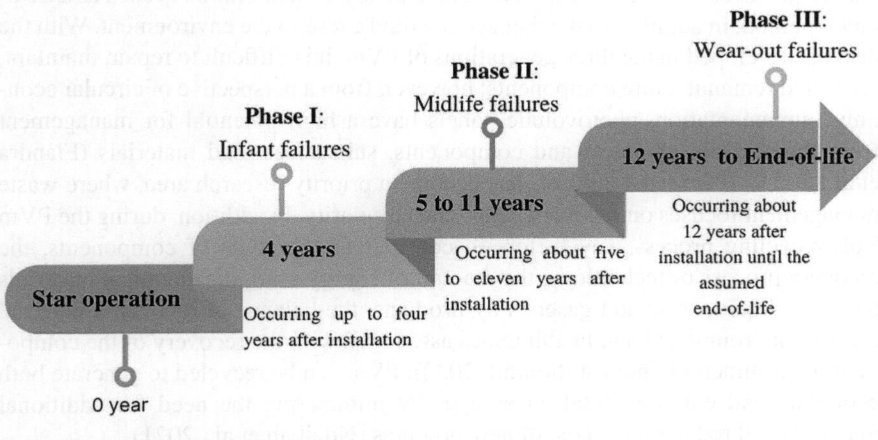

Fig. 8 Three main panel failure phases

Also, the PVm that are not in operation due to any damage suffered during transport or operation can be subjected to a technical review and assess whether it is possible to repair them, in most cases when the damage is to the cables and J-box, which prevents the PVm from reaching the end of its life cycle. Regardless of the technology or generation of PVm, the main failures detected occur in three phases (Mulazzani et al., 2022): Infant (I), mid-life (II) and Wear-out (III), Fig. 8.

The primary opportunity and driving force behind the PV industry's shift to a circular economy is the potential volume of waste (Rabaia et al., 2022). The decommissioning of PVm, when they arrive at EoL, is a business opportunity that unlocks a large volume of raw materials and some valuable components to generate new PVm and ensure the future security of supply through photovoltaics (Ndzibah et al., 2021).

Our whole terrestrial ecosystem will become contaminated if the PVMs EOL are not managed responsibly. This might reduce the net environmental benefit of solar energy collection and indirectly encourage mining and ongoing resource extraction from the Earth. On the other hand, a satisfactory recovery of the photovoltaic modules outside of use may reduce resource extraction and waste generation. Also, it might provide enough economic rentability and value to fund the production of another 2 billion photovoltaic modules from now until 2050. As a result, the photovoltaic materials that are no longer in use need to be properly managed and promoted in the CE (Tan et al., 2022).

4.2 *Managing EoL of PVm*

The management of PVm, when its EoL occurs, is of commercial interest due to the materials that can be recovered, which could be lost if incineration is used as a treatment method, in addition to the damage it would cause to the environment. With the designs developed in the three generations of PVm, it is difficult to repair, maintain, reuse and remanufacture components; however, from a perspective of circular economy implementation, photovoltaic panels have a high potential for management from three levels: products and components, substances, and materials (Fiandra et al., 2023). PVm reaching EoL has become a priority research area, where waste management focuses on recovery, reuse and circularity. In addition, during the PVm EoL recycling process, key factors to consider are the type of components, the recovery process or technology, the amount of energy used, the chemical materials used, the liquid, solid and gaseous by-products, the balance of recovered material and the environmental and health issues associated with the recovery of the components of commercial interest (Senthil, 2022). PVm can be recycled to generate both economic and environmental advantages by minimizing the need for additional materials and reducing the cost of new products (Ndzibah et al., 2021).

Different types of hierarchy in circularity have been proposed in a systematic way, for which the *R*-framework is used, the simplest 3R being: reduce, reuse, and recycle (del Río et al., 2021). For the application of the 3Rs (Fig. 8), the first step of any circular route is the dismantling and collection of PVm; if the modules are not collected, it is possible that they end up in a landfill without proper treatment. Reuse is when the PVm is maintained or returned to the field considering the level of power degradation beyond the nominal threshold. Repair is when failed modules are present and changes are made to the junction box that allows it to remain in use (Ovaitt et al., 2022).

The process of sustainability of a process or system improves from the optimization of the traditional linear (unidirectional) economic model to create a circular economy model. To this end, the 3R, 6R and 10R categories for sustainability must be taken into account, Table 2. Also, it was reported that the PV system's circular economy business model comprises many business stages that fall under the following main categories (Rabaia et al., 2022): (a) collection and processing of raw materials, (b) manufacture, (c) distribution and installation, (d) useful or operation of the system, (e) EoL and decommissioning, (g) sorting of materials and dismantling, and (h) landfill disposal.

Table 2 10R's for PVm considering the circularity

Category or principle	Category or principle	Possible applications with EC
Refuse	Reject the use of hazardous materials, identify the possibility of using non-virgin materials or propose sustainable alternatives	Investigates the main advances in materials associated with the technology developed and evaluates the impact of the material used on people's health and the associated environmental damage. It explores the generations of panels developed
Rethink	Make products for more intensive use or compatible	Optimize the designing of modules and identify compatibility with other similar products to perform life cycle planning of PV modules
Reduce	Increase efficiency in the manufacture of products	Reduce the consumption of natural resources used in fabrication
Reuse	Reuse by another consumer under the principle of correct operation	Use PVm that are still able to generate energy
Repair	Repairs and maintains those products that maintain original functionality	Maintain defective PV modules to their best possible functionality
Refurbish	Refurbishing old products or upgrading them for new use	Refurbishment of old PVm to meet technical functionality specifications
Remanufacture	Uses scrap parts in the manufacture of new products	Identifies scrap parts in new PVm manufacturing
Repurpose	Identify discarded parts in manufacturing new products and put them to a different use	Identify rejected parts in PVm manufacturing and give them new applications
Recycle	Processing to obtain materials of equal or lesser quality	Processing of materials from dismantling to obtain PVm of equal or lower quality
Recovery	Recovery of materials and energy, incineration can be used as a final stage of energy recovery	Optimize processes to recover as much materials and energy as possible

Source: Author's elaboration using Potting et al. (2017), Okorie et al. (2018), Rabaia et al. (2022)

4.3 Treatment of Photovoltaic Modules

As a means to minimize production costs and improve the energy efficiency of photovoltaic modules, different technologies have been developed, where 51% of the panel market contemplates polycrystalline silicon (Si-poly), 41% monocrystalline silicon (Si-mono) and the rest with materials such as Cd, Te, In, Ga and others (Rubino et al., 2021). The potential environmental advantages and material recovery drive the ongoing development of PV recycling to further increase the efficacy and efficiency to achieve the circularity of raw materials used in PVm (Sim et al., 2023). After the end of their useful life, the PVm are transported for recycling or final disposal.

Waste management in the European Union is through Directive 2012/19/EU, where waste PVm at the end of its useful life or that were damaged during the production, transportation, installation, and operation process are categorized as waste electrical and electronic equipment, and the recovery of secondary raw materials from PV panels now has to be 85% efficient. In addition to the Extended Product Responsibility (EPR) where the responsibility of the collection and recycling costs is imposed on the companies that produce the modules, which prevents the PVm from being sent to landfills when they arrive at EoL (Czajkowski et al., 2023).

The recycling process starts with the removal of the aluminum frame, cables and junction box; this process of recycle is known as *downcycling*, with the recovery of materials with lower value or purity. In the recycling process called upcycling, high-value materials or materials with commercial interest that are in greater demand are obtained.

Figure 9 shows the process of crushing, which is carried out in machines that allow the mechanical separation of materials such as glass. The EVA (ethylene vinyl acetate) resin is frequently utilized in PVm as a greater moisture resistant and is typically the second stage of the recycling process from the discarded PVm. Many techniques, including heat breakdown and nitric acid dissolving, were used to remove the EVA (Kang et al., 2012). Also, to break the encapsulation, another treatment commonly used is thermal treatment; however, when the panel is coated with fluoropolymers, the emission of toxic gases negatively affects the environment and human health. Czajkowski et al. (2023) describe that for the process of separation and purification of silicon cells and high-value metals such as silver, tin, lead, and copper, chemicals and electrochemical procedures are used (Fig. 10).

Even if the recycling of solar waste for commercial purposes has already begun, some technologies still need to be developed in order to calculate costs, levels of recuperation and recycling, and also processing efficiency. It still takes a lot of work to advance recycling technology and satisfy the current demand. So, more study and development are required for solar PVm EOL management (Rathore & Panwar, 2021).

It has been reported that around 10% of PVm are recycled in the world and the main reasons are poor regulation and the cost of recycling versus the value of recycled products, with the option of disposal in landfills being evident as the most

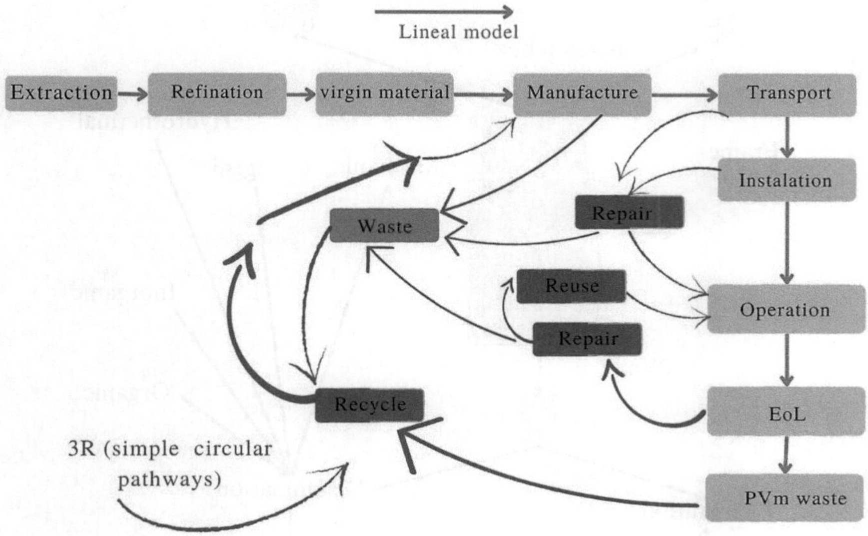

Fig. 9 PVm-EC linear economy and circularity diagram. Light grey boxes are the lineal model

attractive option from the economic point of view. In this context, the advantages of recycling are also visualized, such as guaranteeing long-term supply chain sustainability and therefore the reduction of CO_2 emissions associated with the PVm operation from recycling (Lunardi et al., 2018).

It is a reality that recycling processes today are not economically profitable; it is an area that requires basic and applied research to obtain timely and technically and economically accessible treatment technologies, where the type of PVm generation must be considered to be more specific in material recovery.

In addition, the correct management of PVm has the potential to become a business opportunity for companies, involving producers and users of the technology to reach a global consensus that benefits the photovoltaic sector.

The social and economic benefits of the implementation of CE principles reduce the environmental impact by ceasing to deposit PVm in landfills, with labor being the beneficiary of the recycling process if global regulatory mechanisms and economic incentives for the sector are damaged.

5 Conclusions

In this chapter we conducted a quantitative analysis of the literature on the topic of photovoltaic modules ending their life cycle through the publications contained in the Scopus database in the period from 2012 to 2022. From 2018 to 2022 there was

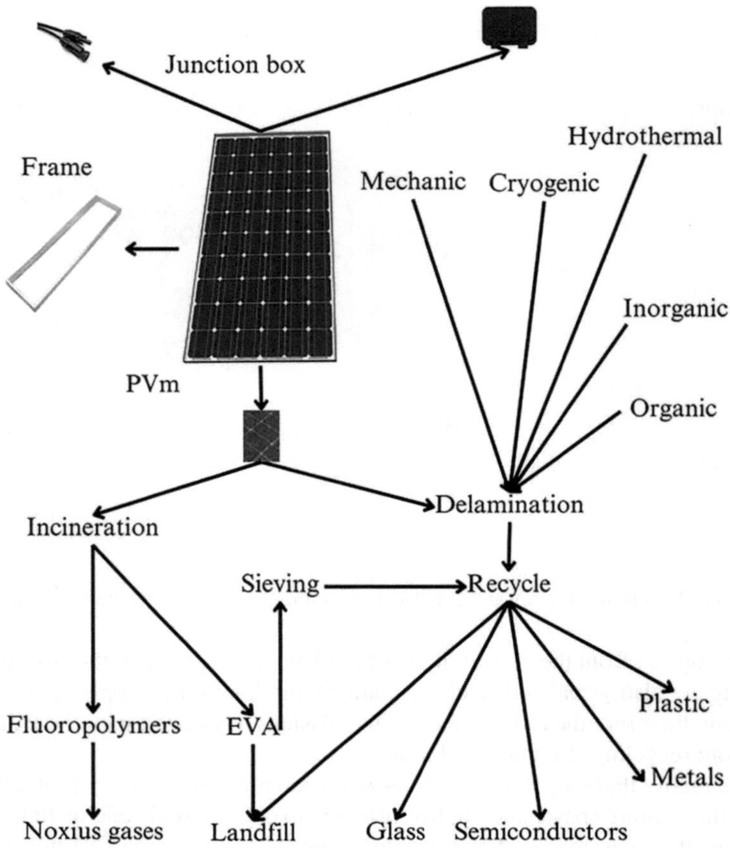

Fig. 10 Recycle of PVm

an exponential growth in the number of publications and citations also coincides with the period of increase in publications; the following findings were found:

In the last 5 years, the number of publications and citations related to the circular economy of the P-V-m that reach the EoL has increased exponentially. From the data of publications and citations, we identified with the gompertz model that there is a wide interest among researchers in finding recycling technologies in the different generations of PVm.

It shows how different countries are focused on complying with international agreements on waste treatment related to PVm that suffer damage in different parts of the production, use or recycling process.

The authors carry out their publications in high-impact journals to contribute to the global community on the topic of study and under the search criteria specified in this work.

Due to the growth of the photovoltaic sector, the need to establish waste treatment conditions for PVm is imminent; regulatory mechanisms are necessary to promote recycling and recover materials that allow maintaining the renewable energy production chain. The need for circularity principles helps to reduce the disposal of PVm waste in landfills, reducing the environmental impact.

For the recycling process there is no technology development for each generation of PVm, and the cost of recycling is not competitive with respect to disposal in landfills; however, the global regulation of producers and consumers of PVm could be the key to successful commercial recycling and circularity of materials.

The disposal in landfills of waste PVm that reach their EoL presents environmental disadvantages associated with the composition of each generation of photovoltaic panel; the environmental impact they would cause is incalculable due to the diversity of materials that would affect the area where they are disposed of. In addition, photovoltaic technologies and renewable energy are currently positioned as environmentally viable options to reduce emissions to the environment and with the lack of policies that regulate the treatment process for panels at the end of their useful life, it is possible that the real benefit of renewables is questioned.

In another sense, the management of materials in the photovoltaic industry should be seen from a circularity perspective to ensure the reduction of the exploitation of virgin raw materials and the use of materials from recycling or reconditioning.FundingThis publication was funded by the Consejería de Economía, Ciencia y Agenda Digital de la Junta de Extremadura and by the European Regional Development Fund of the European Union through the reference grant GR21161.

References

Ansanelli, G., Fiorentino, G., Tammaro, M., & Zucaro, A. (2021). A life cycle assessment of a recovery process from end-of-life photovoltaic panels. *Applied Energy, 290*, 116727. https://doi.org/10.1016/j.apenergy.2021.116727

Chowdhury, M. S., Rahman, K. S., Chowdhury, T., Nuthammachot, N., Techato, K., Akhtaruzzaman, M., Tiong, S. K., Sopian, K., & Amin, N. (2020). An overview of solar photovoltaic panels' end-of-life material recycling. *Energy Strategy Reviews, 27*, 100431. https://doi.org/10.1016/j.esr.2019.100431

Czajkowski, A., Wajda, A., Poranek, N., Bhadoria, S., & Remiorz, L. (2023). Prediction of the market of end-of-life photovoltaic panels in the context of common EU management system. *Energies, 16*, 284. https://doi.org/10.3390/en16010284

del Río, P., Kiefer, C. P., Carrillo-Hermosilla, J., & Könnölä, T. (2021). The micro-level approach to the circular economy. *Green Energy and Technology, 2021*, 73–87. https://doi.org/10.1007/978-3-030-74792-3_4

Deng, R., Chang, N. L., Ouyang, Z., & Chong, C. M. (2019). A techno-economic review of silicon photovoltaic module recycling. *Renewable and Sustainable Energy Reviews, 109*, 532–550. https://doi.org/10.1016/j.rser.2019.04.020

Einhaus, R., Madon, F., Degoulange, J., Wambach, K., Denafas, J., Lorenzo, F. R., Abalde, S. C., Garcia, T. D., & Bollar, A. (2018). Recycling and reuse potential of NICE PV-modules. In *2018 IEEE 7th world conference on photovoltaic energy conversion (WCPEC) (A joint conference of 45th IEEE PVSC, 28th PVSEC & 34th EU PVSEC)*. https://doi.org/10.1109/pvsc.2018.8548307

Farrell, C. C., Osman, A. I., Doherty, R., Saad, M., Zhang, X., Murphy, A., Harrison, J., Vennard, A. S. M., Kumaravel, V., Al-Muhtaseb, A. H., & Rooney, D. W. (2020). Technical challenges and opportunities in realising a circular economy for waste photovoltaic modules. *Renewable and Sustainable Energy Reviews, 128*, 109911. https://doi.org/10.1016/j.rser.2020.109911

Fiandra, V., Sannino, L., & Andreozzi, C. (2023). Photovoltaic waste as source of valuable materials: A new recovery mechanical approach. *Journal of Cleaner Production, 385*, 135702. https://doi.org/10.1016/j.jclepro.2022.135702

Ganesan, K., & Valderrama, C. (2022). Anticipatory life cycle analysis framework for sustainable management of end-of-life crystalline silicon photovoltaic panels. *Energy, 245*, 123207. https://doi.org/10.1016/j.energy.2022.123207

Heath, G. A., Silverman, T. J., Kempe, M., Deceglie, M., Ravikumar, D., Remo, T., Cui, H., Sinha, P., Libby, C., Shaw, S., Komoto, K., Wambach, K., Butler, E., Barnes, T., & Wade, A. (2020). Research and development priorities for silicon photovoltaic module recycling to support a circular economy. *Nature Energy, 5*(7), 502–510. https://doi.org/10.1038/s41560-020-0645-2

Jain, S., Sharma, T., & Gupta, A. K. (2022). End-of-life management of solar PV waste in India: Situation analysis and proposed policy framework. *Renewable and Sustainable Energy Reviews, 153*, 111774. https://doi.org/10.1016/j.rser.2021.111774

Jiménez-Islas, D., Pérez-Romero, M. E., & de la Cruz Del Río-Rama, M. (2022). Mapping research trends in publications related to bio-jet fuel: A Scientometric review. *International Journal of Design & Nature and Ecodynamics, 17*, 1–8. https://doi.org/10.18280/ijdne.170101

Kang, S., Yoo, S., Lee, J., Boo, B., & Ryu, H. (2012). Experimental investigations for recycling of silicon and glass from waste photovoltaic modules. *Renewable Energy, 47*, 152–159. https://doi.org/10.1016/j.renene.2012.04.030

Khoaele, K. K., Gbadeyan, O. J., Chunilall, V., & Sithole, B. (2023). The devastation of waste plastic on the environment and remediation processes: A critical review. *Sustainability, 15*, 5233. https://doi.org/10.3390/su15065233

Li, X., Liu, H., You, J., Diao, H., Zhao, L., & Wang, W. (2022). Back EVA recycling from c-Si photovoltaic module without damaging solar cell via laser irradiation followed by mechanical peeling. *Waste Management, 137*, 312–318. https://doi.org/10.1016/j.wasman.2021.11.024

Lunardi, M. M., Alvarez-Gaitan, J. P., Bilbao, J. I., & Corkish, R. (2018). A review of recycling processes for photovoltaic modules. In *Solar panels and photovoltaic materials*. IntechOpen. https://doi.org/10.5772/intechopen.74390

Mirletz, H. M., Ovaitt, S., Gaulding, A., Sridhar, S., & Barnes, T. (2022). Quantifying energy flows in PV circularity processes. In *2022 IEEE 49th photovoltaics specialists conference (PVSC)*. https://doi.org/10.1109/pvsc48317.2022.9938594

Morante-Carballo, F., Montalván-Burbano, N., Arias-Hidalgo, M., Domínguez-Granda, L., Apolo-Masache, B., & Carrión-Mero, P. (2022). Flood models: An exploratory analysis and research trends. *Water, 14*(16), 2488. https://doi.org/10.3390/w14162488

Mulazzani, A., Eleftheriadis, P., & Leva, S. (2022). Recycling c-Si PV modules: A review, a proposed energy model and a manufacturing comparison. *Energies, 15*, 8419. https://doi.org/10.3390/en15228419

Nain, P., & Kumar, A. (2020). Understanding the possibility of material release from end-of-life solar modules: A study based on literature review and survey analysis. *Renewable Energy, 160*, 903–918. https://doi.org/10.1016/j.renene.2020.07.034

Nain, P., & Kumar, A. (2022). A state-of-art review on end-of-life solar photovoltaics. *Journal of Cleaner Production, 343*, 130978. https://doi.org/10.1016/j.jclepro.2022.130978

Ndzibah, E., Andrea Pinilla-De La Cruz, G., & Shamsuzzoha, A. (2021). End of life analysis of solar photovoltaic panel: Roadmap for developing economies. *International Journal of Energy Sector Management, 16*(1), 112–128. https://doi.org/10.1108/ijesm-11-2020-0005

Nevala, S.-M., Hamuyuni, J., Junnila, T., Sirviö, T., Eisert, S., Wilson, B. P., Serna-Guerrero, R., & Lundström, M. (2019). Electro-hydraulic fragmentation vs conventional crushing of photovoltaic panels—Impact on recycling. *Waste Management, 87*, 43–50. https://doi.org/10.1016/j.wasman.2019.01.039

Okorie, O., Salonitis, K., Charnley, F., Moreno, M., Turner, C., & Tiwari, A. (2018). Digitisation and the circular economy: A review of current research and future trends. *Energies, 11*(11), 3009. https://doi.org/10.3390/en11113009

Oteng, D., Zuo, J., & Sharifi, E. (2021). A scientometric review of trends in solar photovoltaic waste management research. *Solar Energy, 224*, 545–562. https://doi.org/10.1016/j.solener.2021.06.036

Ovaitt, S., Mirletz, H. M., Hegedus, A., Gaulding, A., & Barnes, T. (2021). PV evolution in the light of circular economy. In *2021 IEEE 48th photovoltaic specialists conference (PVSC)*. https://doi.org/10.1109/pvsc43889.2021.9518683

Ovaitt, S., Mirletz, H., Seetharaman, S., & Barnes, T. (2022). PV in the circular economy, a dynamic framework analyzing technology evolution and reliability impacts. *iScience, 25*(1), 103488. https://doi.org/10.1016/j.isci.2021.103488

Pavlopoulos, C., Kelesi, M., Michopoulos, D., Papadopoulou, K., Lymperopoulou, T., Skaropoulou, A., Tsivilis, S., & Lyberatos, G. (2022). Management of end–of–life photovoltaic panels based on stabilization using Portland cement. *Sustainable Chemistry and Pharmacy, 27*, 100687. https://doi.org/10.1016/j.scp.2022.100687

Potting, J., Hekkert, M., Worrell, E., & Hanemaaijer, A. (2017). *Circular economy: Measuring innovation in the policy report*. PBL Netherlands Environmental Assessment Agency.

Prasad, D. S., Sanjana, B., Kiran, D. S., Srinivasa Kumar, P. P., & Ratheesh, R. (2022). Process optimization studies of essential parameters in the organic solvent method for the recycling of waste crystalline silicon photovoltaic modules. *Solar Energy Materials and Solar Cells, 245*, 111850. https://doi.org/10.1016/j.solmat.2022.111850

Rabaia, M. K. H., Semeraro, C., & Olabi, A.-G. (2022). Recent progress towards photovoltaics' circular economy. *Journal of Cleaner Production, 373*, 133864. https://doi.org/10.1016/j.jclepro.2022.133864

Rathore, N., & Panwar, N. L. (2021). Strategic overview of management of future solar photovoltaic panel waste generation in the Indian context. *Waste Management & Research: The Journal for a Sustainable Circular Economy, 40*(5), 504–518. https://doi.org/10.1177/0734242x211003977

Rubino, A., Schiavi, P. G., Altimari, P., & Pagnanelli, F. (2021). Valorization of polymeric fractions and metals from end of life photovoltaic panels. *Waste Management, 122*, 89–99. https://doi.org/10.1016/j.wasman.2020.12.037

Salim, H. K., Stewart, R. A., Sahin, O., & Dudley, M. (2019). Drivers, barriers and enablers to end-of-life management of solar photovoltaic and battery energy storage systems: A systematic literature review. *Journal of Cleaner Production, 211*, 537–554. https://doi.org/10.1016/j.jclepro.2018.11.229

Senthil, R. (2022). A technical note on integrating thermal energy systems into solar photovoltaic panels toward a circular economy. *Renewable Energy Focus, 42*, 97–100. https://doi.org/10.1016/j.ref.2022.06.002

Seo, B., Kim, J. Y., & Chung, J. (2021). Overview of global status and challenges for end-of-life crystalline silicon photovoltaic panels: A focus on environmental impacts. *Waste Management, 128*, 45–54. https://doi.org/10.1016/j.wasman.2021.04.045

Sica, D., Malandrino, O., Supino, S., Testa, M., & Lucchetti, M. C. (2018). Management of end-of-life photovoltaic panels as a step towards a circular economy. *Renewable and Sustainable Energy Reviews, 82*, 2934–2945. https://doi.org/10.1016/j.rser.2017.10.039

Sim, Y., Tay, Y. B., Pham, H. K., & Mathews, N. (2023). A facile crush-and-sieve treatment for recycling end-of-life photovoltaics. *Waste Management, 156*, 97–106. https://doi.org/10.1016/j.wasman.2022.11.023

Tan, J., Jia, S., & Ramakrishna, S. (2022). End-of-life photovoltaic modules. *Energies, 15*(14), 5113. https://doi.org/10.3390/en15145113

Tsanakas, J. A., van der Heide, A., Radavičius, T., Denafas, J., Lemaire, E., Wang, K., Poortmans, J., & Voroshazi, E. (2020). Towards a circular supply chain for PV modules: Review of today's challenges in PV recycling, refurbishment and re-certification. *Progress in Photovoltaics: Research and Applications, 28*, 454–464.

Yafetto, L. (2022). Application of solid-state fermentation by microbial biotechnology for bioprocessing of agro-industrial wastes from 1970 to 2020: A review and bibliometric analysis. *Heliyon, 8*(3), e09173. https://doi.org/10.1016/j.heliyon.2022.e09173

Zhang, Y., Ye, E., Liu, F., Lai, N., You, X., Dong, J., & Dong, J. (2023). The relationship between landscape construction and bird diversity: A bibliometric analysis. *International Journal of Environmental Research and Public Health, 20*, 4551. https://doi.org/10.3390/ijerph20054551

Renewable Energy Production with a Circular Economy Approach: Entrepreneurial Brazilian and Italian Case Studies

Adele Finco, Pery Francisco Assis Shikida, Deborah Bentivoglio, Carlo Eduardo de Freitas Vian, and Giulia Chiaraluce

Abstract Global environmental concerns, depletion of fossil energy resources and the recent energy crisis and security, especially due to the Russia-Ukraine war, are among the main reasons for promoting a shift from fossil fuels to sustainable renewable energy. However, today, to support the sustainable market position of renewable energies, it is necessary to encourage the transition from the traditional linear model to the circular one. In this scenario, the use of waste shows a huge potential to enhance this change. While the implementation of circular principles represents an opportunity for companies, this entails several barriers. Therefore, this chapter aims at identifying and analyzing the main barriers and opportunities to circular economy adoption in European and Brazilian agroindustry based on case studies. In detail, the chapter focuses on the Brazilian case of sugarcane ethanol and on an Italian frozen vegetable company because they are strategic sectors, they produce considerable quantities of waste, and both produce renewable energy from waste. The implementation of circular practices in companies could represent a business opportunity and a source of value creation. However, policymakers need to first better understand the complex challenges faced by small and medium enterprises to develop appropriate policies to encourage a sustainable transition.

A. Finco · D. Bentivoglio (✉) · G. Chiaraluce
Department of Agricultural, Food and Environmental Sciences (D3A), Polytechnic University of Marche, Ancona, Italy
e-mail: a.finco@staff.univpm.it; d.bentivoglio@staff.univpm.it; g.chiaraluce@staff.univpm.it

P. F. A. Shikida
Department of Economics, Universidade Estadual do Oeste do Paraná (UNIOESTE), Toledo, Paraná, Brazil
e-mail: pery.shikida@unioeste.br

Carlo Eduardo de Freitas Vian
Department of Economics, Business and Sociology, ESALQ/USP, Piracicaba, São Paulo, Brazil
e-mail: cefvian@usp.br

C. Rego et al. (eds.), *Transitioning to a Circular Economy*, Studies on Entrepreneurship, Structural Change and Industrial Dynamics,
https://doi.org/10.1007/978-3-031-77661-8_15

Keywords Circular economy · Biomethane · Ethanol · Case study · SMEs · Waste valorization

1 Introduction

At present, the energy supplies from fossil fuels and non-renewable resources cover more than two third of the global energy demand. Nevertheless, global environmental concerns, depletion of fossil energy resources and the recent energy crisis and security, especially due to the Russia-Ukraine war, are among the main reasons for promoting a shift from fossil fuels to sustainable renewable energy (Omer, 2008; Finco et al., 2012; Su et al., 2021). Amongst the renewable energy resources, biomass, such as agricultural and food waste, has always been forecasted as a promising solution to substitute crude oil-based petrol refineries. It is estimated that almost 9% of the global energy supply comes from biomass, which is considered a renewable and sustainable feedstock that is environmentally friendly and available in large quantities. The problem that arose in the last decade was connected to the fact that biofuels like biodiesel, bioethanol, and biogas were mainly derived from biomasses and crops destined for human consumption, the so-called first-generation feedstocks (oil seeds, corn, sugarcane juice and molasses). Nowadays, the debate established between food and fuel forcibly shifted the focus to second-generation feedstocks like lignocellulosic matrices, primarily from agricultural residues (Cherubini, 2010; Bentivoglio et al., 2016; Kapoor et al., 2020). In this perspective, to support the sustainable market position of renewable energies, it is necessary to encourage the transition from the traditional linear model to the circular one. According to the currently most recognized definition of the circular economy (CE) offered by the Ellen MacArthur Foundation (2016), "*A circular economy is one that is restorative and regenerative by design and aims to keep products, components, and materials at their highest utility and value at all times, distinguishing between technical and biological cycles*." Generally, by defining the CE, we refer to a "closed-loop economy" that "does not generate excessive waste and whereby any waste becomes a resource" (Wysokińska, 2016).

The transition from a traditional linear model to a circular one is still on its path, requiring the improvement and implementation of new skills and additional knowledge to establish innovative and sustainable paradigms (Rizos & Bryhn, 2022). There are two opportunities to support this transition, starting from current renewable energy systems: convert existing systems into circular economy models or improve existing models that already use a CE approach. Both these strategies support the provision of sustainable development of renewable energy in the context of a CE.

On these premises, the goal of this chapter is to identify and analyze the main barriers and opportunities to circular economy adoption in Brazil and the European Union (EU) based on case studies: a sugarcane mill in the State of Paraná and a frozen vegetable company in Central Italy. Both selected case studies produce energy from waste: bioelectricity and first-generation ethanol in Brazil and biogas in Italy. The analysis will be based on face-to-face interviews with the managers of the companies. Indeed, few studies have explored the adoption of the circular economy in the sugarcane ethanol sector (Jesus et al., 2021; Vandenberghe et al., 2022); on the other hand, the circular economy concept is much discussed in the EU, but only limited progress has been accomplished so far regarding its implementation (Bonviu, 2014; Domenech & Bahn-Walkowiak, 2019; Mhatre et al., 2021).

The CE in Brazil is still in its early stages (de Oliveira et al., 2018; Mancini et al., 2021). Nonetheless, the sugarcane agroindustry has great potential in the circular economy because circularity is part of the various stages of the production processes. In particular, the reuse of by-products from the sugarcane agroindustry is very intensive. The straw and sugarcane bagasse, vinasse, and filter cake are essential for the generation of valuable products such as bioelectricity, biogas, fertilizers, and second-generation ethanol. Second-generation ethanol, for example, is produced from residues such as straw and sugarcane bagasse that are discarded from the production process of first-generation ethanol (coming this latter from sugarcane sucrose). This reuse means cost rationalization and cash generation for the industries, converting into vital competitive advantages for their survival in the market (Jesus et al., 2021; Nadaleti & Lourenco, 2021).

At the same time, the goal to move to circular economy has been particularly strong in Europe, as the EU embraced the CE not only as an economic concept but also as a social and political goal through the release of dedicated strategies and action plans, referring also to the agri-food sector and the production of waste and by-products (Chiaraluce et al., 2021). When it comes to circular approaches, in Europe biogas production represents an environmentally friendly strategy to successfully recover large amounts of waste and by-products to produce renewable energy (European Commission, 2017). Among the different food supply chains, fruit and vegetables produce a considerable amount of waste and by-products (Kowalska et al., 2017). The recovery and valorization of by-products deriving from the processing and transformation of vegetables, by converting them into biogas, makes it possible to manage a considerable quantity of these substrates on a daily basis. The new frontier consists of converting biogas into biomethane by purification and upgrading (Ryckebosch et al., 2011; Adnan et al., 2019).

The implementation of CE practices in the agroindustry could represent a business opportunity and a source of value creation. However, for companies using commodities as raw materials, the recovery of bio-waste, in line with the CE approach, involves many uncertainties, particularly if the raw material is seasonal. Policymakers need to first better understand the complex challenges faced by small and medium enterprises in order to develop appropriate policies to encourage a sustainable transition (Rizos et al., 2015; Ormazabal et al., 2018). It is expected that the circular business models will be able to help the companies in waste valorization, decreasing management costs, and generating new added value. However, the transition from the linear to the circular model cannot be materialized without sufficient capital investment, making the government's role more crucial in this transition.

The chapter is organized as follows: first, the state-of-art circular economy in Italy is introduced (Sect. 2). Second, the context of the circular economy in Brazil is presented (Sect. 3). For both paragraphs, the political context, the case study, and the barriers and opportunities are described. Finally, we draw a key conclusion.

2 The State-of-Art of Circular Economy in Italy

In this section, the authors will present the current Italian situation about circular economy, with a focus on renewable energy production (Sect. 2.1). Probably, the generation of energy from waste and by-products represents one of the most common ways to implement circular practices. For this purpose, we present a case study based on a frozen vegetable company, the owner of a biogas plant since 2011. In light of the current policies, the authors would like to disclose the possibility behind the conversion of the current plant into a biomethane-producing one (Sect. 2.2). The selected case study was useful in highlighting current barriers and opportunities to the implementation of the circular economy model in Italy.

2.1 Circular Economy and Renewable Energy in Italy

The European Commission decided to address the future of the Union to circularity by launching its first Circular Economy Action Plan in 2015. Since then, the EU has always stressed the high ambitions and broad scope of this new policy initiative (Kovacic et al., 2019; Pantzar & Suljada, 2020), reflected also in the fact that one of the strategies of the European Green Deal is properly the mobilization of the industry for a clean and circular economy. As the agri-food sector is one of the pillars of the European economy and contributes considerably to the production of waste, the Farm to Fork Strategy, at the core of the Green Deal, aims to contribute to achieving a circular economy, reducing the environmental impact of the food processing. The release of the Green Deal in 2019 also entailed the publication of the New Circular Economy Action Plan in 2020 (Chiaraluce, 2021).

Amongst the Member States, Italy stands out as the top in circularity trends among the five major European economies (Circular Economy Network, 2022). In 2017 the document "Towards a circular economy model for Italy—Document of strategic positioning and framing" was published, aiming to provide a general framework of the circular economy and to define the country's strategic positioning on the topic. The national government is adopting and implementing the directives coming from the Circular Economy Package and, in 2022, the Italian National Strategy on the Circular Economy came into force. The strategy, focused on eco-design and eco-efficiency, aims to define new administrative and fiscal instruments to strengthen the market for secondary raw materials, the extended responsibility of the producer and the consumer, the spread of sharing practices and "product as service," supporting the achievement of climate neutrality objectives, and defining a roadmap of measurable actions and targets by 2040.

One of the aspects considered in the circular classification is the renewable energy consumption for electricity, heat, and biofuels. The European stakeholders are conscious that, to reach the climate and energy targets, circular economy can play a major role. The recovery and valorization of waste and by-products open a path towards the production of renewable energy and biofuel, and, in this sense, agricultural and food waste are primary feedstock in bioenergy generation. Among the most promising and highly encouraged is the production of biogas and its consequent upgrade to biomethane. According to the European Biogas Association (EBA), Europe has always been a leader in biogas production, with around 20,000 units in operation in 2021. Around 1000 of them upgrade biogas into biomethane, which is considered the future to phase out the European dependency on fossil fuel resources and to increase the resilience of the EU-wide energy system (Abdalla et al., 2022). Figure 1 shows the evolution of the number of biomethane plants in Europe from 2011 to 2021.

In Italy, in the last decade, the biogas industry was progressively growing, due to dedicated policies and generous government incentives (Benato & Macor 2019; Pirelli et al., 2021). Figure 2 shows the evolution of the number of biogas plants in Italy, growing from 103 in 1999 to 2201 in 2020. Regarding biomethane, Italy currently has 30 active plants in its territory, mainly located in the northern regions.

The history of biogas in Italy dates back to the last decade of the 1990s. The turning point was in 1999, when the green certificate system was established. This was the first effort to improve the production of electricity from biogas, providing subsidies directly from the central government.

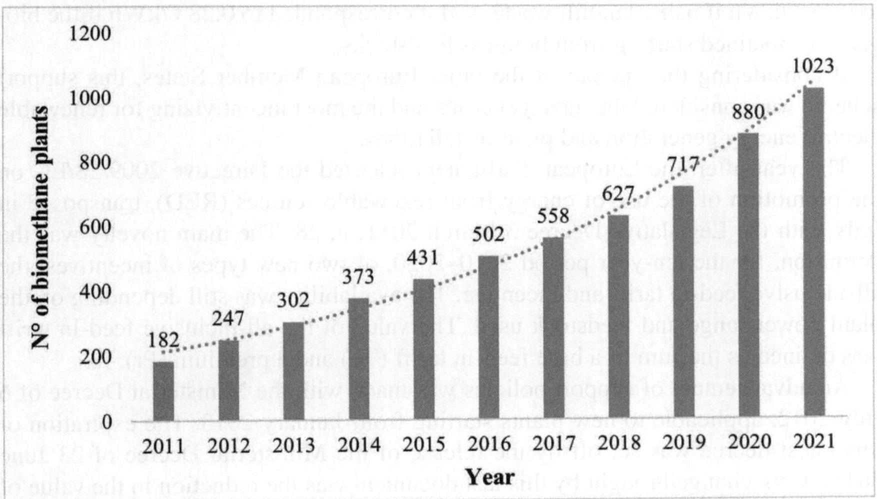

Fig. 1 Evolution of the number of biomethane plants in Europe from 2011 to 2021. *Source:* Authors' elaboration from EBA data

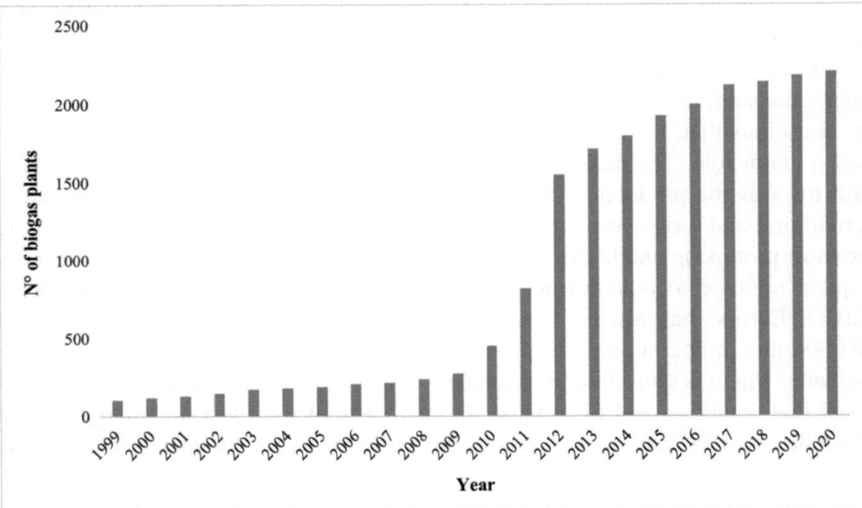

Fig. 2 Evolution of the number of biogas plants in Italy from 1999 to 2020. *Source:* Authors' elaboration from GSE data

Only after more than 10 years, in 2008, the Italian government recognized an all-inclusive feed-in tariff for all those plants with an installed power range between 1 kW and 1 MW. With this system, it was thought to give a shake to the entire compartment. The limitation was that this tariff was guaranteed for only 15 years and only if the plant entered into operation before the endo of December 2012. The monetary value was also different according to the diet of the digestor: it was equal to 0.18 €/kWh if using landfill waste, and it corresponded to 0.28 €/kWh if the biogas was obtained starting from biomass feedstocks.

If considering the operate of the other European Member States, this support scheme was considered the most generous and the most incentivizing for renewable electric energy generation and plant installations.

The year after, the European Parliament released the Directive 2009/28/EC on the promotion of the use of energy from renewable sources (RED), transposed in Italy with the Legislative Decree 3 March 2011, n. 28. The main novelty was the definition, for the ten-year period 2010–2020, of two new types of incentives, the all-inclusive feed-in tariff and incentive. The availability was still depending on the plant power range and feedstock used. The value of the all-inclusive feed-in tariff was defined as the sum of a base feed-in tariff (Tb) and a premium (Pr).

An advancement of support policies was made with the Ministerial Decree of 6 July 2012, applicable to new plants starting from January 2013. The expiration of this latest decree was set off by the release of the Ministerial Decree of 23 June 2016. A big change brought by this last document was the reduction in the value of Tb and Pr. A sort of compensation was provided by some tax exemptions. Additionally, starting in 2017, the feed-in-tariff was dispensed only to biogas plants with a nominal power comprised between 1 and 500 kW, for 20 years instead of 15, and a monetary value between 0.085 to 0.233 €/kWh.

The most actual commitment of the EU is reflected in the new Renewable Energy Directive (RED II) of 2018. An important aspect connected to sustainable development requires EU Member States to provide binding proof of sustainability for the generation of electricity and heat from solid and gaseous biomass fuels (like biogas and biomethane), by July 2021 at the latest. In addition to setting the targets for sustainable energy production, the RED II has provided clarifications on the concept of renewable energy itself, redefining what can be considered a renewable energy source or a biofuel and providing additional explanations related to biogas and biomass production. Furthermore, in RED II, biomethane falls under the category of biogas and is no longer limited to the transport sector, while it is strongly incentivized in its production. RED II also expanded the Guarantee of Origin (GO) scheme to cover not only renewable electricity but also renewable gases, with the objective to prove to the end-consumer that a specific share of the energy consumed origins from renewable resources. Italy has implemented the RED II with the Legislative Decree 8 November 2021 no. 199. This decree contains provisions on energy from renewable sources and defines the instruments, mechanisms, incentives, and the institutional, financial, and legal framework necessary to achieve the objectives of increasing the share of energy from renewable sources to 2030, also in line with the objectives of the Italian PNRR (*Piano Nazionale di Ripresa e Resilienza*). Only 1 year later, in September 2022, the Ministerial Decree 15 September 2022 was released, allowing access to the resources provided by the PNRR for the development of biomethane production facilities. The decree aims to promote biomethane production and injection into the natural gas network recognizing:

(i) A capital support of a maximum of 40% of the eligible expenses incurred. The eligible expenses include the costs for the construction and efficiency improvement of the plant, the equipment for monitoring of biomethane, the costs of connection to the natural gas grid, the costs for the purchase or acquisition of software for the management of the plant, the costs of design, construction management, testing, consultancy, feasibility studies activities, and purchase of licenses. The payment of the capital contribution is resumed in Table 1.
(ii) An incentive tariff applied to the net production of biomethane for a maximum of 15 years. The decree provides for two different incentive mechanisms, depending on the production capacity of the plant: an all-inclusive feed-in tariff (in the case of production capacity up to 250 Smc/h) and a premium tariff (in the case of plants with a production capacity exceeding 250 Smc/h).

After participating in a competitive public procedure (discount auction), both new biomethane production plants, from agricultural biomass or Organic Fraction of Municipal Solid Waste (OFMSW), and biogas plants converted (totally or partially) to biomethane will benefit from the incentives. In this last case, only biogas produced from agricultural residues will be considered, as the conversion of plants fed with OFMSW is not allowed. This latest decree was expressly designed by the

Table 1 Capital grant on the eligible expenses of the investment occurred for a biomethane plant in Italy

Production plant	Production capacity (Cp)	Maximum specific investment cost [€/Smc/h] New plants	Maximum specific investment cost [€/Smc/h] Reconversion	Capital contribution percentage (%)	Maximum specific values of the capital contribution [€/Smc/h] New plants	Maximum specific values of the capital contribution [€/Smc/h] Reconversion
Plants powered by agricultural residues	$Cp \leq 100$ Smc/h	33,000	12,600	40	13,200	5400
	$100 < Cp \leq 500$ Smc/h	29,000	12,600	40	11,600	5400
	$Cp > 500$ Smc/h	13,000	11,600	40	5200	4640
Plants powered by OFMSW	Whatever	50,000	/	40	20,000	/

Source: Author's elaboration

Italian government as an incentive for the spread of the circular model in the production of renewable energy.

In addition, in March 2022, the decree related to urgent measures to counter the economic and humanitarian effects of the Ukrainian crisis promoted the replacement of synthetic chemical fertilizers with digestate equivalent, to favor the transition to a circular economy. Digestate can be considered a secondary product of anaerobic digestion, rich in nitrogen, potassium, and phosphorus, highly nutritious for soils (Baştabak & Koçar, 2020; Głowacka et al., 2020). Digestate is defined as equivalent when obtained by anaerobic digestion of substances and materials alone or in a mixture. The aim is to promote the spread of ecological practices in biogas-biomethane production and reduce the use of chemical fertilizers, also limiting production costs. The purpose of equating digestate is to allow biogas-biomethane plants the possibility of obtaining an extra income from the sale of such digestate as a fertilizer product.

2.2 *Case 1: The Frozen Vegetable Industry (Central Italy)*

The agri-food sector is responsible for the production of a considerable amount of waste and by-products (Lin et al., 2013; Girotto et al., 2015). When considering different supply chains, fruits and vegetables account for 40–50% of the produced waste, followed by cereals (30%), fish (30%), and oilseed, meat, and dairy (20%) (Mor et al., 2021). The fruit and vegetable supply chain also has considerable economic importance in the EU, accounting for almost 14% of the entire European agricultural production value, thus accounting as a central sector in the economy of many Member States, including Italy (Rossi, 2019). The seasonality and geographical localization of vegetable waste pose a limit on how to manage them in a sustainable way from a technical, economic, and environmental point of view. If using, thus valorizing, the vegetable waste deriving from the industry through anaerobic digestion, it will be possible to daily manage high quantities of these matrices, reducing their impact on the supply chain (Bentivoglio et al., 2022). For this purpose, the selected case study is a frozen vegetable company located in Central Italy, the owner of a biogas plant. All the data reported in this chapter were retrieved from face-to-face interviews with the owner of the company.

The company has a cultivated area of 6.6 thousand hectares under organic and integrated schemes. The production of raw material is, on average, 21,000 t/year of peas, green beans, leafy vegetables (spinach, chard, chicory), tomatoes and cherry tomatoes, soup ingredients (celery, cabbage, leek, pumpkin), and herbs (basil, parsley, garlic, onion). On average, yearly, 17% (3480.36 t/year) of the produced raw material is wasted during industrial processing to produce frozen vegetable products. The by-products are composed mainly of leaves, pods, seeds, skins, stems, and products unsuitable for the transformation. Figure 3 shows the amount (tons) of raw material, frozen products, and by-products obtained for each crop in 1 year (data from 2021).

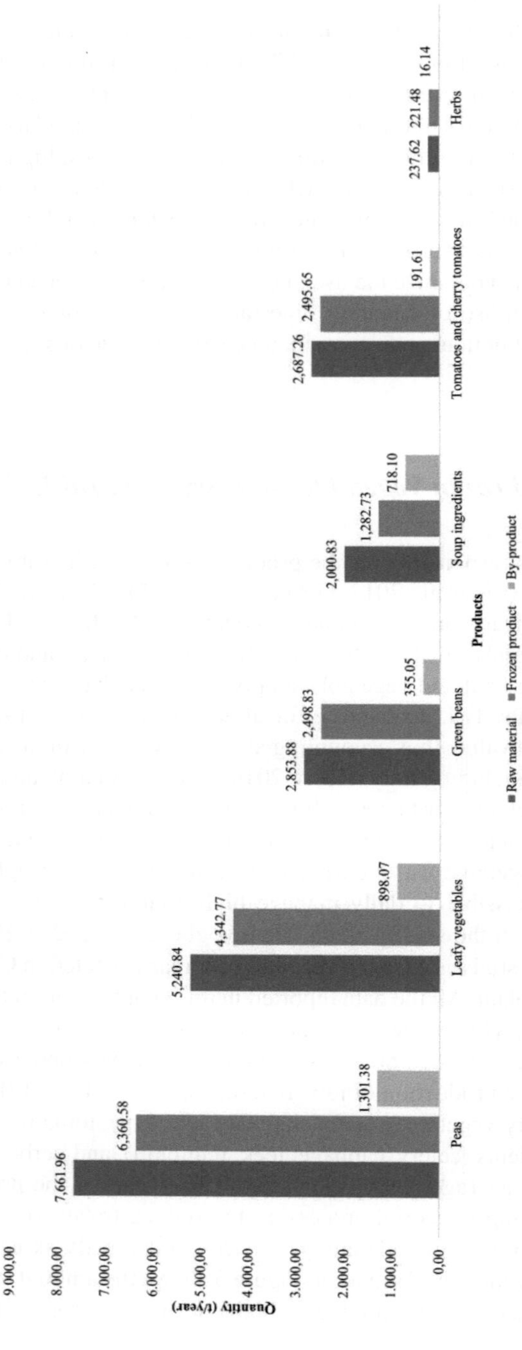

Fig. 3 Quantity (t/year) of raw material, frozen product, and by-product produced for each crop. *Source:* research data (2021)

In 2011, the company started the biogas production plant according to the dictates defined by the Italian government in 2008. In this regard, the plant, fed with vegetable waste (20%), chicken manure (10%) and maize silage (70%), had access to the all-inclusive feed-in tariff equal to 0.28 €/kWh, guaranteed for 15 years. The electric capacity is 1 MWe, producing almost 4Mio m^3/year of biogas. The reasons behind the decision of the company to the realization of the biogas system are both of environmental and economic nature. From the environmental point of view, particular attention is paid to the recovery of waste and, therefore, to the consequent reduction of pollution; from the economic side, this involves the diversification of the agricultural activity, hence of the profit deriving from it. The market for energy products obtained from waste biomass represents an opportunity for the stable integration of farm income and the implementation of the circular model in agriculture.

However, it is important to highlight some limitations of the current energy production model. First, the plant to operate annually is fed not only by vegetable waste and chicken manure but also by silo maize. This is because the quantity of vegetable waste produced solely by the company, combined with its seasonality, is not sufficient for the operation of a plant of this size in a standardized and continuous way. This involves the use of a dedicated crop, with the consequent purchase costs (50 €/t). In addition, the current cogeneration engine has energy losses. As the 15-year incentive period is coming to an end (by 2026), and because of the new Ministerial Decree favoring the production of biomethane from a circular economy perspective, the company is planning to invest in upgrading the biogas plant.

From the interview with the owner, the planned investment will be 1,445,500.00€ for a plant producing 250 Smc/h. Following the Ministerial Decree 15 September 2022, a capital contribution on the eligible costs of the investment incurred, equal to 40% of the expenses, will be recognized, as the conversion will be on a plant fed with agricultural residues. Moreover, as the plant will have a capacity of 250 Smc/h, it will be awarded with an all-inclusive feed-in tariff, calculated based on the rules of application to the biomethane decree promulgated by the Italian Energy Services Operator (GSE). These two incentives will be granted only if the company wins the competitive public procedure and has access to the available funds. Since the plant will produce biomethane for its own consumption, the tariff will be calculated by GSE according to the following algorithm:

$$I_n = \max\left((T - P_{\text{GME}}) * \min\left(M2_n; M_s; M_{\max}\right) * \text{PCS2}_n * (1 - \%\text{SA}) - \text{GO}_\text{T} * P_{\text{GO},\,T}; 0\right)$$

where

I_n = incentive for the monthly production of biomethane.

T = fee indicated in the accepted provision relating to the installation.

P_{GME} = average monthly price of natural gas.

$M2_n$ = monthly quantity of biomethane (in Smc).

M_s = quantity of sustainable biomethane for month n (in Smc). Biomethane respecting the greenhouse gas emission reduction value is considered sustainable.

$M_{\max}$ = maximum monthly producibility.

$PCS2_n$ = average monthly value of the higher calorific value determined based on the chemical composition of biomethane (in kWh/Smc).

%SA = percentage related to energy consumption attributable to auxiliary plant services.

GO = number of Guarantees of Origin of biomethane.

P_{GO} = value of the average monthly price recorded on the market platform for the exchange of Guarantees of Origin.

Up to now, it is not possible to know the real amount (in euros) of the tariff. As it can be seen from the parameters, the amount will depend on several factors changing periodically.

To accomplish full circularity in the digester's diet, silo maize should be replaced. For this purpose, the owner of the company aims to put in place a cooperative system, like an industrial symbiosis, where the nearby farms and/or food industry commit their agricultural and food waste/by-products to the plant. In this way, the cost of the diet will be dramatically reduced: from 750,000.00 €/year to purchase the silo maize to 115,000.00 €/year to collect the waste/by-products from the neighboring farms (if considering 5.50 €/t for the transportation costs, charged to the biomethane plant owner). Nevertheless, it is worth remembering that the rebalancing of a digester diet is not a simple process. It requires several analyses and attempts to find the correct amount of each feed to ensure a correct, stable, and standardized functioning of the plant, so of the energy production. The company is currently working to find the right balance between circularity, economic profitability, and production of renewable energy only from waste.

Lastly, a secondary product of anaerobic digestion is digestate. Digestate has always been considered a highly valuable biofertilizer (Lukehurst et al., 2010). However, in the past years, it was not sold for use in agriculture as official political directives, as well as a reference market, were missing. For this reason, the case study company was giving it away for free. In the last year, to try to reduce the amount of imported Russian fertilizers, the Italian government, with the Legislative Decree 21 March 2022 n. 21, has officially recognized digestate as equivalent to chemical fertilizers. From now on, digestate comparable to chemical fertilizers can be used agronomically, with a nitrogen use efficiency level of at least 90%, an average dry matter content of 1.5–8% and an ammoniacal nitrogen value of up to 75–90% of the total nitrogen present. If the company decides, in the future, to sell the digestate at an average price of 10 €/t (a price lower compared to chemical fertilizers), it will gain an extra income of roughly 150,000.00 €/year (deriving from the production of 15,600 t/year of digestate from the anaerobic digestion). This represents another solution for the diversification of agricultural income, and another way to improve the resilience of the company in global and changing markets.

Upgrading of biogas to biomethane, substitution of silo maize with other agri-food by-products, and digestate sale and use are the basis for a successful roadmap to a circular economy model in agriculture. From the interview with the entrepreneur, more could be done. From a forward-looking perspective, it would be desirable to completely close the circle by recovering the wastewater used in the company to wash the vegetables (to be used for cooling the refrigerators in the freezing process) and the CO_2. By implementing an innovative circular business model, it will be possible to create wealth from an environmental, economic, and social point of view.

2.3 *Barriers and Opportunities for Circular Economy in Italy*

Considering the barriers related to the implementation of a circular model in the agri-food sector in Italy, the entrepreneur has repeatedly stressed the fact that the current legislation on the circular economy is unclear and that there are no incentives to support the implementation of a circular model. Moreover, the public administration and the bureaucracy do not facilitate the enactment of circular practices. Nowadays, policy instruments are proposed more to advertise circularity, rather than reduce the impact of the linear economy. We are faced with action plans and strategies that are not really limiting the traditional model to the detriment of a new regenerative economy (Kirchherr et al., 2018; Corvellec et al., 2022). From an economic point of view, these actions at this time would require investments that are not feasible for small-medium enterprises. Setting-up a biogas plant, and then a biomethane upgrading system, is far from being cheap. It is therefore hoped that the intervention of the European and national government bodies to support the realization and maintenance of the circular model, through an adequate system of fair incentives.

As for the opportunities, the entrepreneur has projected the reduction of the amount of waste and their recovery through valorization, with a consequent creation of added value. In this way, the circular economy can be a valuable asset for income diversification. Thus, economic profitability could be reached if innovative circular business models are developed (Daou et al., 2020). Anaerobic digestion could act as a key model in the realization of sustainable circular business models; however, despite the increasing awareness of scholars, entrepreneurs, and stakeholders, circular business models are still scarce (De Keyser & Mathijs, 2023).

The production of biomethane can provide new opportunities for spreading the circular model. Biomethane could represent a prospect for biogas producers, especially for those plants whose incentivization period is about to finish. Using vegetable waste as a raw material for bioenergy production is an advantageous alternative to dumping processes in soil or landfills, thus reducing the environmental impacts of biomass disposal, reducing the demand for fossil fuels and the emission of greenhouse gases, and contributing to the circular economy (Scapini et al., 2023). However, a limiting aspect is the availability of these wastes: vegetable by-products are produced greatly during the specific period of the year, according to the seasonality of each crop. This will limit the operation of a biomethane plant, which should produce energy constantly all year round. The problem could be solved by integrating the diet whit other agricultural and food waste, like animal manure, straw, pruning residue, and food processing by-products, ensuring a sufficient supply for the digestor. Gathering such a considerable amount of product will require the establishment of an industrial symbiosis system, where the waste of one stream becomes the resource of another, or of a cooperative structure, where more than one farm or food industry commits their own waste to a single producer (Bentivoglio et al., 2022). It is not utopian to think that, in a cooperative form of support, the different actors can become shareholders of the biomethane plant, benefiting not only from the provision of waste but also from any gains deriving from the production of circular renewable energy.

Biomethane produced from upgrading the biogas obtained from the anaerobic digestion of food and agricultural by-products, like vegetable ones, could potentially reduce the carbon footprint of the agri-food sector. Europe, and therefore Italy, is strongly aiming at sustainable energy production, regulating some aspects of environmental protection. The latest Italian biomethane decree provides for the promotion of the so-called "sustainable biomethane," and production plants should comply with specific limits for the reduction of greenhouse gas emissions (at least 65% reduction in GHG for biomethane production facilities for the transport sector; at least 80% GHG reduction for biomethane production facilities for other uses). Moreover, the digestate could further improve the sustainability of this system through its use as fertilizer (Murano et al., 2021; Bose et al., 2022). However, some barriers can be linked to non-technical issues, such as the lack of public acceptance for the biogas-biomethane plants (NIMB—Not In My Backyard syndrome) (D'Adamo et al., 2021).

It is not a secondary aspect that a transition to a bio-based/renewable energy production system, such as biomethane, would be able to better balance the energy supply situation around the world. In this way, more countries, like Italy, would be able to become self-sufficient (Fagerström et al., 2018).

3 The State-of-Art of Circular Economy in Brazil

This topic deals with the case study of the implementation of circular economy strategies in a sugar and ethanol plant in the state of Paraná in Brazil. Section 3.1 presents the possibilities of implementing circularity in the sugarcane agroindustry, highlighting which productive resources can be reused for other activities and productive processes.

Next, Sect. 3.2 discusses the case of Usiban, highlighting the current situation of adopting a circular economy and the benefits and challenges that the company has faced. Finally, Sect. 3.3 examines the barriers and opportunities for adopting circular economy strategies in the Brazilian sugarcane agroindustry.

3.1 Circular Economy in the Sugarcane Agroindustry

Nowadays, the production of goods and services involves issues related to the economy, sustainability, logistics, reverse logistics, etc. In the process of producing goods and services, it is common to use resources extracted from nature in which waste is not always properly reused. Maximizing, as much as possible, the reuse of these residues in other production processes allows significant changes in production, changing not only its forms but also the relationships with the supply chain and the business itself. This context defines the circular economy (Manavalan & Jayakrishna, 2019).

In Brazil, the National Solid Waste Policy (*Política Nacional de Resíduos Sólidos*), instituted in 2010, is the institutional arrangement that serves to guide the circular economy, making mention of reverse logistics. In article 3, item XII of this Law, the instrument of economic and social development can be "[...] characterized by a set of actions, procedures and means aimed at enabling the collection and return of solid waste to the business sector, to reuse, in its cycle or in other production cycles, or other environmentally appropriate final destination" (Brasil, 2010, n.p.).

Before handling the issue of the circular economy and the sugarcane agroindustry, it is important to say that the term sugarcane agroindustry encompasses the agricultural segment and the industrial segment of sugarcane processing, which generates three important products: sugar (for food security), fuel ethanol (for energy security), bioenergy (biogas and bioelectricity) (Shikida & Cattelan, 2020). Furthermore, the main by-products derived from the sugarcane agroindustry are sugarcane bagasse and straw, vinasse (a pasty and malodorous residue derived from the distillation of fermented sugarcane juice, also known as "*vinhoto,*" "*tiborna,*" or "*restilo*"), filter cake (an organic compound from the mixture of crushed sugarcane bagasse and decanting sludge), ash, pellet (solid biomass produced from the compacted residue of sugarcane bagasse), and yeast (Raízen, 2023).

Currently, Brazil is the world's largest producer of sugarcane, whose harvest in 2021/2022 was 578.8 million tons, which generated 26.4 billion liters of ethanol (the second largest producer of this commodity in the world) and 34.9 million tons of sugar (the world's largest producer and exporter of this commodity) (Companhia Nacional de Abastecimento—CONAB, 2023). Therefore, the sugarcane agroindustry has had the circular economy as an important ally for a long time. Based on a compilation from de Fravet et al. (2010), Pacheco (2011), Cezarino et al. (2020) and Raízen (2023), the reuse of by-products from the sugarcane agroindustry is very intensive. Sugarcane straw and bagasse, vinasse and filter cake are essential for generating valuable products such as bioelectricity, biogas, fertilizers, and second-generation ethanol. Second-generation ethanol, for example, is produced from residues such as sugarcane straw and bagasse that are discarded from the production process of first-generation ethanol (first-generation ethanol is derived from sugarcane sucrose). This reuse means rationalization of costs and cash generation for industries, becoming vital competitive advantages for their survival in the market. Figure 4 shows how circularity is part of the various stages of the productive processes of the sugarcane agroindustry, from the production of second-generation ethanol to the production of bioelectricity, fertilizers, etc.

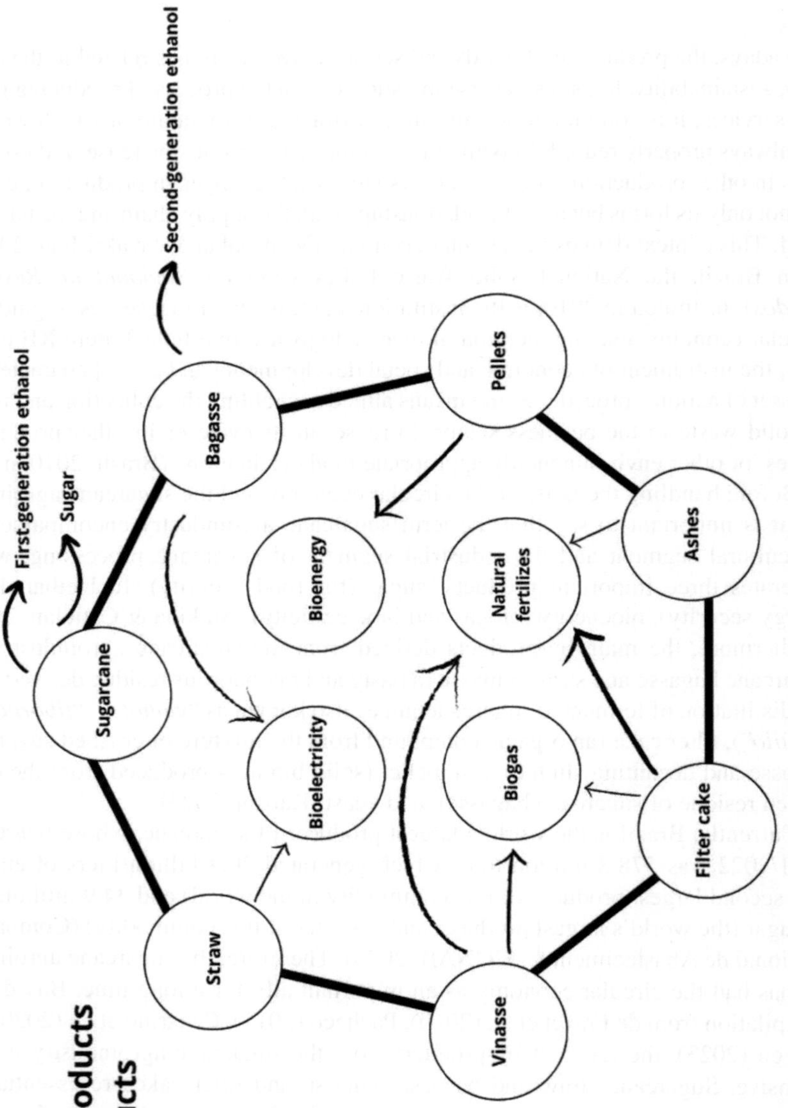

Fig. 4 Sugarcane products, by-products, and the circular economy. Source: Research data (2023)

Authors such as Chagas (2009), Cardoso (2016), and Shikida and Cattelan (2020) unanimously state that the sugar-energy sector has two outstanding products, sugar and ethanol—E1G (first-generation ethanol, formerly called fuel alcohol), which can be anhydrous (for an additive in gasoline, with a water content of around 0.5% by volume) or hydrous (for use as exclusive fuel, containing about 5% of water). Figure 4 points out this highlight in the upper right of the sugarcane circle. Ethanol emerged in Brazil with the National Alcohol Program in 1975. Its main objective was to reduce the use of petroleum derivatives from a renewable and clean alternative, since ethanol emits fewer polluting gases compared to its equivalent fossil procedure counterparts. Currently, Brazil stands out on the world stage for its proficiency in advanced technology, whether in the manufacture or use of ethanol, contributing to the mitigation of greenhouse gases (Cardoso, 2016; Shikida & Cattelan, 2020).

The development of flex-fuel vehicles (from 2003 onwards), which allowed consumers to choose any proportion of gasoline and hydrous ethanol in the fuel supply, was the latest innovation in the sector to the point that where flex-fuel technology was present in 96% of new cars sold domestically in Brazil (Rodrigues & de Belon, 2019). According to de Sousa (2020, n.p.), "about 515 million tons of greenhouse gases have stopped being released into the atmosphere since the beginning of flex-fuel cars in Brazil, in 2003."

More recently, with Law n. 13,576 (Brasil, 2017), the federal government acted on the National Biofuels Policy (RenovaBio), defining decarbonization targets based on the increased use of biofuels, involving the purchase of CBIOs (Decarbonization Credits) by fuel distributors. The definition of the number of CBIOs issued will consider "the volume of biofuel produced, imported and sold by the primary issuer, observing the respective Energy-Environmental Efficiency Index included in the Efficient Biofuel Production Certificate of the primary issuer" (Brazil, 2017, n.p.). Thus, circularity is necessary to produce two important commodities (sugar and first-generation ethanol). Taking Pacheco (2011), Associação dos Fornecedores de Cana de Araraquara (2021) and Raízen (2023) references as a guideline, the by-product sugarcane bagasse (fibrous residue obtained after grinding the sugarcane to extract the juice and produce sugar and ethanol) can be used to generate electricity (bioelectricity), which is essential for energy self-sufficiency in mills and distilleries, and its surplus is sold to energy concessionaires. Second-generation ethanol (also known as E2G) can also be produced from sugarcane bagasse (Fig. 4). Mercante (2020) points out that, for each ton of clean sugarcane, 276 kg of bagasse is generated, and in energy terms this is equivalent to approximately 598,000 kcal (from bagasse alone). According to União da Indústria de Cana-de-Açúcar—Unica (2023), in 2021, energy generation by the sugar-energy sector, which includes not only bagasse but also straw and pellets, represented 79.5% of the total production of bioelectricity supplied to the electricity grid, with a production of 20.2 thousand GWh, equivalent to 30% of the electricity generated by Itaipu Hydroelectric Dam in 2021. The lignocellulosic materials contained not only in sugarcane bagasse but also in sugarcane straw (cellulose, hemicellulose and lignin) can produce second-generation ethanol, increase production efficiency by extracting more ethanol from the same raw material and helping to reduce waste formation without implying the expansion of the sugarcane cultivated area (Pacheco, 2011; Calil Neto, 2017).

However, Brazil is still not producing second-generation ethanol on a large scale "[...] due to the high cost of cellulosic bioethanol production technology, mainly due to the high cost of the enzymes used in the process" (Calil Neto, 2017, p. 151). That is, intensive investments in RD&I are needed to maximize the most appropriate technological routes for the production of ethanol from lignocellulosic material (Pacheco, 2011; Calil Neto, 2017). Even with this context, Raízen is producing on a small scale and is already making progress in investing in a new unit. The pellet, as it also contains lignocellulosic components, can be used both to produce second-generation ethanol and bioenergy by combustion in large furnaces and boilers (Almeida, 2015). Oliveira (2016) highlights that although the process of producing sugarcane bagasse pellets already exists in Brazil, there are few studies on its process. Despite this, more recently Novacana (2020) announced that Raízen and one of the largest electricity generators in Europe, the German company RWE, are studying the use of sugarcane bagasse pellets for energy generation on the European continent. Vinasse is another important sugarcane by-product, relevant to the sector's circular economy. According to Cortez et al. (1992), Serafim et al. (2021) and Novacana (2023), vinasse is the main residue of the production process that remains after the distillation of the fermented sugarcane juice. Considered highly polluting, the direct release of this by-product into water bodies is not allowed in Brazil. However, vinasse can generate biogas and, as it contains high concentrations of organic matter, suspended solids, calcium, potassium, nitrogen, and phosphorus, among other nutrients, its application in agricultural soil, especially in the sugarcane crop itself, enables excellent results in terms of fertirrigation, thus increasing agricultural productivity. It is emphasized that technical criteria and procedures must be followed to avoid negative environmental impacts on the soil, groundwater, and river sources. Estimates indicate that for each liter of alcohol produced, 10 to 18 liters of vinasse are generated as waste. According to Piracicaba Engenharia Sucroalcooleira (2015, n.p.), "[...] productivity can be increased by up to 10% thanks to the use of vinasse as a fertilizer."

In this circular economy, the filter cake is also used to produce biogas and, due to its richness in terms of organic matter, it can be used as a natural fertilizer. Each ton of crushed sugarcane produces 30 to 40 kg of filter cake. If the ash resulting from the burning of bagasse is recovered and the technical criteria established for its use in soils and plants are fully respected, it can serve as a fertilizer due to its richness in potassium, allowing it to reach areas where vinasse may not reach, contributing to close a nutritive circle (Cortez et al., 1992; Bega, 2014; Raízen, 2023). Last but not least, there is yeast, a microorganism that ferments sugarcane juice, responsible for producing ethanol in the mills/distilleries; that is, it transforms the sugarcane sucrose into ethanol (Grangeiro et al., 2001). In this process "the yeast is centrifuged, isolating the wine from the yeast. Afterwards, the yeast returns to the reservoir, where it is treated to restart, as fermentation is continuous," and for each liter of ethanol processed, an excess of 25 g to 30 g of yeast is produced (Faleiro &

Carvalho, 2021, p. 78). The surplus yeast is dried to be sold as a nutritional supplement for animal feed, suitable for cattle, goats, horses, swine, poultry, and other things, as it contains amino acids, B vitamins and proteins (Associação dos Fornecedores de Cana de Araraquara, 2021).

3.2 Case 2: the Sugarcane Mill (State of Paraná)

The first sugarcane agroindustry installed in the State of Paraná was "Usina Bandeirantes" (Usiban). It was created in 1942 by the founders Luiz Meneghel and three brothers from the Ometto and Dedini group who came from Piracicaba (São Paulo), bringing with them the necessary inputs and machinery for sugarcane exploration. Usiban is in the municipality of Bandeirantes, belonging to the "Norte Pioneiro" mesoregion, in Northern Paraná. This locality has the third largest representativeness in the sugarcane sector in Paraná (Bray & Teixeira, 1985; Ribeiro & Endlich, 2010).

As this is a case study at Usiban, the data was collected in loco, through field research by the team, whereby the administrative and manufacturing structure was visited at the end of the harvest period (October 2022), covering the various sugar and ethanol steps of the production process, and verifying the occurrence of the circular economy in it. Managers of the sectors participating in the entire process (plantation, industry, quality, human resources, legal and environment) were interviewed face to face, through focused interviews, with the circular economy theme in the studied agroindustry. Thus, the entire schematic process was drawn up in terms of data on planting, milling, and whether the by-products of the cycle are obtained to understand how Usiban's circular economy occurs.

Usiban has adopted alternatives to improve its economic performance. Thus, according to reports obtained directly from Usiban managers, the company's circular processes are more representative of the reuse/recovery of this waste and residues. In this scenario, in addition to cost savings, reducing investments in industrialized inputs and improving the energy efficiency of machinery, such measures also have environmental sustainability features. It was found from statements of the managers that the proportion adopted by Usiban in managing the production of sugar and/or ethanol depends on the conditions of the external and internal markets of these commodities. Currently, the mixing ratio is 60% for sugar and 40% for ethanol. According to the data obtained, the harvests conducted over 10 months of the 2022 agricultural calendar in an area of 10,000 hectares produced, on average, 10,000 tons of sugarcane per day. For every 10,000 tons, approximately 17,600 50-kg bags of sugar are produced, while the same measure produces 280,000 liters of hydrous ethanol (non-anhydrous).

In this context, of the total sugarcane harvested daily, according to Usiban, 25% or about 2500 tons are residual, mainly sugarcane bagasse. The plant fully reuses it for its energy self-sufficiency. Although there is an autonomous generator, it is noted that this mill does not have a projected utilization structure for the subsequent transfer of energy to the concessionaires. The reuse of waste in production is a reverse flow practice. It is important to mention that the sugarcane agroindustry can also produce bioenergy from straw and pellets, not only from bagasse. On the other hand, Usiban does not produce energy from these other residues. The sector's focus is on transforming bagasse into electricity for its own consumption.

Second-generation ethanol is also a product related to the circular economy. For its production, lignocellulosic materials contained in sugarcane bagasse and straw (cellulose, lignin, and hemicellulose) are used. Usiban does not produce second-generation ethanol. All the residue from sugarcane production is used as a form of bioenergy, as mentioned, or as a natural fertilizer. The latter is the case of vinasse and filter cake, by-products that are reallocated for the fertirrigation of sugarcane fields. After distillation of the already fermented sugarcane juice, the main residue and/or by-product in the production stage is vinasse. It is extremely polluting, but it is also rich in phosphorus and potassium, which are commonly used as a natural fertilizer. Usiban notes that it obtains cost savings of approximately R$ 2500.00 (2000 and 5000 Reais) per hectare with the management of vinasse and filter cake fertilization. In this sense, studies carried out by the company employees in its research laboratory show that approximately 41 kg of filter cake and ash are produced from 1000 kg of sugarcane, which is the largest source of fertilizer for the plant. Usiban also uses industrialized nitrogen fertilizer as a supplementary fertilizer. The good profitability of using vinasse, filter cake and ash as natural fertilizers is demonstrated by the productivity and quality of the sugarcane harvested in Usiban. According to the production data collected in October 2022, the average so far recorded was 86 tons of sugarcane harvested per hectare. The average sugarcane harvest production cycle at the mill is 6 to 7 cuts (it should be noted that after planting, the sugarcane is cut several times before being replanted).

"The sugarcane is paid according to the amount of sugar (ATR in Portuguese: Total Recoverable Sugars) contained in this raw material destined for the production of sugar and ethanol" (de Sousa, 2011, p. 25). The Usiban's ATR rate is 129 kg/tons of sugarcane, which represents a considerable average for the sector, especially in the region. According to the mill, this optimistic data comes from the correct management of the crop, with cultural treatment, circular economy, pest control and especially with fertirrigation residues. For the next few agricultural years, Usiban intends to continue using the circular economy methods and improve them, as they are economically profitable and facilitate the management of waste that, in a linear economy, also demands costs for correct disposal due to environmental legislation. In this sense, they intend to optimize production to reach the goal of harvesting 100 tons of sugarcane per hectare, a record in the Paraná State.

Usiban also has a distillation efficiency of 99.59%, fermentation efficiency of 92.21% and industrial efficiency of 93.7%. It was found in the field research that human capital is a group that works and thinks collectively. The mill also values its employees and invests in assertive communication. It points out that the integration of the team, in general, is a great competitive differential, enabling savings in internal transaction costs. Due to the shortage of qualified labor and courses in the region, the mill itself trains its staff and offers professional training courses aimed exclusively at the sugar and energy sector. Currently, the number of employees has decreased due to the mechanization of the sector. Usiban has approximately 120 permanent employees. However, although there is a reduction in the number of employees, the quality of the service provided has improved with investments in the qualification of human capital and professionalization in specific positions, quality control and input analysis. Figure 5 presents a visualization of the circular economy at Usiban, summarizing what was exposed in the results and discussion.

Comparing Figs. 4 and 5, Usiban does not produce second-generation ethanol and pellets, by the decision of the board of directors, which argues the preference for first-generation ethanol and the use of bagasse exclusively for the production of bioelectricity. It is not produced biogas either, with vinasse and/or filter cake being used as fertirrigation.

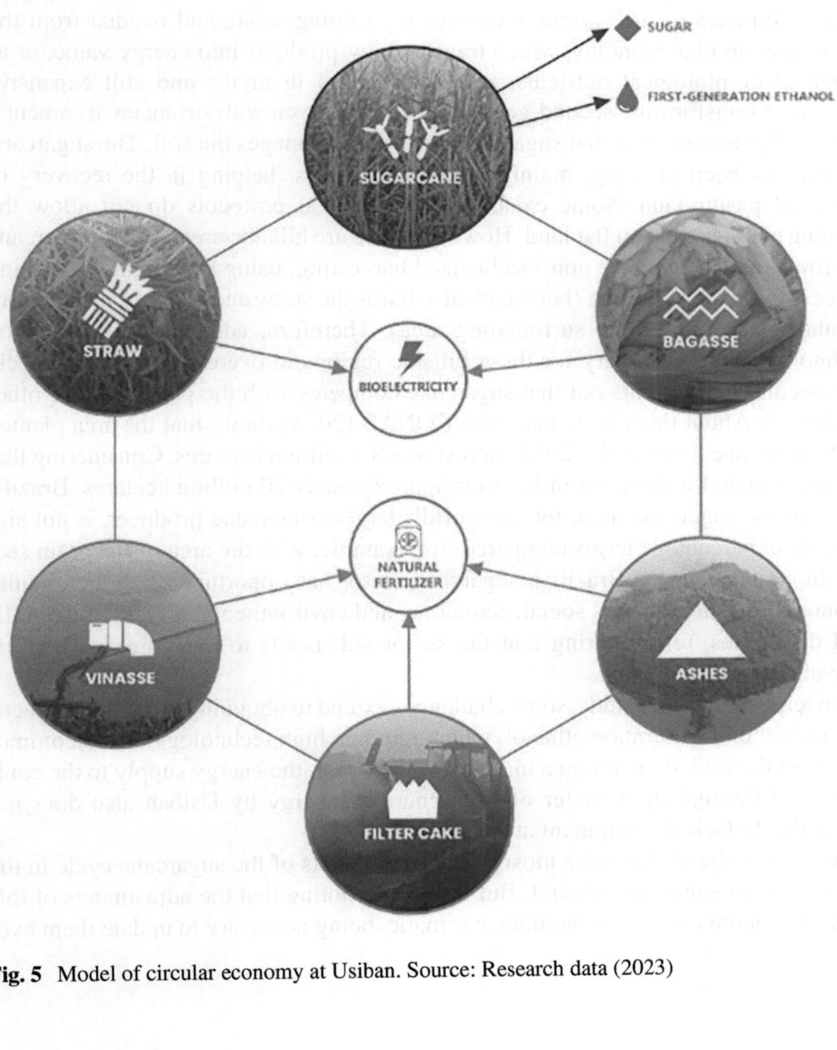

Fig. 5 Model of circular economy at Usiban. Source: Research data (2023)

3.3 Barriers and Opportunities for Circular Economy in Brazil

As mentioned in Brito (2020), “nothing is lost from sugarcane” (popular expression). Valuing the sugarcane economy’s waste and residues makes it possible not only to reduce the demand for natural resources, reverberating in environmental sustainability, but also economic and social advantages. Therefore, circular economy opportunities for the sugarcane economy consist of sustainable practices and the use of clean technologies that add economic, environmental, and social value to the sector, ethanol is an example.

According to Brito (2020), taking the case of a Brazilian state, 99% of the productive units of the sector in the Minas Gerais State have mechanized harvesting processes. Thus, the straw is not burned, as it is used to protect the soil in the off-season so that the plants regrow, and also serves as an input for bioelectricity generation. Barriers to the various processes for valuing waste and residue from the sugarcane circular economy, which transform by-products into energy value, or as technical or biological nutrients, are concentrated in myths and still expensive process of transforming second-generation ethanol, even with strong environmental appeal. The first myth is that sugarcane cultivation damages the soil. The sugarcane culture has been growing, mainly in degraded areas, helping in the recovery of degraded pastureland. Some existing environmental protocols do not allow the burning of sugarcane on flat land. However, there are hillside areas where sugarcane is grown that still require non-mechanized harvesting, using burning as a cleaning process to facilitate cutting (but which also burns the straw and worsens the environmental conditions of the surrounding area). Therefore, adequate mechanization technologies are necessary for these hillside regions to overcome this bottleneck. The second myth points out that sugarcane competes with the production of other foodstuffs. About this myth, data from CONAB (2023) shows that the area planted with sugarcane in the 2021–2022 harvest was 8.3 million hectares. Considering that the area planted with grains in Brazil is approximately 70 million hectares, Brazil’s share in the sugarcane area, for the world’s largest sugarcane producer, is not significant in percentage terms compared, for example, with the area of the grain sector. In conclusion, the Brazilian sugarcane sector has opportunities in the circular economy, which improves social, economic, and environmental drivers of the mills and distilleries, remembering that the sector still needs to overcome its barriers as well.

In terms of the case study, some challenges extend to obtaining more by-products, such as second-generation ethanol, which requires high technology and accommodation in the mill, demanding a high investment cost; the energy supply to the common grid through the transfer of self-generated energy by Usiban also does not occur due to lack of equipment at the plant.

However, the Usiban uses most of the by-products of the sugarcane cycle in the production of sugar and ethanol. But it is worth noting that the adjustments of this type of economy are not something automatic, being necessary to update them over time. Thus, further research is indicated with the scope of understanding the various effects that the functioning of the circular economy brings to Usiban.

4 Conclusions

The concept of circular economy is wide, involving a systemic change in production and consumption patterns that encompasses innovation and technology systems but also policies and business models. The basis is to create a system regenerative by itself where products and materials are maintained for as long as possible, and resources can be recovered and reintegrated into the economy or provide nutrients to natural systems.

Today, both Italy and Brazil are trying to apply circular models for the production of renewable energy.

Table 2 summarises the main aspects and differences highlighted in this study for both Italy and Brazil.

Table 2 Comparison between Italian and Brazilian case study in applying the circular economy model for bioenergy/biofuel production

	Italy	Brazil
Case study	Frozen vegetable company (Central Italy)	Sugarcane mill (state of Paraná)
Principal waste/ by-product	Vegetable by-products	Vegetable and agro-industrial by products: Bagasse; ash; pellet straw; vinasse; filter cake; yeast
Policies for circular economy currently in force	Green Deal; New Circular Economy Action Plan; Italian National Strategy on the Circular Economy; RED II; MD 15 September 2022	National Solid Waste Policy; National Biofuels Policy (RenovaBio)
Circular approach	Renewable energy production, sustainable agriculture (organic fertilizers)	Renewable energy, sustainable agriculture (fertilizers)
Circular products	Biogas/biomethane; digestate	Biogas; bioelectricity; biomethane; fertilizers; second-generation ethanol
Barriers	Legislation; economic (incentives and investment); public administration; government strategies; waste availability (seasonality); public acceptance (NIMB)	Myths: Sugarcane cultivation damages soil and sugarcane competes with foodstuffs Investments in efficiencies to produce more by-product Credit for investments in innovation
Opportunities	Waste valorization; decreasing management costs (profitability); creating added value; industrial symbiosis	Cost reduction; energy efficiency; waste reduction

Source: Author's elaboration

Brazil is a relevant player in the global sugar-energy sector, being the main producer and exporter of sugar, in addition to standing out in the field of advanced technology for the manufacture and use of fuel ethanol. All this is due not only to its continental extension and favorable soil and climate conditions but also to the role played by the circular economy throughout the sugarcane agroindustry, with the reuse of various by-products in the production of its main commodities, sugar and ethanol.

On the other hand, in Italy, the biogas-biomethane chain permits the recovery of by-products from the agri-food sector, which can be converted into energy, allowing the implementation of a full circular model. This is particularly true for the vegetable industry, which is among the food sector with the highest loss rates in the production chain (mainly in agricultural production) generating economic, environmental, and social impacts.

In both case studies, the application of a CE strategy implies a shift from "cradle to grave" to "cradle to cradle" waste management with an increasing decrease in waste generation using new business models. The circular business models will help the companies in waste valorization, decreasing management costs, and generating new added value.

However, the transition from a linear to a CE approach cannot be materialized without sufficient support from the policy. Compared to Brazil, the EU has support from policymakers, as showcased by a dedicated European policy mix in place since 2015, and various national government strategies.

Despite this, the transition to a CE is still at an early stage and only limited progress has been achieved. In fact, the policy can motivate and stimulate firms to adopt circular economy innovation activities. However, because different types of barriers have been highlighted based on the Italian and Brazilian contexts that could impede firms from adopting this model, it is necessary to define ad hoc policy instruments.

The implementation of CE strategies by companies is often contingent on economic factors. Prompted by increasing concerns over resource depletion and commodity price volatility, companies are motivated to search for new solutions that can help them reduce waste and energy costs as well as improve their competitive advantages. CE processes can provide opportunities for new revenue streams and have a positive impact on profitability. Finally, further research is needed to study the circular economy so that policymakers and entrepreneurs can have a better understanding of how to preserve nature, maximizing scarce resources.

References

Abdalla, N., Bürck, S., Fehrenbach, H., Köppen, S. & Janosch Staig, T. (2022). *Biomethane in Europe*. ifeu – Institut für Energie - und Umweltforschung Heidelberg. Accessed March 10, 2023, from https://www.ifeu.de/fileadmin/uploads/ifeu_ECF_biomethane_EU_final_01.pdf

Adnan, A. I., Ong, M. Y., Nomanbhay, S., Chew, K. W., & Show, P. L. (2019). Technologies for biogas upgrading to biomethane: A review. *Bioengineering, 6*(4), 92.

Almeida, L. F. P. D. (2015). *Pelletização do bagaço de cana-de-açúcar: estudo das propriedades do produto e rendimento do processo* (Master's thesis, UTFPR/Brasil).

Associação dos Fornecedores de Cana de Araraquara. (2021). *Como funciona a economia circular da cana-de-açúcar*. Accessed February 21, 2023, from https://canasol.com.br/como-funciona-a-economia-circular-da-cana-de-acucar/

Baştabak, B., & Koçar, G. (2020). A review of the biogas digestate in agricultural framework. *Journal of Material Cycles and Waste Management, 22*, 1318–1327.

Bega, R. M. (2014). *Aplicação de cinza do bagaço de cana-de-açúcar em latossolo cultivado com cana-de-açúcar* (PhD thesis, UNESP/Brasil).

Benato, A., & Macor, A. (2019). Italian biogas plants: Trend, subsidies, cost, biogas composition and engine emissions. *Energies, 12*(6), 979.

Bentivoglio, D., Finco, A., & Bacchi, M. R. P. (2016). Interdependencies between biofuel, fuel and food prices: The case of the Brazilian ethanol market. *Energies, 9*(6), 464.

Bentivoglio, D., Chiaraluce, G., & Finco, A. (2022). Economic assessment for vegetable waste valorization through the biogas-biomethane chain in Italy with a circular economy approach. *Frontiers in Sustainable Food Systems, 6*, 1035357.

Bonviu, F. (2014). The European economy: From a linear to a circular economy. *Romanian Journal of European Affairs, 14*, 78.

Bose, A., O'Shea, R., Lin, R., Long, A., Rajendran, K., Wall, D., Sudipta De, & Murphy, J. D. (2022). The marginal abatement cost of co-producing biomethane, food and biofertiliser in a circular economy system. *Renewable and Sustainable Energy Reviews, 169*, 112946.

Brasil. (2010). *Lei n. 12.305, de 02 de agosto de 2010. Institui a Política Nacional de Resíduos Sólidos; altera a Lei no 9.605, de 12 de fevereiro de 1998; e dá outras providências. Diário Oficial da União, Brasília, DF, 03 de agosto de 2010*. Accessed February 21, 2023, from https://www.planalto.gov.br/ccivil_03/_ato2007-2010/2010/lei/l12305.htm

Brasil. (2017). *Lei n. 13.576, de 26 de dezembro de 2017. Dispõe sobre a Política Nacional de Biocombustíveis (RenovaBio) e dá outras providências. Diário Oficial da União, Brasília, DF, 27 de dezembro de 2017*. Accessed March 23, 2023, from https://legis.senado.leg.br/norma/26330795

Bray, S. C., & Teixeira, W. A. (1985). O processo de implantação e expansão do complexo canavieiro, açucareiro e alcooleiro no Estado do Paraná. *Revista Boletim de Geografia - UEM, 3*(3), 17–30.

Brito, V. (2020). Revolução nos canaviais. *Revista Ecológico, 130*, 50–55.

Calil Neto, A. (2017). *Análise de modelos de negócios para a produção brasileira de bioetanol de segunda geração em escala comercial* (PhD thesis, UFRJ/Brasil).

Cardoso, B. F. (2016). *Análise do sistema agroindustrial do biodiesel no Brasil e na União Europeia* (PhD thesis, UNIOESTE/Brasil).

Cezarino, L. O., de Araújo, G. J. F. & Ronquim Filho, A. (2020). O licenciamento ambiental e a produção mais limpa sob ótica da economia circular nas usinas sucroenergéticas. In *Anais do Simpósio de Engenharia de Produção, Bauru, SP, Brasil*, p. 27.

Chagas, A. L. S. (2009). *Três ensaios sobre o setor produtor de cana-de-açúcar no Brasil*. (PhD thesis, FEA/USP-Brasil).

Cherubini, F. (2010). The biorefinery concept: Using biomass instead of oil for producing energy and chemicals. *Energy Conversion and Management, 51*(7), 1412–1421.

Chiaraluce, G. (2021). Circular economy in the agri-food sector: A policy overview. *Italian Review of Agricultural Economics, 76*(3), 53–60.

Chiaraluce, G., Bentivoglio, D., & Finco, A. (2021). Circular economy for a sustainable agri-food supply chain: A review for current trends and future pathways. *Sustainability, 13*(16), 9294.

Circular Economy Network. (2022). *4° Rapporto Sull'economia Circolare in Italia*. Accessed March 10, 2023, from https://circulareconomynetwork.it/wp-content/uploads/2022/04/Rapporto-sulleconomia-circolare-2022-CEN.pdf

Companhia Nacional de Abastecimento – CONAB. (2023). *Cana-de-Açúcar – Indústria e Cana-de-Açúcar – Agrícola e Industrial*. Accessed February 09, 2023, from https://www.conab.gov.br/info-agro/safras/serie-historica-das-safras

Cortez, L., Magalhães, P., & Happi, J. (1992). Principais subprodutos da agroindústria canavieira e sua valorização. *Revista Brasileira de Energia, 2*(2), 1–17.

Corvellec, H., Stowell, A. F., & Johansson, N. (2022). Critiques of the circular economy. *Journal of Industrial Ecology, 26*(2), 421–432.

D'Adamo, I., Falcone, P. M., Huisingh, D., & Morone, P. (2021). A circular economy model based on biomethane: What are the opportunities for the municipality of Rome and beyond? *Renewable Energy, 163*, 1660–1672.

Daou, A., Mallat, C., Chammas, G., Cerantola, N., Kayed, S., & Saliba, N. A. (2020). The eco-canvas as a business model canvas for a circular economy. *Journal of Cleaner Production, 258*, 120938.

de Fravet, P. R. F., Soares, R. A. B., Lana, R. M. Q., Lana, A. M. Q., & Korndörfer, G. H. (2010). Efeito de doses de torta de filtro e modo de aplicação sobre a produtividade e qualidade tecnológica da soqueira de cana-de-açúcar. *Ciência e Agrotecnologia, 34*(3), 618–624.

De Keyser, E., & Mathijs, E. (2023). A typology of sustainable circular business models with applications in the bioeconomy. *Frontiers in Sustainable Food Systems, 6*.

de Oliveira, F. R., França, S. L. B., & Rangel, L. A. D. (2018). Challenges and opportunities in a circular economy for a local productive arrangement of furniture in Brazil. *Resources, Conservation and Recycling, 135*, 202–209.

de Sousa, I. C. (2011, April–June). What is the cost of a ton of cane? *Revista Opiniões*, 25. Accessed February 21, 2023, from https://issuu.com/opinioesbr/docs/opaa28?fr=sNzRjMzEzOTgzMDA

de Sousa, E. L. (2020). *Cem milhões de elefantes*. Accessed February 21, 2023, from https://unica.com.br/noticias/opiniao-cem-milhoes-de-elefantes/

Domenech, T., & Bahn-Walkowiak, B. (2019). Transition towards a resource efficient circular economy in Europe: Policy lessons from the EU and the member states. *Ecological Economics, 155*, 7–19.

Ellen MacArthur Foundation. (2016). *Circularity indicators*. Accessed February 20, 2023, from https://www.ellenmacarthurfoundation.org/programmes/insight/circularity-indicators

European Commission. (2017). *Communication from the Commission to the European Parliament, the Council, the European Economic and Social Committee and the Committee of the Regions - The role of waste-to-energy in the circular economy*. European Commission. Accessed March 16, 2023, from https://eur-lex.europa.eu/legal-content/EN/TXT/PDF/?uri=CELEX:52017DC0034&from=en

Fagerström, A., Al Seadi, T., Rasi, S., & Briseid, T. (2018). The role of anaerobic digestion and biogas in the circular economy. *IEA Bioenergy Task, 37*, 8.

Faleiro, C. P., & Carvalho, K. A. (2021). Implantação de sistema de beneficiamento de levedura residual para produção de ração animal. *Revista S&G, 16*(1), 77–83.

Finco, A., Bentivoglio, D., & Nijkamp, P. (2012). Integrated evaluation of biofuel production options in agriculture: An exploration of sustainable policy scenarios. *International Journal of Foresight and Innovation Policy, 8*(2–3), 173–188.

Girotto, F., Alibardi, L., & Cossu, R. (2015). Food waste generation and industrial uses: A review. *Waste Management, 45*, 32–41.

Głowacka, A., Szostak, B., & Klebaniuk, R. (2020). Effect of biogas digestate and mineral fertilisation on the soil properties and yield and nutritional value of switchgrass forage. *Agronomy, 10*(4), 490.

Grangeiro, M. G. A., Fuentes, M. D. F. F., Freitas, E. R., Espíndola, G. B., & Souza, F. M. D. (2001). Inclusão da levedura de cana-de-açúcar (Saccharomyces cerevisiae) em dietas para frangos de corte. *Revista Brasileira de Zootecnia, 30*, 766–773.

Jesus, G. M. K., Jugend, D., Paes, L. A. B., Siqueira, R. M., & Leandrin, M. A. (2021). Barriers to the adoption of the circular economy in the Brazilian sugarcane ethanol sector. *Clean Technologies and Environmental Policy, 2021*, 1–15.

Kapoor, R., Ghosh, P., Kumar, M., Sengupta, S., Gupta, A., Kumar, S. S., Vijay, V., Kumar, V., Vijay, V. K., & Pant, D. (2020). Valorization of agricultural waste for biogas based circular economy in India: A research outlook. *Bioresource Technology, 304*, 123036.

Kirchherr, J., Piscicelli, L., Bour, R., Kostense-Smit, E., Muller, J., Huibrechtse-Truijens, A., & Hekkert, M. (2018). Barriers to the circular economy: Evidence from the European Union (EU). *Ecological Economics, 150*, 264–272.

Kovacic, Z., Strand, R., & Völker, T. (2019). *The circular economy in Europe: Critical perspectives on policies and imaginaries*. Routledge.

Kowalska, H., Czajkowska, K., Cichowska, J., & Lenart, A. (2017). What's new in biopotential of fruit and vegetable by-products applied in the food processing industry. *Trends in Food Science & Technology, 67*, 150–159.

Lin, C. S. K., Pfaltzgraff, L. A., Herrero-Davila, L., Mubofu, E. B., Abderrahim, S., Clark, J. H., Koutinas, A. A., Kopsahelis, N., Stamatelatou, K., Dickson, F., Thankappan, S., Mohamed, Z., Brocklesby, R., & Luque, R. (2013). Food waste as a valuable resource for the production of chemicals, materials and fuels. Current situation and global perspective. *Energy & Environmental Science, 6*(2), 426–464.

Lukehurst, C. T., Frost, P., & Al Seadi, T. (2010). Utilisation of digestate from biogas plants as biofertiliser. *IEA Bioenergy, 2010*, 1–36.

Manavalan, E., & Jayakrishna, K. (2019). An analysis on sustainable supply chain for circular economy. *Procedia Manufacturing, 33*, 477–484.

Mancini, S. D., de Medeiros, G. A., Paes, M. X., de Oliveira, B. O. S., Antunes, M. L. P., de Souza, R. G., Ferraz, J. L., Bortoleto, A. P., & de Oliveira, J. A. P. (2021). Circular economy and solid waste management: Challenges and opportunities in Brazil. *Circular Economy and Sustainability, 1*(1), 261–282.

Mercante, A. L. P. (2020) *Geração de bioeletricidade, através do bagaço e da palha da cana-de-açúcar* (Monograph-Agronomic Engineering Course, UFSCar).

Mhatre, P., Panchal, R., Singh, A., & Bibyan, S. (2021). A systematic literature review on the circular economy initiatives in the European Union. *Sustainable Production and Consumption, 26*, 187–202.

Mor, R. S., Panghal, A., & Kumar, V. (2021). Circular economy in the agri-food sector: An introduction. In *Challenges and opportunities of circular economy in agri-food sector: Rethinking waste* (pp. 1–14). Springer.

Murano, R., Maisano, N., Selvaggi, R., Pappalardo, G., & Pecorino, B. (2021). Critical issues and opportunities for producing biomethane in Italy. *Energies, 14*(9), 2431.

Nadaleti, W. C., & Lourenco, V. A. (2021). A mathematical, economic and energetic appraisal of biomethane and biohydrogen production from Brazilian ethanol plants' waste: Towards a circular and renewable energy development. *International Journal of Hydrogen Energy, 46*(54), 27268–27281.

Novacana. (2020). *Raízen e RWE estudam uso de pellets de bagaço para geração de energia na Europa*. Accessed February 21, 2023, from https://www.novacana.com/n/industria/raizen-rwe-estudam-uso-pellets-bagaco-geracao-energia-europa-301020

Novacana. (2023). *Uso da vinhaça na cultura da cana-de-açúcar*. Accessed February 21, 2023, from https://www.novacana.com/cana/uso-vinhaca-cultura

Oliveira, B. M. (2016). Qualidade de pellets de bagaço de cana-de-açúcar e serragem de *eucalipto em diferentes composições* (Master's thesis, UFES/Brasil).

Omer, A. M. (2008). Energy, environment and sustainable development. *Renewable and Sustainable Energy Reviews, 12*(9), 2265–2300.

Ormazabal, M., Prieto-Sandoval, V., Puga-Leal, R., & Jaca, C. (2018). Circular economy in Spanish SMEs: Challenges and opportunities. *Journal of Cleaner Production, 185*, 157–167.

Pacheco, T. F. (2011). *Produção de etanol: primeira ou segunda geração? Circular técnica, 004 Embrapa Agroenergia*. Embrapa Agroenergia.

Pantzar, M., & Suljada, T. (2020). *Delivering a circular economy within the planet's boundaries: An analysis of the new EU circular economy action plan* (pp. 1–30). Institute for European Environmental Policy (IEEP) and Stockholm Environment Institute (SEI).

Piracicaba Engenharia Sucroalcooleira. (2015). *Vinhaça – principais técnicas de utilização*. Accessed February 21, 2023, from https://www.piracicabaengenharia.com.br/vinhaca-principais-tecnicas-de-utilizacao/

Pirelli, T., Chiumenti, A., Morese, M. M., Bonati, G., Fabiani, S., & Pulighe, G. (2021). Environmental sustainability of the biogas pathway in Italy through the methodology of the global bioenergy partnership. *Journal of Cleaner Production, 318*, 128483.

Raízen. (2023). *Economia circular*. Accessed February 21, 2023, from https://www.raizen.com.br/agenda-esg/economia-circular

Ribeiro, V. H., & Endlich, A. M. (2010). O avanço da agroindústria canavieira na Mesorregião Noroeste Paranaense. *Revista Percurso Nemo, 2*(1), 73–92.

Rizos, V., & Bryhn, J. (2022). Implementation of circular economy approaches in the electrical and electronic equipment (EEE) sector: Barriers, enablers and policy insights. *Journal of Cleaner Production, 338*, 130617.

Rizos, V., Behrens, A., Kafyeke, T., Hirschnitz-Garbers, M., & Ioannou, A. (2015). *The circular economy: Barriers and opportunities for SMEs. CEPS Working Documents*.

Rodrigues, L., & de Belon, J. G. O. (2019). *Etanol e veículos flex-fuel: possibilidade de escolha e a economia efetiva aos consumidores*. Accessed February 21, 2023, from https://www.udop.com.br/noticia/2019/02/06/etanol-e-veiculos-flex-fuel-possibilidade-de-escolha-e-a-economia-efetiva-aos-consumidores.html

Rossi, R. (2019). *The EU fruit and vegetable sector. Main features, challenges and prospects*. EPRS. European Parliamentary Research Service, 2019. PE 635.563. Accessed March 13, 2023, from https://www.europarl.europa.eu/RegData/etudes/BRIE/2019/635563/EPRS_BRI(2019)635563_EN.pdf

Ryckebosch, E., Drouillon, M., & Vervaeren, H. (2011). Techniques for transformation of biogas to biomethane. *Biomass and Bioenergy, 35*(5), 1633–1645.

Scapini, T., Bonatto, C., Dalastra, C., Bazoti, S. F., Camargo, A. F., Júnior, S. L. A., Venturin, B., Steinmetz, R. L. R., Kunz, A., Fongaro, G., & Treichel, H. (2023). Bioethanol and biomethane production from watermelon waste: A circular economy strategy. *Biomass and Bioenergy, 170*, 106719.

Serafim, R. F., Yabuki, L. N. M., Queluz, J. G. T., Galdeano, L. R., & Garcia, M. L. (2021). Efeitos da aplicação de vinhaça na fertilidade do solo. *Irriga, 26*(2), 439–459.

Shikida, P. F. A., & Cattelan, R. (2020). Evolução e dinâmica no setor sucroenergético brasileiro. In *Uma jornada pelos contrastes no Brasil: cem anos do Censo Agropecuário* (pp. 361–374). José Eustáquio Ribeiro Vieira Filho, José Garcia Gasques.

Su, C. W., Khan, K., Umar, M., & Zhang, W. (2021). Does renewable energy redefine geopolitical risks? *Energy Policy, 158*, 112566.

União da Indústria de Cana-de-Açúcar – UNICA. (2023). *Cana gerou 79% da bioeletricidade ofertada à rede*. Accessed February 21, 2023, from https://unica.com.br/noticias/setor-sucroenergetico-gerou-79-da-bioeletricidade-ofertada-a-rede/

Vandenberghe, L. P. S., Valladares-Diestra, K. K., Bittencourt, G. A., Torres, L. Z., Vieira, S., Karp, S. G., et al. (2022). Beyond sugar and ethanol: The future of sugarcane biorefineries in Brazil. *Renewable and Sustainable Energy Reviews, 167*, 112721.

Wysokińska, Z. (2016). The "new" environmental policy of the European Union: A path to development of a circular economy and mitigation of the negative effects of climate change. *Comparative Economic Research, 19*(2), 57–73.

MIX
Papier aus verantwortungsvollen Quellen
Paper from responsible sources
FSC® C105338

If you have any concerns about our products,
you can contact us on
ProductSafety@springernature.com

In case Publisher is established outside the EU,
the EU authorized representative is:
Springer Nature Customer Service Center GmbH
Europaplatz 3, 69115 Heidelberg, Germany

Printed by Libri Plureos GmbH
in Hamburg, Germany